Springer
Berlin
Heidelberg
New York
Hong Kong
London
Milan
Paris
Tokyo

Alain Meunier • Bruce Velde

Illite

Origins, Evolution and Metamorphism

With 138 Figures

Springer

Professor Dr. Alain Meunier
University of Poitiers
UMR 6532 CNRS
HYDRASA Laboratory
40, Avenue du Recteur Pineau
86022 Poitiers, France
E-mail *alain.meunier@hydrasa.univ-poitiers.fr*

Dr. Bruce Velde
Research Director CNRS
Geology Laboratory
UMR 8538 CNRS
Ecole Normale Supérieure
24, rue Lhomond
75231 Paris, France
E-mail *papa@geologie.ens.fr*

ISBN 3-540-20486-5 Springer-Verlag Berlin Heidelberg New York

Library of Congress Control Number: 13502272

Bibliographic information published by Die Deutsche Bibliothek
Die Deutsche Bibliothek lists this publication in die Deutsche Nationalbibliography; detailed bibliographic data is available in the Internet at <http://dnb.ddb.de>.

Springer-Verlag Berlin Heidelberg New York
Springer-Verlag is a part of Springer Science+Business Media

springeronline.com

Cover Desing: Erich Kirchner, Heidelberg
Typesetting: LE-TeX, Leipzig

Printed on acid free paper 32/3141/LT – 5 4 3 2 1 0

To the memory of John Hower
and to Bruno Lanson

This book has taken some time to assemble, bringing together much data resulting from approximately 50 years of scientific endeavour by many researchers. It has been inspired by many people, although some seem more important to us than others, mostly for personal reasons.

One of us (BV) began his research career in trying to answer the question "What is Illite?". We hope that in a few years we can answer it more fully than in 1965. John Hower posed this question to one of his students, one among many. John was an inspiration to many people, not only his students, and for this reason we would like to dedicate this book to him. In fact, one can observe that much of the work on clay mineralogy has been influenced by the ideas and discoveries of John Hower. One needs only to scan the pages of major clay mineral journals.

The present authors, finding their way towards the ends of their careers, one faster than the other, have followed the paths laid out by John Hower and other eminent clay scientists. The question of illites, however, is not yet answered completely, and the importance of using them to solve geological and environmental problems persists. Because of this we must look to the young active scientists who are better equipped than we were to solve the problems of the future. We would like then to dedicate this book also to the younger generation, especially one member, Bruno Lanson, who has shown his capacities in the recent past, and whom we would like to wish well in future endeavours. As with John Hower, we have worked with Bruno and appreciate his qualities as we appreciated those of John.

Our book is therefore dedicated to the memory of John Hower and to Bruno Lanson, generations and worlds apart but both passionately engaged in clay science.

Preface

It is our pleasure to present this large body of information and thoughts of many mineral scientists which has accumulated over decades. Illite is a mineral that has been discovered relatively recently, even though it has great importance in the geological cycles of weathering, sedimentation and burial. Illite is the major potassium mineral among silicates in the surface environment. Potassium represents the only alkaline metal, to be bound in silicate structures during the great chemical reshuffling called weathering. The weathering environment is one of strong chemical segregation, where Si and Al become the resistant, residual elements of silicate rocks. Iron forms an oxide and potassium forms the stable clay illite. Then Si and Al form smectites and kaolinite. Sodium, calcium and to a large extent magnesium are extracted from the solids as dissolved ionic species of the altering fluids. Ca and Mg are reintroduced into solid minerals via carbonate precipitation, and Na remains to make the sea saline.

This mineral has been difficult to study because it is of fine grain size, as are all clays: 2 µm in diameter. Illite, along with other clays, had to wait to be discovered until a useful method of X-ray detection became available. With such a tool clays, whose definition was initially based upon the resolving power of an optical microscope (2 µm), could be efficiently investigated. In fact the study of illite parallels the use and development of X-ray diffraction techniques. Initial powder amounts giving rise to linear photographic film records gave rather rough indications as to mineral phase identifications, especially in cases of multiphase assemblages so common to clay mineral occurrence. Analog recording methods developed in the 1950s provided graphical representations of diffracted beam intensities which made identification much more precise. For the next three decades, X-ray diffractograms were the major tool of clay mineralogy. This was a period of enormous progress in the identification, interpretation and modelling of clay mineral interaction using X-ray radiation. The geology of clays became a true discipline. It was possible to study large series of samples in order to gain a general idea of phenomena based upon spatial distribution and variations of clays as they reacted to chemical changes and temperature changes over different periods of time. These are the fundamental variables of geology, time-temperature-composition; it takes a large amount of information in order to understand their functions.

Of course, Ralf Grim is the pioneer of illite research; John Hower and Jack Burst outlined the evolution of clays in sedimentary burial processes which produce illite-rich clays and Bob Reynolds provided us with the tools of calculation which were used to determine the physical reality and unmask some wishful thinking in the interpretation of X-ray diagrams. These new tools kept some of us busy for several decades, until it became apparent that more information of a more precise nature was necessary in order to interpret the X-ray signal detected by diffractometers. Bruno Lanson gave us a solid method of decomposing the different diffraction components in an X-ray diagram so that much more precise and numerically exact comparisons could be made with the modeled diffraction diagrams. Use of digital output from a diffraction spectrum and comparison with calculated X-ray diffraction spectra give us the possibility to quantify clay mineral occurrence and give viable identifications to mineral change. This is the great leap forward of the last years of the 20th century.

These were the tools. Along with them we had the inspiration of work using the concepts of crystal growth; elegantly explained by Alain Baronnet, which caused us to think in another dimension, the third instead of the classical phyllosilicate two-dimensioned sheet structures. Some of the first applications of these ideas to clay series were performed by Atsuyuki Inoue, which propelled us to our present state of discovery.

The present authors have participated in this rather fascinating development and have enjoyed the fruits of the research of our friends, colleagues and students. Here we have tried to focus this effort on the problem of illite: illite is a mineral that occurs over the whole range of clay mineral stability – time, temperature and chemical activity – and often in great abundance. It is frequently but not always present. At times it plays a critical role in the development of a geological system, either by its presence or absence. Illite is one of the most common clay minerals in soil clays developed under temperate climates, in the most fertile of soils. It is the end-product of burial diagenetic processes, supplanting the diagenetically formed smectites in pelitic sedimentary materials. It is the common clay mineral associated with many types of ore deposits. The correct identification and determination of the origin of illite can provide important information concerning many processes which lead to economic deposits. Illite, by virtue of its potassium content, is a key mineral concerning soil usage. Its presence, absence or instability determine much of what is called soil fertility. Thus illite is not just a nice mineral to study but an important one. We hope to convey some of this information and a sense of the significance of illite in natural and man-made environments to our readers.

Quite abruptly, we decided to write this book with "four hands" (if not two full heads) in November 2002 in order to put a "milestone" in the long path (more than 30 years) which we have walked. Since a memorable afternoon of 2002 during which BV visited the "Pédologie" lab of the University of Poitiers, we worked and discussed (sometimes with more passion than necessary) together. Science is above all a debate! Our greatest satisfaction is undoubtedly to have learned so much from our students. Some of them are now our colleagues and masters in their specialities (D. Beaufort, B. Lanson, A. Bauer among others). Perhaps it is time to say, after 30 years, that one of the greatest benefits of our common scientific adventure was to make friends all over the world. This could be, at least, encouraging for new student generations to learn that science offers great possibilities, to meet exceptional people and to maintain enthusiasm which is more or less the equivalent of youth.

Contents

List of Abbreviations

CEC	cation exchange capacity (meq/100 g or cmol kg^{-1})
FWHM	full width at half maximum intensity (°2θ Cu or Co Kα)
HIS	hydroxy-Al interlayer smectite
HIV	hydroxy-Al interlayer vermiculite
I/S MLM	illite-smectite mixed layer mineral. In case of coexistence of several I/S MLM, they are distinguished as follows:
I/Sil	illite-rich I/S MLM
I/Ssm	smectite-rich I/S MLM
PCI	poorly crystalline illite
R=0 or R0	randomly ordered mixed layer mineral (see glossary)
R=1 or R1	ordered mixed layer mineral (see glossary)
WCI	well-crystallized illite
XRD	X-ray diffraction
$\Delta G°$	free energy of formation (kcal mol^{-1} or kJ mol^{-1})

Introduction

The study of illite may seem rather narrow in scope, but we hope to demonstrate that this is not completely true. In fact this assemblage of published and unpublished research tries to analyze the accumulated knowledge and interpretations in order to make a summary of illite occurrence, to define more precisely what illite is and then to establish its place in a geological-environmental context. The information gathered since the 1950s has made proper identification and interpretation much easier as time has passed and more work has been done in fields where little information was previously available. However, research on the identification and diagnostic of illite occurrence and its related clay mineral suites stagnated somewhat in the 1980s. It seemed that further investigation would bring little new information. This was especially true in studies of soil clay mineralogy. The research activity in agronomy departments in institutions over the world has almost come to a halt. This is largely due to the difficulty in further pursuing clay mineral identification with the tools at hand, as well as a shift of interest from soil to plants in agronomic research. The statics of clay mineral identification and the slow mineral change observed in studies of sedimentary series, occurring on the scale of millions of years, seems to have discouraged people from observing the changes possible in active soil sequences. Silicates were assumed to be stable, or at least of such slow kinetic reaction that their existence was a given in soil clay problems. Interest in clay minerals shifted to problems related to sedimentary basins subjected to burial diagenesis and related problems of petroleum generation. Here the general scheme was established and one had only to apply the recipe to a given problem. The 1990s seemed to be devoted to refining a classical model and applying it to natural occurrence.

However, in a small corner of geological-mineralogical research, a new method was being developed and combined with a more well-known corpus of theoretical information. This method of numerical analysis of X-ray diffraction recordings (curve decomposition) was being coupled with the physics of X-ray diffraction to provide a new, powerful tool for mineral identification. Not only a correct identification can be made but one can also follow the changes in relative mineral content in natural samples. This allows one to follow changes in composition of clay minerals as well as their relative abundance. With this tool it is possible to determine the reaction kinetics of clay change.

The chapters which follow are based largely on observations of curve decomposition and comparison of these observations with calculated X-ray spectra.

This is our tool for determining what illite is. We have based these observations on our own samples and on spectra taken from publications (using a scanner and manual translation of positions on a rectangular grid). The results obtained are surprisingly consistent, indicating that most workers have done a good job, and thus that the principles of scientific publication are validated. However, there is much work to be done because the method is new and has just begun to be applied systematically by different researchers in different fields of clay mineral investigation.

The strategy of this investigation of illite mineralogy was to assemble the occurrences reported for various environments where illite is known to exist: weathering, soils, sediments, sedimentary rocks and geothermal-hydrothermal occurrences. First of all a general assessment of what is generally known about illite is made and a provisional definition based upon the judgement of previous workers. Then we investigate more intensively the large body of data available, applying our criteria for mineral identification (decomposition and comparison with calculated curves). Identification of illite and the chemical-physical conditions of its formation are used to define illite in a full mineralogical sense. Minerals do not exist in vacuum nor only in mineral collections.

We feel also that the dynamic aspect of illite occurrence, its formation in response to chemical and temperature changes, is fundamental to an understanding of how illite interacts or is a part of a geological system. This concept of dynamic attainment of an equilibrium is essential for an understanding of the effect of human activity on geological materials. How much change can a system sustain without being modified? This is in fact the crucial point in human-related problems in the environment. The duration of confinement barriers for nuclear waste is critical for all strategies of construction of waste structures. Kinetics is an all-important consideration. How much chemical stress (inorganic and organic) can a soil undergo before it changes, losing illite, the reservoir of potassium which is a fundamental component of soil fertility? The dynamics of soil mineral change will determine farming practice in the future if one wishes some approach to sustainable agriculture. These are the challenges of the 21st century. Yet more methods with higher precision are needed to correctly follow the very rapid mineral change observed in some agricultural systems (illite loss in as little as 15 years). There is a great need for the co-ordination of scientific endeavor in the fields of organic chemistry and plant physiology with clay mineralogy in order to solve these problems of biosphere interaction.

We hope that the following chapters might provide an inspiration to continue and expand the study of clay minerals.

CHAPTER 1

The Mineralogy of Illite – What is Illite?

The purpose of this chapter is to try to establish a guide line among the numerous uses of the term "illite", considering first historical definitions, then recent advances in the fields of mineralogy and thermodynamics. The goal is to describe the variability of the crystal structure of illite in its different states of crystallization through X-ray diffraction, particle morphology and HERTM observations. If illite is "a phase", we must understand how it can grow. From the examination of our present knowledge we hope to propose a "working" definition of illite before considering the way in which illite forms in natural environments.

1.1 Illite Definitions

1.1.1 Definitions of the Past

Mineral definitions of fine-grained materials have posed a problem and still do. After using optical microscopes for many years as the main tool of mineral identification, geologists were liberated from the constraints of grain size by using X-ray diffraction techniques on powder (fine-grained) samples. In this way great progress was made in the study of soil and other clay (less than 2 micrometer diameter) materials. Illite was identified among the new materials.

The term "illite" was proposed by Grim et al. (1937) as a general term, not as a specific clay mineral name, for the mica-like clay minerals. The name was derived from the name of the state of Illinois where it was first described near the town of Fithian. It has been widely used as a name for clay minerals with a 10 Å $d_{(001)}$ spacing which show substantially no expanding-lattice characteristics, i.e. no swelling clay components. Grim et al. (1937) gave illite a general unit formula which in fact includes both di- and trioctahedral structures. Hence illite is a mineral group in this conception and not a specific mineral. Although this definition is generally accepted by soil scientists, it is not completely satisfying because in other geological situations (diagenesis, supergene and hydrothermal alterations), illite is often considered as a mineral species. In these other environments, the formation of illite crystals has been shown to respect the classical rules of nucleation and growth processes.

Since then, the illite definition was progressively improved. Gaudette et al. (1966) claimed that illite is not necessarily a mixed-layer mineral. In 1982–3 the Clay Minerals Society (USA) Nomenclature Committee made a more precise definition (Clays Clay Minerals 32:239, 1984) of illite where the mineral illite is:

1. non-expandable
2. a dioctahedral mineral
3. ionic substitutions occur in both the octahedral and tetrahedral sites of the mica structure. There is a phengite component in illite.
4. interlayer substitutions can include not only potassium but also hydronium (H_3O^+) ions.
5. interlayer charge ranges between 0.8 and 0.6.
6. mineral structure (polytype) is not considered as a valid criteria.

An even more accurate definition was given by Eberl and Srodon (1984) who restricted the term "illite" to the non-expanding dioctahedral aluminous, potassium mica-like minerals occurring in the clay-size (< 4 µm) fraction of a geological sample.

However, even though Eberl and Srodon were more precise in their definition than were Grim et al. (1937), this later definition is not fully satisfying either since it is strictly based on XRD analysis of a given size fraction. Indeed, according to these authors, any crystal the size of which is greater than 4 µm is not an illite but a mica. Crystal size should not be a criterion for a mineral group definition even though clay minerals are normally relegated to the clay sized fraction of a geologic material (< 2 µm).

Thus, considering the above definitions we can consider that material called illite should respect the following conditions:

- it is restricted to dioctahedral series (trioctahedral illites were shown to be debris of trioctahedral micas);
- the 10 Å non expandable criterion (and non-contracting) conditions (XRD from oriented preparations) is necessary;
- the interlayer charge is lower than that of micas, < 1 per $O_{10}(OH)_2$;

Several questions should be addressed which are not fully answered in the classical definitions of illite:

- what is the extension of the chemical composition range in the dioctahedral phyllosilicate group? What substitutions occur in the tetrahedral, octahedral and interlayer sites?
- are the models of the crystal structure coherent with the crystal growth processes?
- what are the theoretical possibilities of phyllosilicate structures in 3D space and those which are typical of illite?

1.1.2
Definition Based on XRD Examination

1.1.2.1
The Illite Problem

Considering the questions mentioned above, it should be useful to revisit the XRD definition of illite. One important aspect is the non-expanding or mica-like properties seen by XRD which are currently used in order to distinguish illite from smectite-vermiculite and dioctahedral micas. The major ambiguity at present is the simplistic assimilation of illite layers to non-expandable ones. This creates several sources of error:

- illite XRD patterns from oriented preparations in the air-dried and ethylene glycol saturated states must be perfectly superimposable. However, even if this condition is fulfilled, the detection limits of powder diffractometer analyses are between 0 and 5% of the expandable layers;
- high-charge smectite or vermiculite layers may collapse irreversibly when K-saturated. Their detection needs to compare XRD patterns from preparations in different cationic saturation states (Ca, K and K-Ca, for instance).

It is also necessary to investigate the 3-D crystal structure. Smectites are characterized by *turbostratic* stacking modes showing only wide, non-defined peaks in the regions of 4.5 and 2.06 Å while illite is more ordered ($1M_d$, 1M, $2M_1$ polytypes), showing clear peaks in these regions. Potassium saturation alone is not able to significantly increase the stacking order of smectites; additional energy has to be provided through drying-wetting cycles (Mamy and Gaultier 1975; Eberl et al. 1986). However, the maximum degree of ordering attainable remains largely *turbostratic*: a progressive transition toward the 1M polytype was never observed in smectite or vermiculite having experienced K-saturation and drying-wetting cycles (Plançon et al. 1978).

If the stacking order of the non-expandable layers (polytypism) could be considered as a criterion to separate illite from K-bearing smectite or vermiculite (high-charge smectite), what can be used to help to distinguish illite from dioctahedral mica? Polytypes of the two species are similar even if the 1M one is more common for illite.

Thus, from the XRD point of view, we must choose another criterion: the difference between illite and dioctahedral micas could be related to the size of the coherent scattering domains (CSDS) in the c* direction. The narrower the $d_{(00l)}$ peaks, the higher the CSDS. Thus, a simple measure of the full width at half height (FWHM) of these peaks may provide a useful guide: FWHM is lower than 0.25 °2θ Cu Kα for micas (Lanson and Champion 1991; Jaboyedoff 1999). This is not, of course, a rigorous definition but simply an empirical method which is quite satisfying when XRD patterns are decomposed into Gaussian and/or Lorentzian elementary curves (Lanson 1997).

The approach of distinguishing illite from other minerals by diffracting domain size has been employed for some time, using different terminology, in the illite crystallinity index method developed in the 1960s by Kübler (see the review by Kübler and Jaboyedoff 2000).

The physical boundary of crystallographic differences between illite and micas has been distinguished using peak width criteria. A second method is by peak decomposition (Gharrabi et al. 1998). In this study it is observed that the illite recrystallizes (using XRD criteria) to smaller-grained micas at its upper thermal stability limit. Thus, large illites, having grown under diagenetic conditions, are not stable, forming smaller mica minerals as metamorphic conditions increase. In this study there is no information concerning the chemistry of illite compared to micas, which should be part of a mineral definition.

However, an illite definition based only on XRD properties cannot be complete since a mineral is not only characterized by a crystal structure but also by its chemical composition.

1.1.2.2
X-Ray Diffraction of Illite and Illitic Mixed-Layer (I/S MLM) Materials: Theory

Moore and Reynolds (1989) give a well-documented discussion of the effects of X-ray diffraction of mixed-layer minerals. We are initially interested in the characteristics of mixed-layer mica and smectite or expanding layers in a mixed-layer structure. Such determinations can be used to distinguish between illite, mixed-layer minerals and high-temperature micas. The variables of a mixed-layer (I/S MLM) crystallite are composition, i. e. the proportion of mica (10 Å) and smectite layers, and n, the number of coherently diffracting layers in the crystallites. The effects of changes in these parameters on the XRD diagram are roughly as follows:

1. The number of coherently diffracting layers (Coherent Scattering Domain Size, CSDS) in a crystallite is designated as N. This value determines the width of a diffraction peak and, to a certain extent, its XRD position. In general, the smaller the value of N, the wider the diffraction peak. Also, the smaller N is, the larger the apparent d spacing of the diffraction peak. Hence, if we begin with a mica crystallite with 20 or more coherently diffracting layers and we reduce the number of these layers, the peak will gradually (and non-linearly) widen and change position to greater than 10 Å.
2. The presence of smectite layers in an illitic I/S MLM will tend to widen the basal spacing peak and shift it to higher values. This is true if the smectite is hydrated or has only one layer of glycol per smectite layer. If the smectite layer is fully glycol-saturated (17 Å basal spacing), and the crystallites have a very high illite content (> 95%), the peak position near 10 Å will be shifted to a spacing of less than 10 Å. Thus glycol treatment can allow the distinction between the effect of diffracting domain size and the presence of smectite.

In general, a wide peak of slightly greater than 10 Å will represent an illitic mineral (high proportion of illite) with either small diffracting domains or the presence of a small number of smectite layers in an I/S mixed layer mineral. If the peak shifts to less than 10 Å upon glycol saturation, the mineral contains smectite layers. If the peak does not significantly change position, the mineral contains only a few smectite layers which do not expand fully (two glycol layers) or it is composed of very small crystalline domains. The initial criterion for the XRD identification of illite then, according to XRD methods of identification, is that the material investigated is composed uniquely of 10 Å layers. Thus the presence of smectite will disqualify a material from being an illite.

As a starting point, one can define wide peaks (> 0.4 °2θ Full Width at Half Maximum intensity, FWHM) at slightly greater than 10 Å (> 10.2 Å) which do not change peak position with glycol saturation as being poorly crystallized illite, PCI; the wide peak is due to small grain size or small amounts of interstratified smectite in the structure which does not fully expand under glycol saturation. WCI will be a mica layer structure of small peak width, < 0.4 °2θ Cu Kα. We will test these initial criteria on natural samples.

1.1.2.3 The Decomposition of XRD Patterns

Illite rarely occurs as a single mineral deposit in nature. In general, when a mineral is found as a pure phase it is formed due to extraordinary conditions and these mineral phases tend not to represent the more common mineral found in nature. Geologic and surface materials reflect the multicomponent nature of the Earth, with usually seven or eight major chemical components. Such conditions rarely give rise to a single phase. However, in order to define a mineral, according to the rules of science, one needs to have a sufficient amount of the mineral to do a certain number of diagnostic tests. Hence, pure-mineral illite, when found, does not often represent the illites found in soils, sediments and sedimentary rocks.

In order to introduce the reader to the problem of illite, as perceived by X-ray diffraction, we show several examples of nominally pure illite, assumed so in the past, and examples of illites in natural samples of different origins.

1.1.2.4 PCI and WCI

It has been observed for some time that the 10 Å mica peak in sediments and sedimentary rocks is asymmetric (Kübler 1967). The major intensity is at or near 10 Å and a lower shoulder is seen which represents another or several diffraction maxima. This shape is not the same in each sample, and its differences have been used to determine the evolution of illite in sedimentary rocks under the influence of diagenesis and early metamorphism. Using new methods of numerical analysis, Lanson (1997) has defined the peak at 10 Å as being composed of two major components, one wide and displaced to a position greater than 10 Å and another more narrow and centered on 10 Å. Both are dominantly Gaussian in shape. The first wide peak is called in the present work poorly crystallized illite (PCI) while the second is well-crystallized illite (WCI). This terminology is based upon the fact that wider peaks generally represent phases with smaller diffracting domain sizes (smaller crystallites for the most part) while more narrow peaks indicate the presence of larger diffracting domains. It should be remembered that it is not possible to establish the presence of illite (whose composition shows a low potassium content, and high silica content) solely using XRD methods. However, some data of the chemistry of individual clay crystallites in sedimentary rocks (Lanson and Champion 1991) indicate that such an interpretation might be valid. Whether or not this is true in all cases of illite and mica occurrence remains to be proven. It is possible that the narrow peak in a given sample can be formed by a contribution of illite of large diffraction domain size and by that of a true, detrital or high-temperature muscovite. For the moment, we can only use the description given by decomposing XRD spectra as a first approximation to the fine and more coarse clay fraction micaceous material. PCI and WCI are used as provisional descriptive terms.

The concept of PCI and WCI which is based upon observations of XRD diagrams of natural mineral suites can be referred to calculated diagrams using the NEWMOD program of Reynolds (1985). Figure 1.1 shows characteristics of diagrams the calculated for mica structures with different grain sizes or

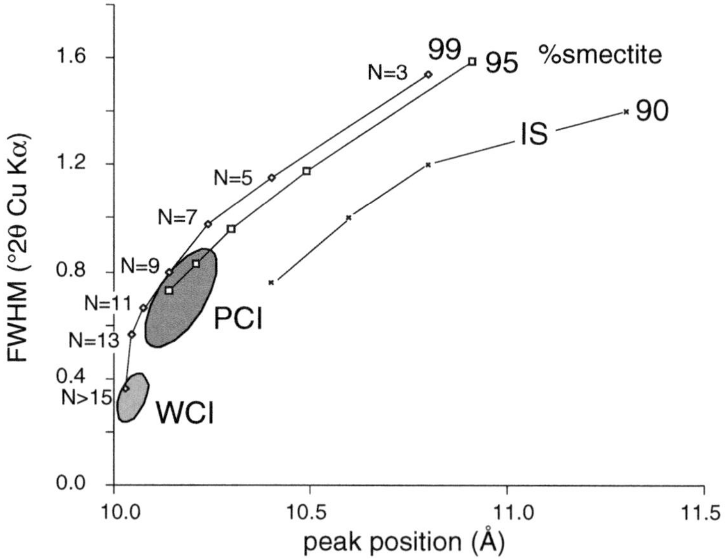

Fig. 1.1. Peak position and widths for calculated spectra of illite minerals as a function of the number of layers (N) in the coherent diffracting domains present. N indicate numbers of layers diffracting. A shift from the calculated line to higher peak positions will indicate the presence of smectite layers in the structure. The shaded areas indicate the PCI and WCI values obtained from sequences of buried sedimentary rocks reported by Lanson et al. (1998)

more precisely coherent diffracting domains. In this diagram the width of the peak at half height (FWHM) and the position of the peak maximum are plotted. The numbers (*N*) on the diagram refer to the coherent diffracting domains in the minerals. The greyed zones show the characteristics of the PCI and WCI minerals as observed by Lanson et al. (1998) for different series of clays in sedimentary rocks. In general, as the number of coherent diffracting domains becomes smaller, the diffraction peak becomes wider and it shifts to higher *d* values. WCI represents crystallites of greater than about 13 coherent diffracting domains. PCI lies approximately between 12 and 10. It should be noted that the greyed areas lie only partially on the line of pure micas. Shifts to the high peak position side usually indicate interlayering with very small amounts of smectite. Thus the PCI can contain some smectite layers (usually less than 5%) whereas the WCI is almost exactly on the mica line. It should also be observed that, as pure mica changes its diffracting domain dimension, it will be in the PCI region and, as the particle size domain increases, it will come into the WCI domain.

Figure 1.2 indicates the effect of particle size or coherent diffracting domain as it changes the width of the diffraction peak. In Fig. 1.2a crystals are thin, i. e. one can see through them under the transition electron microscope beam.

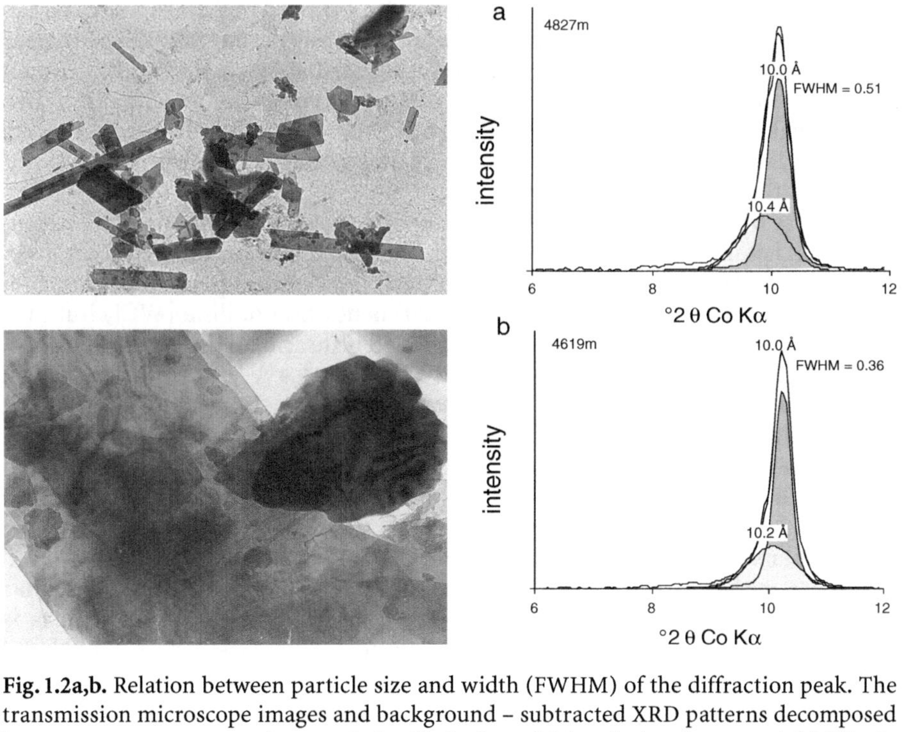

Fig. 1.2a,b. Relation between particle size and width (FWHM) of the diffraction peak. The transmission microscope images and background – subtracted XRD patterns decomposed into two components are characteristic of lath-shaped (**a**) and platy hexagonal (**b**) illite in the Rotliegendes sandstones

The peak width is 0.51 °2θ Cu Kα. In Fig. 1.2b thicker crystals are present, and the grains are larger. The thickness is indicated by the wavy diffraction fringes in the particles. Small grains are also seen among the larger particles. The peak width is only 0.36 °2θ Cu Kα. These samples are illites from sandstones in the Rotliegendes unit in Germany. Both are then mica in structure (10 Å) but the different crystal thickness, affecting the coherent diffracting domains, has resulted in different peak widths.

Illites should show peak position – width relations similar to those shown in Fig. 1.1 along the calculated line.

1.1.3
Examples of Pure Illites

Several samples of pure illite minerals have been reported, providing XRD and chemical data. These examples can serve as the starting point for an investigation of the XRD properties of illitic materials. Three of the pure illites studied by Gaudette et al. (1966) show a lack of expandable material which indicates the absence of a mixed-layer mineral phase association. They are the Beavers Bend, Marblehead and Rock Island samples. They are from midcontinent Silurian and Ordovician age sedimentary rocks in the Central United States. In these rocks the conditions of formation were those of diagenesis at relatively low temperatures over long periods of time. The XRD spectra which are presented in the publication indicate an apparently sharp single peak near 10 Å. However decomposition of these published spectra shows that in fact the apparent single peak is composed of several components (Fig. 1.3). Two major peaks are present, one near 10 Å, but usually less than 10 Å when glycol saturated, and another at higher spacings 10.2 to 10.4 Å. The proportions and widths of the peaks vary.

Because of a peak position significantly below 10 Å (9.65 Å) the Marblehead and Rock Island samples should probably be considered to contain an ordered, mixed layer I/S MLM of low smectite content. The second peak in the Marblehead illite at 10.16 Å which is wide (1.4 °2θ) is not that of illite (WCI) but PCI. This sample is a mixture of poorly crystallized illite and an illite-rich I/S MLM.

The Beaversbend sample has a narrow peak at 10.4 Å which is at a position for PCI but its narrow width suggests that in fact it is well crystallized. Hence it is a very illite-rich I/S (> 95% illite) but of large crystal domain size. The 9.96 Å peak is that of an illite but its width is 0.6 °2θ indicates a less well crystallized mineral and possibly an I/S MLM of high illite content due to the position below 10 Å.

Another sample was presented by Moe et al. (1996) which shows a two-band structure with a smaller peak at less than 10 Å (9.8 Å) indicating a regular mixed-layer I/S mineral of low smectite content (Fig. 1.3). The major peak is at 10.6 Å with a width of 0.6 °2θ Cu Kα radiation. This peak can be considered to be that of a PCI mineral, being at greater than 10 Å and of a width of greater than 0.4 °2θ Cu Kα.

A fourth example assumed to be pure illite is given by Kodama and Dean (1980). This mineral has characteristics similar to those of the Moe et al. sample.

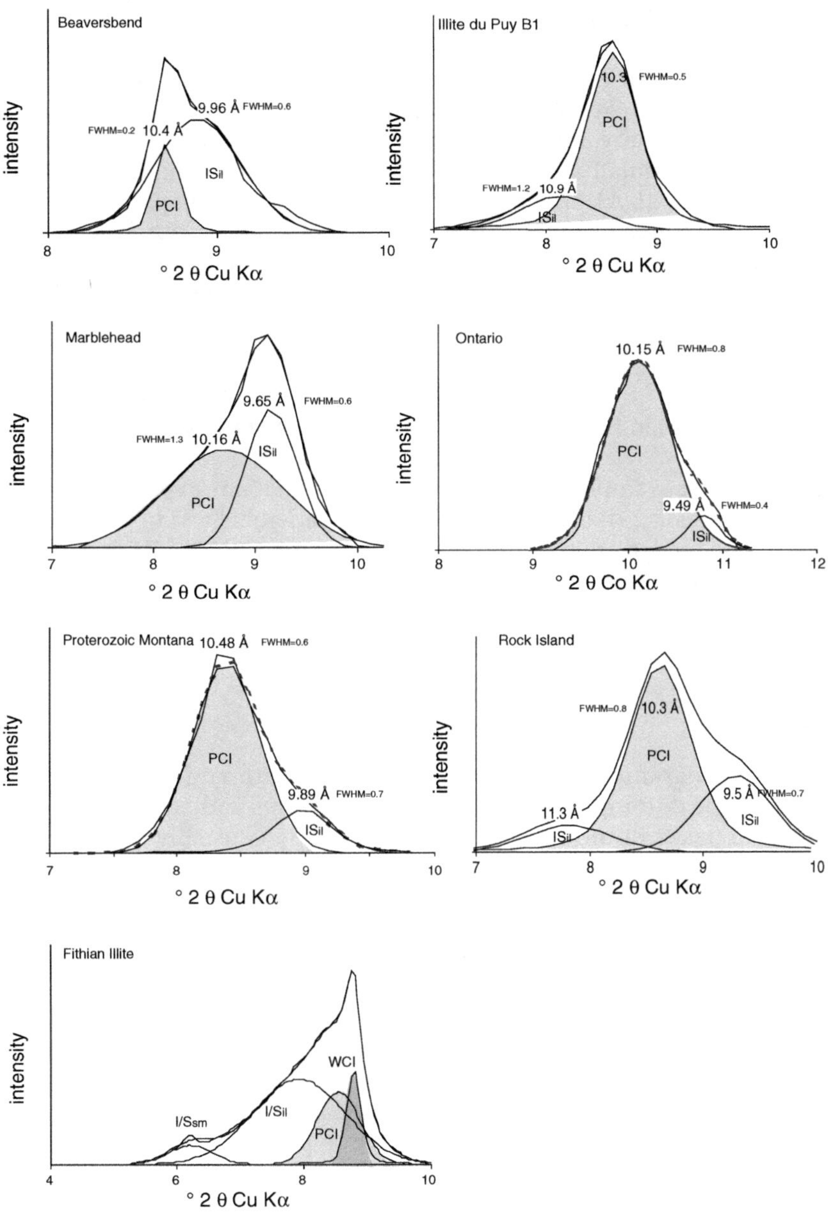

Fig. 1.3. Decomposition of background - subtracted XRD diagrams for illite samples: Beaversbend and Marblehead (from Gaudette et al. 1966), Le Puy (France), Proterozoic disturbed belt, Montana (from Moe et al. 1996), Rock Island and Ontario, Canada (from Kodama and Dean 1980), Fithian illite (collected by author). Samples were observed in the glycol-saturated state. Hence peaks at less than 10 Å are due to I/S MLMs of a high illite content. Peaks at more than 10 Å with a large FWHM (> 0.5 °2θ) are considered to be PCI phases

However, the peak position is closer to 10 Å (10.15 Å), but wider (0.8 °2θ Cu Kα). These two illite samples are then mostly PCI materials. For reference we show a spectrum for the Fithian illite (unpublished data of the author) which is decomposed into PCI, WCI, and two I/S MLMs.

The assumed pure illites discussed above show multiple peaks for the crystallites with small and large diffracting domains. This combined with non-illite (WCI, 10 Å) peak positions suggests that perhaps these minerals are in fact mixtures of highly illitic I/S MLMs (small proportion of smectite (< 5%) interstratified in the structure) and PCI minerals. If this is true, the materials are not pure illite according to the initial definitions of this mineral.

1.1.4 Examples of Illites in Natural Soils, Sediments and Sedimentary Rocks

If the pure illite examples do not correspond to what one expects of illite, what are the occurrences of illite in mixed-mineral assemblages? In order to fix the ideas for the reader, we present several typical examples of XRD patterns for materials common in different geological environments. These examples can serve as a basis for the discussion which follows in this chapter.

1.1.4.1 Soils

X-ray diffractograms of multiphase soil materials with a micaceous component can usually be decomposed into several components with, usually, two clearly observable peaks in the 10 Å region (Fig. 1.4). Soils usually show the peaks, PCI and WCI. Examples are given from prairie and agricultural soils from midcontinent (Velde et al. 2003) and an unpublished Italian Vertisol, where one or two types of mixed layer minerals, illite/smectite are frequently present. A forest soil from France (unpublished) shows a strong vermiculite peak with subordinate I/S and with PCI and WCI minerals.

The mixed layer minerals are both disordered I/S MLMs, one of high smectite content (I/S_{sm}) and the other more illitic (I/S_{il}). The PCI illite band (found between 10.1 and 10.5 Å) usually has a greater surface area than that of the WCI, being significantly wider than the second peak. The WCI, or well-crystallized phase, is sharper, having a peak width of 0.2–0.4 °2θ Cu Kα while that of the illite (poorly crystallized illite) is wider, ranging from about 0.8 to 1.6 °2θ Cu Kα. Most often pedologists and soil chemists will attribute the WCI to a detrital, mica-precursor phase which has been little altered chemically and hence is structurally the same as a high-temperature phase.In fact, published data for forest soils often show relatively more sharp illite-mica peaks, WCI suggesting the presence of old detrital mica. However there is little precise information (decomposition of peaks) for these materials at present.

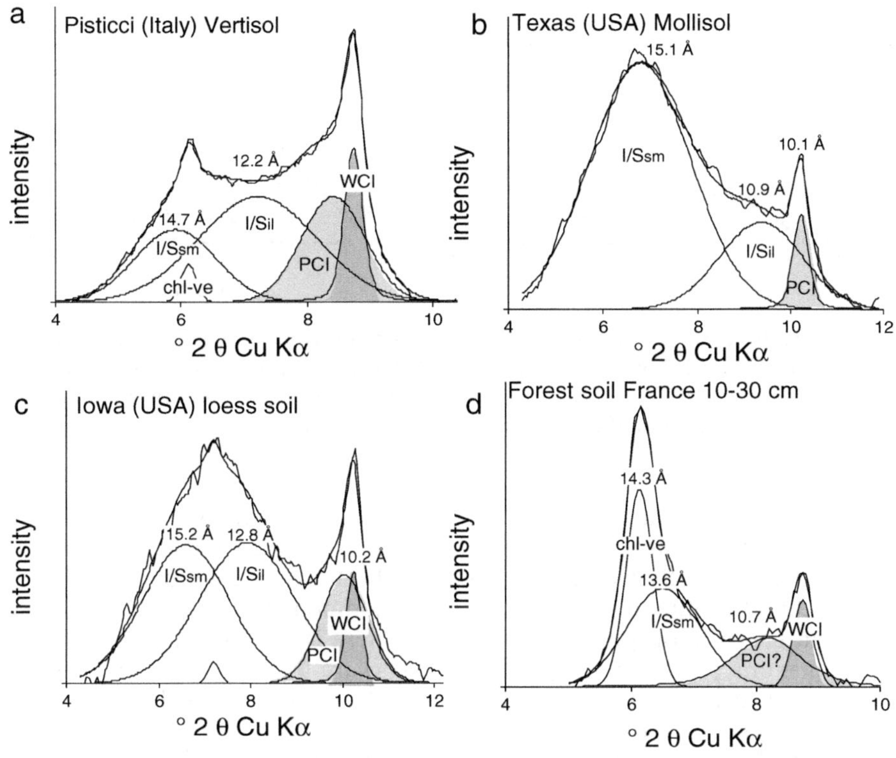

Fig. 1.4a–d. Decomposition of background – subtracted XRD patterns of various soils. Pisticci A-horizon soil based upon a Tertiary marine sediment. Texas Mollisol agricultural horizon based upon Tertiary sediments. Iowa agricultural horizon Mollisol based upon smectitic loess deposits. A-horizon of a forest soil from French Alps

1.1.4.2
Marine Sediments

Two examples of sediment XRD diagrams are given in the Fig. 1.5. It is clear that such materials should indicate the types of clays in the soils which find their way into the river-sediment cycle. An unusual example is given in the figure for Delaware Bay sediments (Velde and Church 1999); the material is essentially glacial flour deposited on the continental shelf which had experienced little interaction at the Earth's surface (weathering). Here the sediment is not derived from soils but essentially finely powdered sedimentary and metamorphic rocks. Hence one sees strong chlorite and illite peaks. However, the PCI–WCI peak relation is similar to what one sees in soil materials. For the most part near-shore sediments can be expected to be similar to the soil materials of the drainage basin of the rivers flowing into the ocean.

An example of evolved soil material forming sediments is that from the Marais at Brouage salt marshes (France) where input from the large Gironde

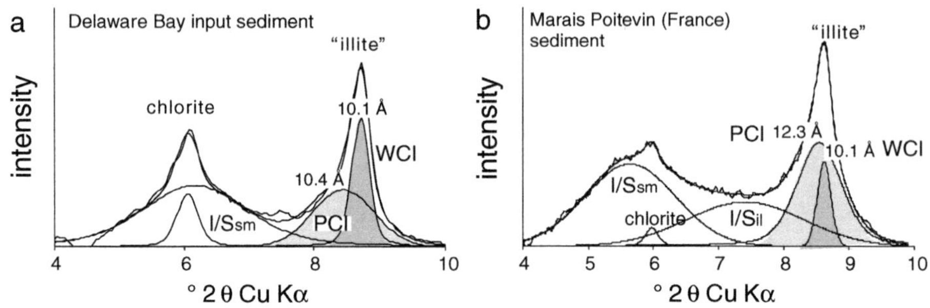

Fig. 1.5a,b. Background – subtracted XRD diagrams for two salt marsh sediments. Comparison of Delaware Bay (USA) and Marais Poitevin (Atlantic French coast) salt marsh sediments showing different proportions of I/S MLMs but a peak intensity maximum near 14 and 10 Å which is most often taken to indicate the presence of illite-chlorite assemblages

and Loire river basins provide material which is essentially brought into the marshes from the near-shore Atlantic Ocean. This material shows strong I/S components, usually two different phases, both disordered; one is smectitic (I/S_{sm}) and the other illitic (I/S_{il}). The illite component has both PCI and WCI present, based upon peak surface areas. There is a strong similarity in the peak maxima positions comparing the Delaware Bay sediments and those of the Marais at Brouage. A sharp peak at 10 Å and a small peak at 14.2 Å can be distinguished. In using just these criteria (sharp peak maxima) one would designate the material as being composed of illite (mica) and chlorite which was the case in the past for most of the Deep Ocean Drilling Project determinations of ocean sediment. However, closer inspection (decomposition) shows that a great majority of material is in fact I/S MLM and the mica component is largely PCI. Hence the soil component of these sediments is expressed in wide, overlapping peaks for the I/S MLMs (14.8 Å I/S_{sm} and 12.2 Å I/S_{il}) and a strong PCI component. Superficial peak recognition would equate the Delaware bay material and that of the Brouage salt marsh sediments as being of the same mineralogy.

1.1.4.3
Sedimentary Rocks

Usually the sedimentary rocks studied by petroleum geologists (hence from burial depths greater than 1000 m), show only one I/S band and the classic PCI-WCI peak pair. Figure 1.6 shows examples from deep wells in Paleozoic rocks from eastern France and a well in Tertiary sediments in the Gulf Coast USA. In all wells the I/S MLM is the ordered, R=1 form with approximately 75% illite. PCI illite peaks are relatively wide and the WCI bands narrow. The proportion of WCI (peak at 10.0 Å) is lower in the Tertiary sedimentary rocks (10% of combined WCI+PCI peaks) than in the Permian rocks (near 30%). In general the width of the PCI band is reduced as the peak position decreases

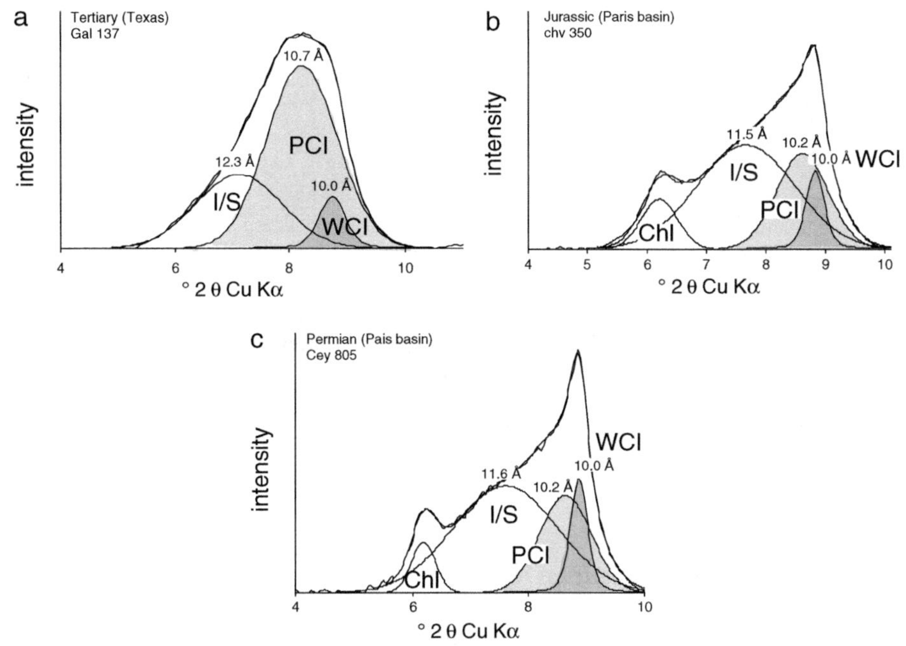

Fig. 1.6a–c. Decomposed background XRD patterns for deep well examples from the Tertiary beds in the Texas Gulf Coast (Gal 137) Jurassic (chv350) and Permian (Cey 805) layers in the Paris basin. The sequence shows the change in relative surface area (proportion of the phases) for PCI and WCI as a function of the age of the sedimentary rocks. All have experienced overall geothermal gradients of near 30 °C/km. The sharpness of the 10 Å part of the peak envelope as a function of age indicates the increasing importance of WCI in older sediments

(Lanson et al. 1998; Gharrabi et al. 1998). This gradual change is typical of the diagenesis trends. In much of the diagenetic materials the PCI has a more important surface area than does the WCI component. The proportion of WCI tends to increase with the age of the sediments which have experienced low temperature burials (near 30 °C/km). Figure 1.7 shows this relationship for several different wells from the Gulf Coast and the Paris Basin.

In all of these examples of natural mineral assemblages, the PCI (wider, higher spacing peak) is of greater surface area than the WCI peak. Eventually this relationship changes as the rocks approach the conditions of metamorphism and the WCI peak becomes more important. However, it is important to remember that in most environments the poorly crystallized fraction of illite-bearing assemblages, PCI, is dominant, at least by the criteria of X-ray diffraction.

In soils, sediments and sedimentary rocks one finds very similar XRD patterns of a double-peak illite or illite-mica material. A PCI and WCI peak are present in different proportions depending upon the material investigated.

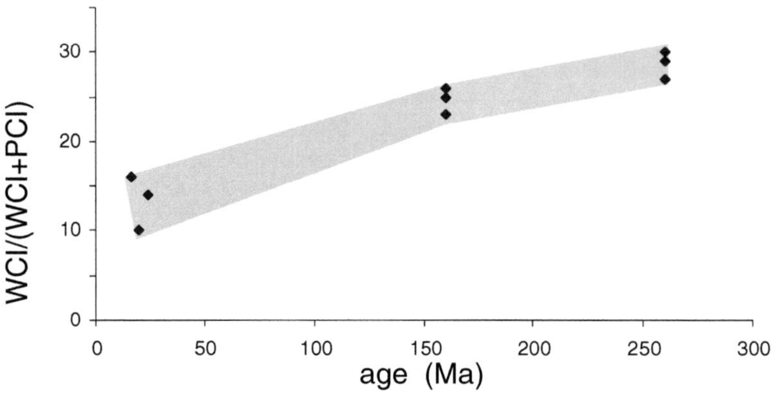

Fig. 1.7. Plot of peak surface areas of WCI and PCI for Tertiary, Triassic and Permian age sedimentary rocks of the Texas Gulf Coast and Paris Basin

1.1.4.4
Hydrothermal Alteration: Sericites

Sericite is a name given to fine-grained mica which is not quite mica. The potassium content is low and the silica content high (Grim 1985). In most circumstances this would be called illite, but the geological occurrence of sericite gives it its name. This is a hydrothermal mineral which has largely been observed under the optical microscope. Hydrothermal alteration is most often observed in granular rocks which are amenable to microscopic study. Eberl et al. (1987) have gone the furthest to identify and describe this mineral in a series of occurrences in Colorado, USA. Sericite is of course illite with another name. This mineral seems to have a pure mica XRD diagram (spacing less than 10.3 Å) and a rather narrow peak (FWHM less than 0.6 °2θ Cu Kα.), as observed from scans of published diagrams in Srodon et al. (1987) and Parry et al. (2002) which would identify it as a WCI mineral.

1.1.4.5
Bentonites

Bentonites are layers of volcanic ash which have been buried and experienced diagenesis (i.e. a gradual increase of temperature with time due to burial in a sedimentary basin). Initially the bentonites are almost pure smectites, with a low potassium content and a low overall charge. Burial diagenesis results in a mineral change through diffusion of potassium into the beds and a loss of silica from the bed. The conversion of smectite to illite is very clearly pronounced in these materials. In the later stages of transformation clays in bentonites show different relations of WCI and PCI peaks depending upon the thermal and the diffusion history of the bed. Figure 1.8 shows some different patterns of bentonite beds of different ages. The older samples probably have not experienced

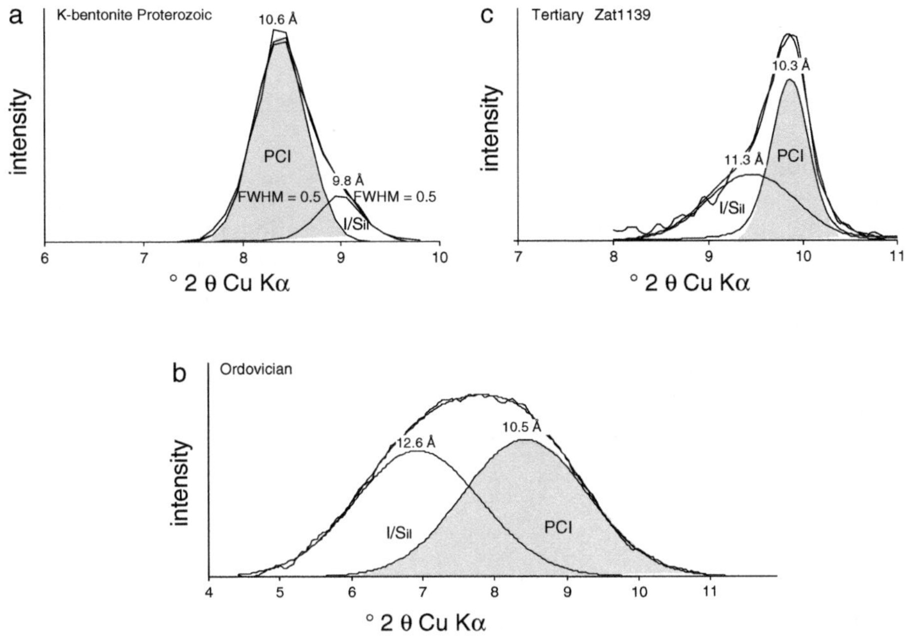

Fig. 1.8a–c. Decomposed background – subtracted XRD spectra for several bentonite samples. Proterozoic (data treated from Moe et al. 2000), Ordovician Swedish bentonite (data of author BV), Tertiary Panonian basin well sample (sample provided by V. Sucha)

high temperatures, i. e. those above normal geothermal gradients of 30 °C/km. However, the Tertiary sample was subjected to temperatures above 200 °C. In all cases the most intense peak is a PCI. This distinguishes bentonites from diagenetic, pelitic mineral assemblages where both PCI and WCI are present.

1.1.5 Summary of One-Dimensional Analysis of Natural Minerals by XRD

From the examples given above, it is clear that there are normally two populations of crystallites present in materials called illite. This result is in a way part of the original observations and definitions of illite. The presence of an asymmetric diffraction peak near 10 Å was always part of the initial observations on illites made using XRD techniques. The asymmetry is caused by the presence of PCI and WCI components in the diffraction spectra and at times I/S MLMs. In most cases, the PCI appears to be less intense than the WCI peak (estimation of the peak height) but the overall peak surface of PCI is generally greater than that of the WCI component. Thus visually one sees a 10 Å peak with a tailing to high *d* spacings. The relative proportions of each component need to be measured by peak surface area in order to have an idea of the relative abundance of the two components of the illite diffraction spectra.

It can be said that illite diffraction spectra are generally composed of two components, one nearer to a well-crystallized mica form than the other.

1.2 Definition Based on Chemical Composition

The illite mineral is generally considered to be a dioctahedral, potassium-deficient mica-like mineral. In this definition the potassium deficiency arises from two possibilities, that which affects the interlayer charge site in the tetrahedral or dioctahedral site.

1.2.1 Solid Solutions of Illite and Glauconite

1.2.1.1 General Statements

Illite is part of a series of mica-like minerals which occur at or near the surface of the Earth, in the realm of sedimentation and diagenesis. The only micas stable or forming in these environments are potassic, and dioctahedral. Other mica minerals such as biotites, paragonite and clintonite are not formed under these conditions. All of the potassic low-temperature mica-like minerals show a deficit of charge which results in a less than full or normal occupancy in the interlayer site by potassium ions. A common question in the past addressed the continuity of composition between these minerals, the possibility of a solid solution gap. Such a gap is of course a prerequisite for naming or distinguishing two minerals. Several groupings have been proposed over time, and the present authors participated in these propositions (Velde and Odin 1975; Courbe et al. 1981 for example). However, looking at the sum of older and more recent data, (Weaver and Pollard 1973; Thompson and Hower 1975; Berg-Madsen 1983; Li et al. 1997; Newman and Brown 1987, Alt et al. 1992; Clayton and Pearce 2000) it seems clear that the range of compositions of these minerals often overlaps.

Since chemistry must be accompanied by X-ray diffraction analysis to verify the homogeneity and purity of a given sample, most studies cannot be used with certainty. The use of XRD as a diagnostic has decreased directly as a function of the use of electron microprobe microanalysis. It is now very rare to find an X-ray diagram accompanying a chemical determination of a clay mineral. This is a great problem for the potassic mica-like minerals because they are often accompanied or preceded by interstratified mica/smectite phases. Hence it is extremely important to establish that the mineral identified as a non-expanding, mica like mineral is just that and not a mixture of mica and other minerals. Given that complete data sets are rare, one can only base arguments on the limits of compositions observed, i.e. the highest potassium content representing a pure mica, and so forth. The following discussion is based upon such information.

1.2.1.2
Illite-Glauconite Chemical Compositions/Overlap in Tetrahedral-Octahedral Substitutions

Illite is reputed to be the most aluminous of the low-temperature micaceous minerals. Its next most aluminous neighbor is glauconite. In order to compare the mineral groups, analyses have been selected on the basis of potassium content, indicating an approach to mica composition. Such criteria are necessary because it is rare to find a chemical analysis presented with XRD data which could be used distinguish if the sample is mono- or multi-mineral in nature. In selecting the data of Berg-Madsen (1983) and Thompson and Hower (1976) for glauconite and Weaver and Pollard (1973), Newman (1987) and Gabis (1963) for illite with more than 7% K_2O it is clearly shown that there is an overlap in the compositional range of Al and Fe content for low to high potassium content minerals called illite and glauconite.

The aluminous, potassic glauconites all originate from sedimentary rocks. Since glauconite is known to form on the shallow ocean bottom, the temperature of origin is clearly low, in fact lower than the terrestrial surface. From the Fig. 1.9 it appears that there is a definite overlap of glauconite and illite compositions, using the classical iron and aluminium content as distinguishing elements. However, the glauconite micas seem to be uniquely of the 1M polytype, which might shed light on the problem as we will understand later.

Then the overlap in composition demonstrates that minerals in glauconite occurrence, i.e. small pellets formed on the ocean bottom where the mineral change to form the micaceous phase is accomplished through very local diffusion processes. It can have the same mineral characteristics as those from sedimentary pelitic environments. If the chemistry and mineral structure is the same, or nearly so, then the only distinction between the two minerals is on

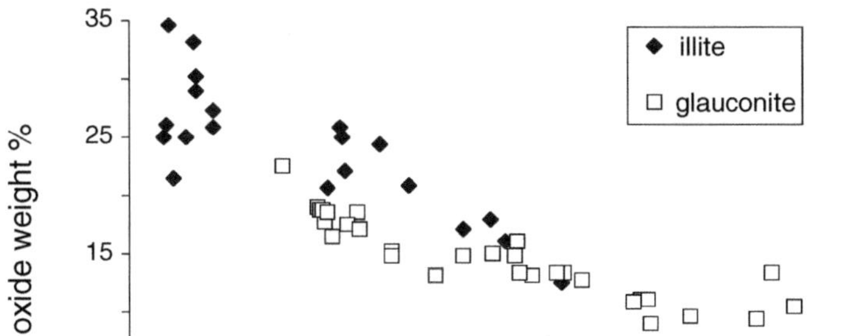

Fig. 1.9. Al_2O_3 vs Fe_2O_3 diagram for illites and glauconites. Data from Berg-Madsen (1983), Thompson and Hower (1975) Weaver and Pollard (1973), Gabis (1963) and Newman (1987). Squares are glauconites and diamonds are illites

the basis of geological occurrence. We will leave this discussion at this point, considering only those minerals which are formed in pelitic environments in a bedded occurrence. What one can say is that illite is not a green mineral localized in a pellet.

1.2.2 Charge-Lowering Substitutions

1.2.2.1 The Chemical Composition of Some Standard Illite Samples

It was shown in the early 1960s that illites contain less potassium than well-crystallized micas. Gaudette et al. (1966) gave the chemical composition of a series of samples considered as standard for illite. Except in samples containing expandable mixed layers (Fithian illite and Grundite, Marblehead and Rock Island), the Beaversbend illite is of the $2M_1$ polytype and exhibits intense 00*l* peaks. This is also the case for the K-bentonite described by Moe et al. (1996) and Kodama and Dean (1980). In spite of the fact that these standard illites have an ordered 3D crystal structure similar to that of micas, their chemical compositions are different: tetrahedral and interlayer charges are significantly lower (Table 1.1). Even if the unit formulae proposed by these authors are uncertain because of the presence of impurities, it is remarkable that the $2M_1$ spatial organization is compatible with such low interlayer charges. All three of these minerals are of PCI peak types.

Table 1.1. Layer charge characteristics of standard illite samples (from Gaudette et al. 1966; Moe et al. 1996; Kodama and Dean 1980)

Layer charges	Beavers Bend	Moe et al	Kodama
Tetrahedral	−0.66	−0.52	−0.62
Octahedral	0	−0.42	−0.06
Interlayer	+0.65	+0.89	+0.67

1.2.2.2 Illite End-Member Chemical Compositions in Mixed Layer I/S MLMs

Classically, illites are considered to be similar to phengites (micas), except that their layer charge is lower than that of a mica, i.e., 1 per $O_{10}(OH)_2$. The illite chemical composition can be fully described in a three-component system (Fig. 1.10): muscovite and celadonite "molecules" (micas) which explain the tetrahedral and octahedral substitutions respectively in the context of a mica substitutional scheme, and pyrophyllite, a non-charged 10 Å mineral, which explains the lowering of the layer charge observed in illites (Hower and Mowatt 1966). Using extrapolation to 0% expandable layers of I/S chemical trends, these authors concluded that illite has a layer charge of 0.75 per $O_{10}(OH)_2$. Illite seems to contain more water than micas do. Norrish and Pickering (1983) showed

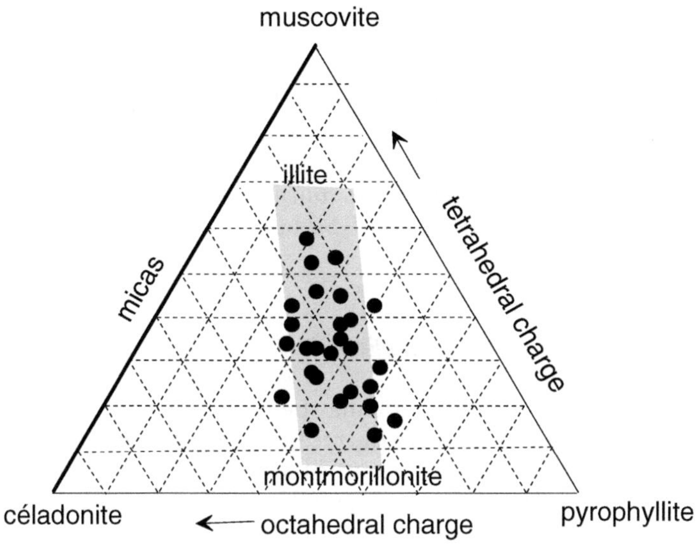

Fig. 1.10. The three-component chemiographic system muscovite – celadonite – pyrophyllite defined with data points from Hower and Mowatt (1966)

that H_2O contents are inversely related to K_2O in illites which apparently do not contain any smectite layers. Even if the presence of H_3O^+ cations replacing K^+ in the interlayers is still controversial, it remains a fact that the DTA curves of illites are different from those of micas.

Nevertheless, the range of the layer charge from 0.75 to 1 per $O_{10}(OH)_2$ is large. Why is this so? Two alternative solutions may be envisaged: a mixture illite+mica (biphase field) or a solid solution domain of illite (monophase field). The answer is not as simple as it sounds because, in natural environments, the "illite" phase is not homogeneous, neither for the habits and sizes of crystals nor for the chemical composition (pure illite or I/S). Using different techniques, several studies showed that the illite end-member in I/S series has a higher charge than that defined by Hower and Mowatt (1966): 0.9 per $O_{10}(OH)_2$ (Meunier and Velde 1989, Lanson and Champion 1991; Srodon et al. 1992). The chemical compositions for the illite layers in I/S MLMs obtained by extrapolation are quite similar (Table 1.2).

Table 1.2. Chemical composition of the illite end-member from I/S series

Reference	Si	^{IV}Al	^{VI}Al	^{VI}Fe	Mg	K
Meunier and Velde (1989)	3.30	0.70	1.78	0.05	0.17	0.87
Lanson and Champion (1991)	3.25	0.75	1.73	0.09	0.19	0.90
Srodon et al. (1992)	3.20	0.80	1.85	0.05	0.10	0.89

It is thus rather evident that the illite in mixed-layer minerals has a composition with an interlayer charge compensated by potassium of near 0.9. The substitutions causing this charge are found in both octahedral and tetrahedral sites of the 2:1 structure. Pure illites, or samples assumed to be pure illites, have variable charges originating in one site or both. The variations in composition of these samples is most likely due to impurities in the sample (essentially the presence of I/S phases). Therefore it is more prudent to take the extrapolated values of illite composition from the I/S MLMs as a model for illite composition.

1.2.2.3
Why is Illite not Simply a K-Deficient Mica?

K-Deficiency The end-member illite may be considered as a K-deficient mica. Thus it is a muscovite according to the recent IMA report on the nomenclature of micas (Rieder et al. 1998) where the mica solid solution domain was extended from 1 to 0.85 interlayer cations per $O_{10}(OH)_2$. This definition ignores the fact that muscovite is not stable with respect to illite, neither in nature nor in the laboratory (McDowell and Elders 1980; Yates and Rosenberg 1996; Giorgetti et al. 2000). The transformation of illite into a mica phase (muscovite or phengite) is observed in the prograde transition from diagenesis to low-grade metamorphism (Gharrabi et al. 1998; Merriman and Frey 1999). The differences between K-deficient micas and illite have been summarized by Rosenberg (2002). For our purpose, the fact that the maximum illite layer charge is about 9/10 of a mica implies that 1 hexagonal cavity in 10 in the arrangement of oxygens in the basal layer of silica tetrahedra is electrically neutral and remains vacant. Then a question arises: since illite must have a neutral hexagonal vacancy (pyrophyllite-type), how are these vacancies accommodated in a mica-type crystal structure? In other words, are they totally empty or do water molecules enter in the free interlayer space of the vacant hexagonal cavity of the surface oxygen array of the silica tetrahedra?

For this question it is important to carefully examine the thermodynamic status of water in the illite crystal structure: hydronium ions (H_3O^+) or neutral H_2O molecules. If the negative charge is usually fully compensated by $K^+$and Na^+ cations, in some cases an additional K, Na-deficiency is observed because the negative layer charge is partly compensated by NH_4^+ or H_3O^+ ions (Norrish 1952; Hower and Mowatt 1966; Hervig and Peacock 1989; Loucks 1991). Tobellite, a dioctahedral ammonium-hydronium mica (Higashi 1982), is a compositional end-member. Using statistical treatments of a set of high-quality mica and illite chemical analyses, Loucks (1991) showed that the hydronium ion is commonly found in the interlayer space of micas formed in acidic to neutral pH conditions below 450 °C. Most often, the calculated small excess of octahedral occupancy is an artifact which can be corrected if hydronium ions are taken into account in the unit formula calculation. These ions are strongly bonded to the 2/1 layer units; they are extracted from the crystal structure at high temperature conditions: 250–500 °C. The OH crystalline water sites in the octahedral layer dehydrate above 500 °C.

Water Content The negative charge of the 2/1 layer unit being totally compensated by alkali, ammonium or hydronium ions, the water molecules that remain in the pyrophyllite-like hexagonal cavities is 1 interlayer site per 10. These sites cannot be occupied by ionic species, only by neutral molecules because all of the charged sites have been satisfied by monovalent cations. The presence of H_2O molecules in these vacant interlayer sites has been deduced from chemical analyses by Loucks (1991) and confirmed using IR spectroscopy by the 3420 and 3260 cm^{-1} absorption bands (Besson and Drits 1997a). The water molecules are structurally connected to the adjacent 2/1 layer units by weak hydrogen bonds. Thus, they are extracted from the crystal structure at low temperature conditions in the 100–200 °C interval.

The key conclusion from this discussion is that at certain points in the illite structure, a neutral, pyrophyllite-like configuration exists. This is a type of solid solution between mica and pyrophyllite. The amount seems to be one tenth of the possible sites. Since the structure is not completely closed by potassium or other monovalent cations, some water molecules enter the illite crystallites on these sites. This solid solution is unstable at higher temperatures, and true micas form.

1.2.3 The Crystal Structure of Illite and Solid Solution

1.2.3.1 Fundamentals for the Illite Crystal Structure

It has been observed that most of the samples called illites of the 1M and $2M_1$ polytypes have a layer charge from 0.75 to 0.90 which refers to the composition of a half unit cell based upon $O_{10}(OH)_2$ (Fig. 1.11a). Because of the structure of the octahedral sheet (the layer "monoclinicity"), the facing tetrahedral sheets are shifted by $a/3$ in the positive or negative direction along 1 of the 3 symmetry axes (Fig. 1.11b). Thus, as shown by Smith and Yoder (1956), there are 6 possible shift vectors (Fig. 1.11c)

Because of the strong K^+–O^{2-} bonds, the facing tetrahedral layers are stacked without any shift in spite of the Si^{4+}–Si^{4+} repulsion (Fig. 1.12a). Thus, the respective orientation of facing 2:1 layers depends on the monoclinic direction of their octahedral sheets. In a stacking sequence of 2:1 layers, the orientation of the shift within each layer can be repeated from layer to layer by the three symmetry hexagonal axes in the positive or negative direction. These shifts in the monoclinic direction from 2:1 layer to 2:1 layer stacking which form polytypes. Smith and Yoder (1956) showed theoretically that there are only 6 possible polytypes (1M, $2M_1$, $2M_2$, 2Or, 3T, 6H) when layers are stacked one on the other to form a larger crystal. Any polytype can be symbolized by a sequence of the corresponding shift vectors (Fig. 1.12a).

Among the 6 possible polytypes, the 1M and $2M_1$ stacking (either one 2:1 layer unit or a coupled two layer unit) are the most frequently encountered in illite species:

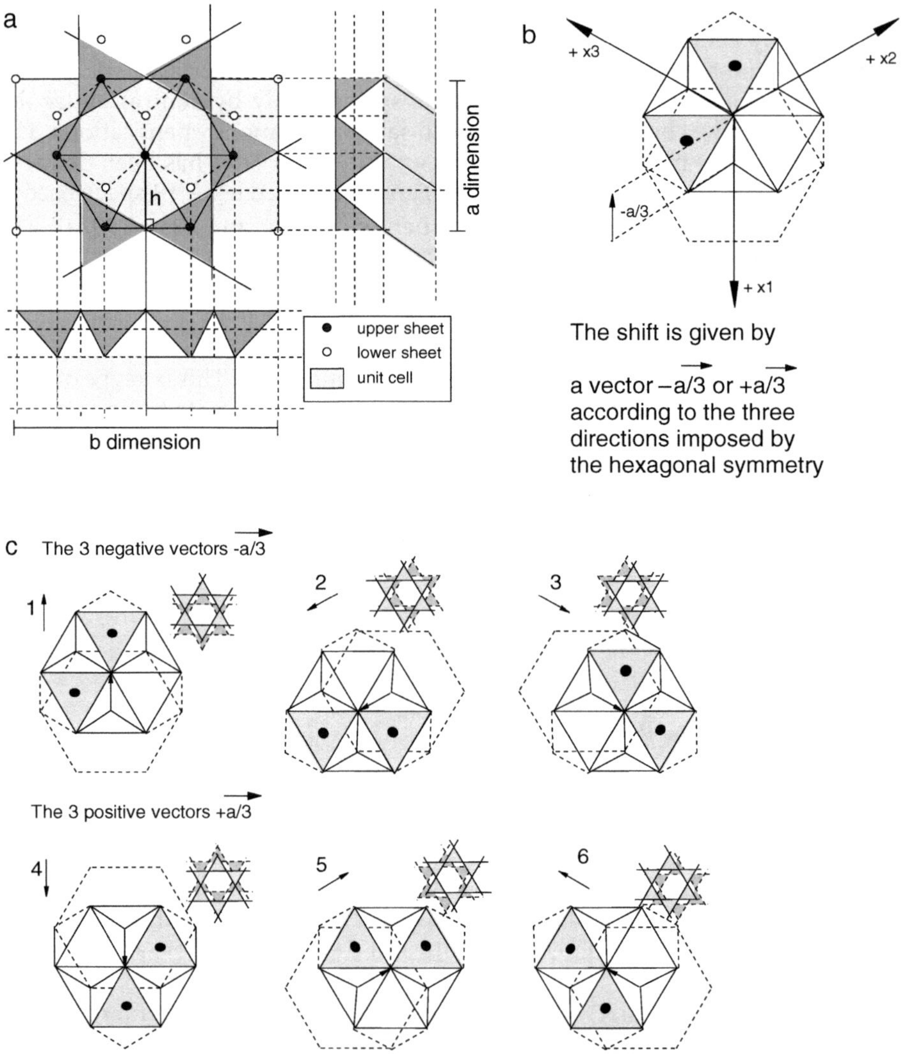

Fig. 1.11a–c. The crystal structure of illite layers. **a** The unit cell with *a* and *b* dimensions. **b** The shift between the two tetrahedral sheets in a given 2:1 layer is *a*/3. **c** The 6 possible shift vectors (from Smith and Yoder 1956)

- the 1M which is characterized by vectors aligned along a single symmetry axis (Fig. 1.12b);
- the $2M_1$ in which the vectors form a zigzag line because of 120° rotations (Fig. 1.12c).

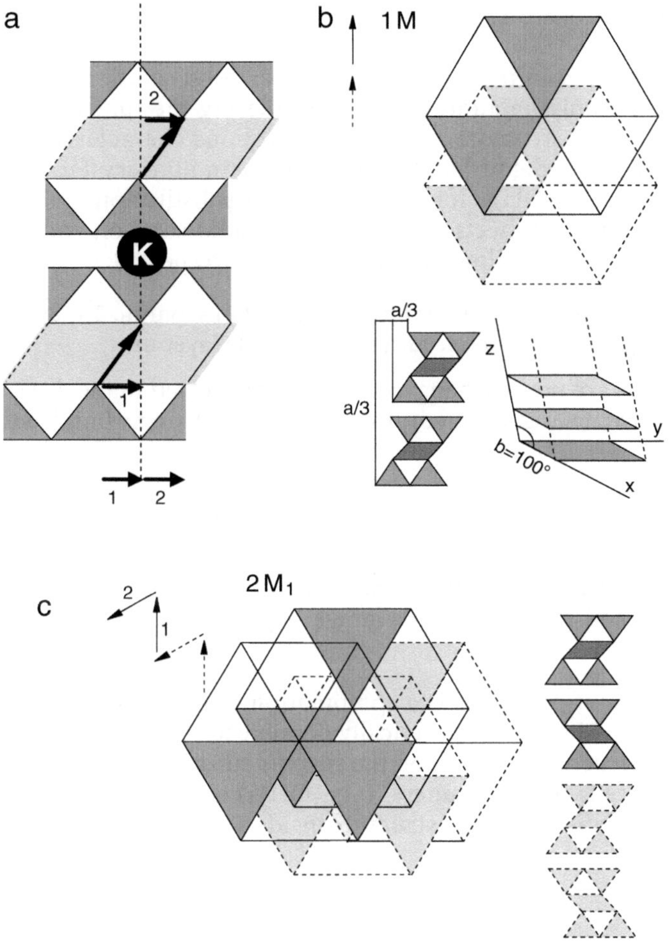

Fig. 1.12a–c. The illite polytypes. **a** Simplified cross section of a two- layer stacking showing how it can be represented by two vectors. **b** The 1M polytype. **c** The $2M_1$ polytype

A third type is frequently observed in illite samples. It derives from the 1M polytypes which exhibit a variable degree of disorder in the stacking sequence: $1M_d$. Disorder derives from the presence of stacking defects in the 1M sequence due to non-rational rotations between two adjacent layers. Illite with lower charges are more likely to have an 1M or $1M_d$ polytype (Eberl et al. 1987). However, this is not a rule since the two polytypes have been identified for illites having a 0.9 layer charge in active hydrothermal systems (Lonker and Fitz Gerald 1990) or in Precambrian diagenetic basins (Laverret 2002). Summarizing the present state of our knowledge, it seems that the layer charges of 1M or $1M_d$ illite vary within the 0.75 to 0.90 per O_{10} $(OH)_2$ range while those of $2M_1$ illites are very near to 0.9. Given that bulk chemical compositions are not conclusive evidence, are $2M_1$ illites really different from dioctahedral aluminous micas?

The problem of polytypes in minerals of nonregular solid solution substitution such as illite can only be addressed by considering the fundamental aspects of the mica crystal structure and the substitutions characteristic of illite. According to the above definition, the layer charge in illite crystals may vary from 0.75 to 0.9 per O_{10} $(OH)_2$. It results from ionic substitutions in both the tetrahedral and octahedral sheets. If illite is a true solid solution and not an interstratified structure, the following rules have to be respected:

- octahedral and tetrahedral charges are not segregated in separate 2:1 layers. If there is any segregation, it must occur inside a single layer unit;
- the interlayer cations are non-exchangeable. They are located in the ditrigonal cavities of the two facing tetrahedral sheets and are strongly bonded to 6 basal oxygens;
- the octahedral and tetrahedral charge distribution inside a single 2:1 layer is similar to that of dioctahedral micas. It can be derived from that of phengites: each hexagonal cavity is negatively charged because of heterovalent ionic substitutions either in the tetrahedra or in octahedra. No position is electrically neutral and all present an average negative charge of −1 per $O_{10}(OH)_2$.

The crystal composition of phengites is a combination of the muscovite and celadonite substitutions. Thus, it is theoretically possible to establish the basic rules for their crystal structure based upon the specific substitutions. Here we will use the "half unit cell" of a 2M mineral ($O_{20}(OH)_4$) or the 1M "half unit cell" ($O_{10}(OH)_2$) to describe the crystal structure inside a single layer whatever the polytype.

1.2.3.2
End-Member Minerals

Muscovite The negative charge originates exclusively from Al for Si substitutions in the tetrahedral sheets (no charge from the octahedral sheet). Considering a half unit cell (Si_3 Al_3 O_{10} $(OH)_2$ K), the average negative charge for each hexagonal cavity is equal to 1. It is located on the O^{2-} anions forming the 4 apices of the Al-bearing tetrahedra (valency is not totally compensated). The induced positive charge deficit is shared by the three neighbouring hexagonal cavities (Fig. 1.13a). Each of them has a negative charge of −1/2. As the facing 2:1 layer is exactly equivalent, the resulting dodecagonal cavity has a total charge of −1 (Fig. 1.13b). Taking into account the Lowenstein rule which states that two Al-tetrahedra cannot be direct neighbours, the theoretical distribution of Al ions in the tetrahedral sheet can be regular, ordered or segregated. These different distributions are shown in Fig. 1.13c where the groups of three neighbouring hexagonal cavities surrounding each Al-bearing tetrahedron are represented by grey triangles.

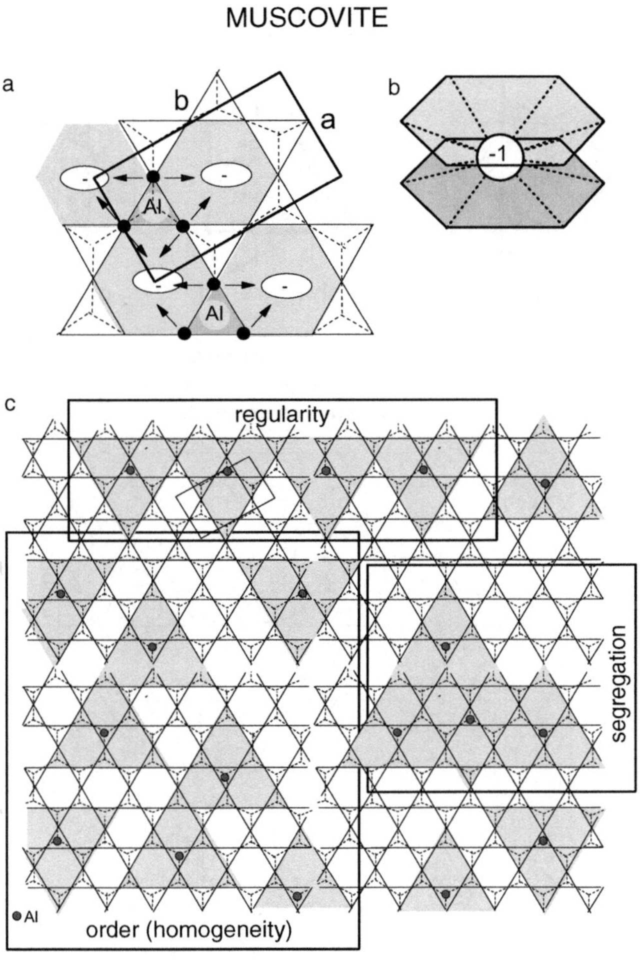

Fig. 1.13a–c. Structure of the tetrahedral sheet in a muscovite. **a** Charge distribution around the Al for Si substituted tetrahedron. **b** A dodecagonal cavity is created when two 2:1 layers are facing each other. The negative charge of the dodecagonal cavity is −1. **c** Theoretical representation of different types of distribution of Al ions in the tetrahedral sheet

Celadonite The negative charge originates exclusively from the ionic substitutions of R^{2+} for R^{3+} in the octahedral sheet (no charge from the tetrahedral sheets). Considering a half unit cell (Si_4 R^{2+} R^{3+} O_{10} $(OH)_2$ K), the average negative charge for each hexagonal cavity is equal to 1. It is located on the apical O^{2-} anions of the tetrahedral sheet and the OH forming the 6 apices of the R^{2+}-bearing octahedra (Fig. 1.14a). The vacant site is in the trans-position. Different types of R^{2+} cation distribution are possible in the three different octahedral sites of the octahedral layer which are just below the hexagonal

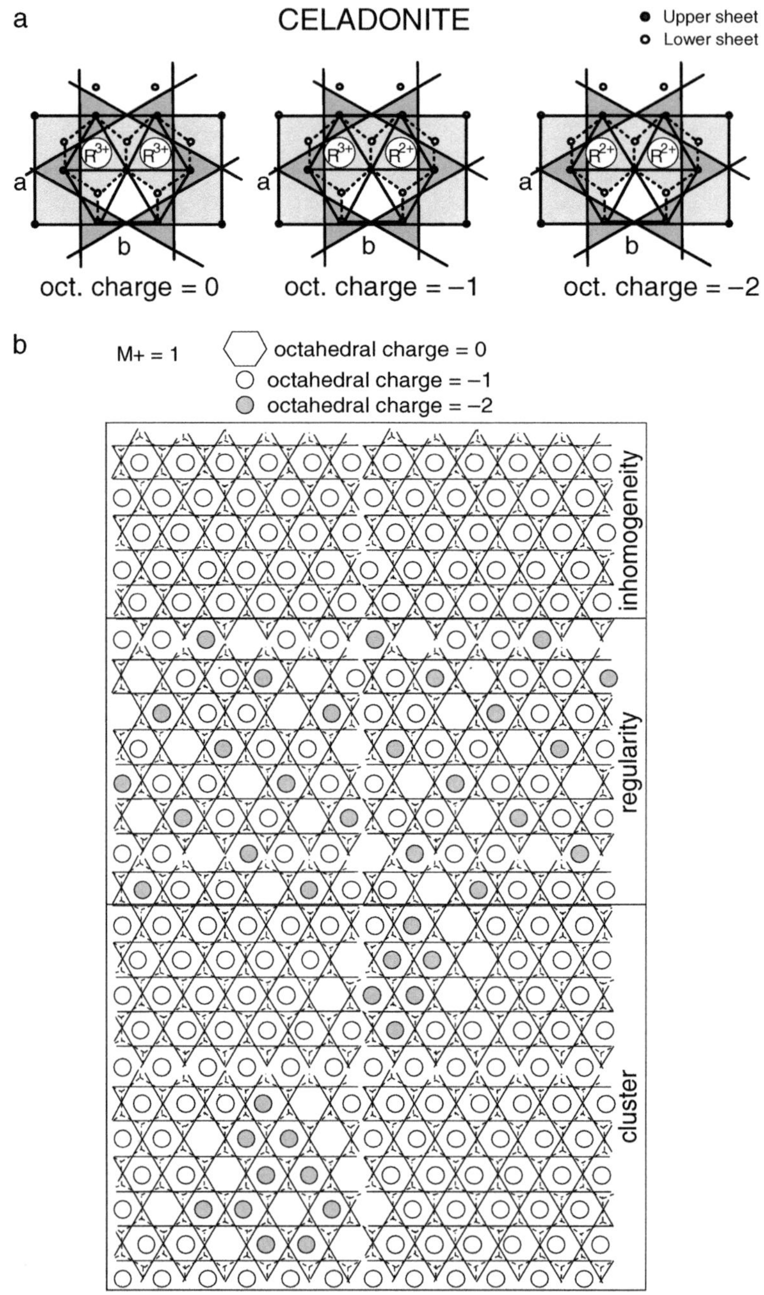

Fig. 1.14a,b. Theoretical possible crystal structure for celadonite (no tetrahedral charge-position of the octahedral vacant site: *trans*). **a** The three possible octahedral charges **b** Three possible distributions of octahedral charges respecting a total layer charge equal to 1 per $O_{10}(OH)_2$

cavity of the basal tetrahedral oxygen of the 2:1 layer:

- R^{3+}–R^{3+}: no octahedral charge;
- R^{3+}–R^{2+}: octahedral charge equal to −1;
- R^{2+}–R^{2+}: octahedral charge equal to −2.

These different types of cation distribution are observed in natural samples which exhibit R^{2+}–OH–R^{2+} and R^{3+}–OH–R^{3+} absorption bands (Besson et al. 1987) indicating that the related hexagonal vacancies have a charge of −2 or 0, respectively.

Then, several theoretical distribution models can be proposed respecting 1) the position of cations in the *cis*1 or *cis*2 octahedral sites and 2) the total layer charge (average on hundred site) is equal to −1 (Fig. 1.14b):

- homogeneous distribution of octahedral charges: each site has a negative charge equal to 1 originating from a R^{3+}–R^{2+} cation pair;
- regular distribution of three types of sites with a charge of 0, −1 and −2, respectively;
- partly clustered: the sites formed of $2R^{2+}$–R^{2+} + trans vacant octahedra may be in directly neighbouring positions forming clusters. Consequently, some pyrophyllite-type zones are formed.

Phengite The negative charge in phengite layers originates both from the tetrahedral and the octahedral sheets. Theoretically, all the hexagonal cavities are equally negatively charged and occupied by a K^+ cation. This means that tetrahedral negative charges are never superimposed to octahedral ones. Thus, several charge distribution models are theoretically possible (Fig. 1.15):

- random distribution of tetrahedral and octahedral charges;
- ordered (regular) distribution of tetrahedral and octahedral charges;
- clustered distribution forming "muscovite" or "celadonite"-type zones.

Pyrophyllite As the 2/1 layers are electrostatically neutral, there is no interlayer cation present. The different layers are linked by van-der-Waals bonding, supplemented by weak ionic attraction between adjacent layers. In order to minimize the repulsion forces of Si^{4+} cations in the facing tetrahedral sheets, adjacent layers are shifted by about 0.3a along one of the three pseudohexagonal symmetry axes (Bailey 1980). As illite has a layer charge less than −1 per $O_{10}(OH)_2$, this implies that some hexagonal cavities are neutral and do not fix any interlayer cation. A pyrophyllite-type structure will be assigned to a pair of neutral hexagonal cavities facing in two adjacent layers. Pyrophyllite is commonly of a 2M polytype.

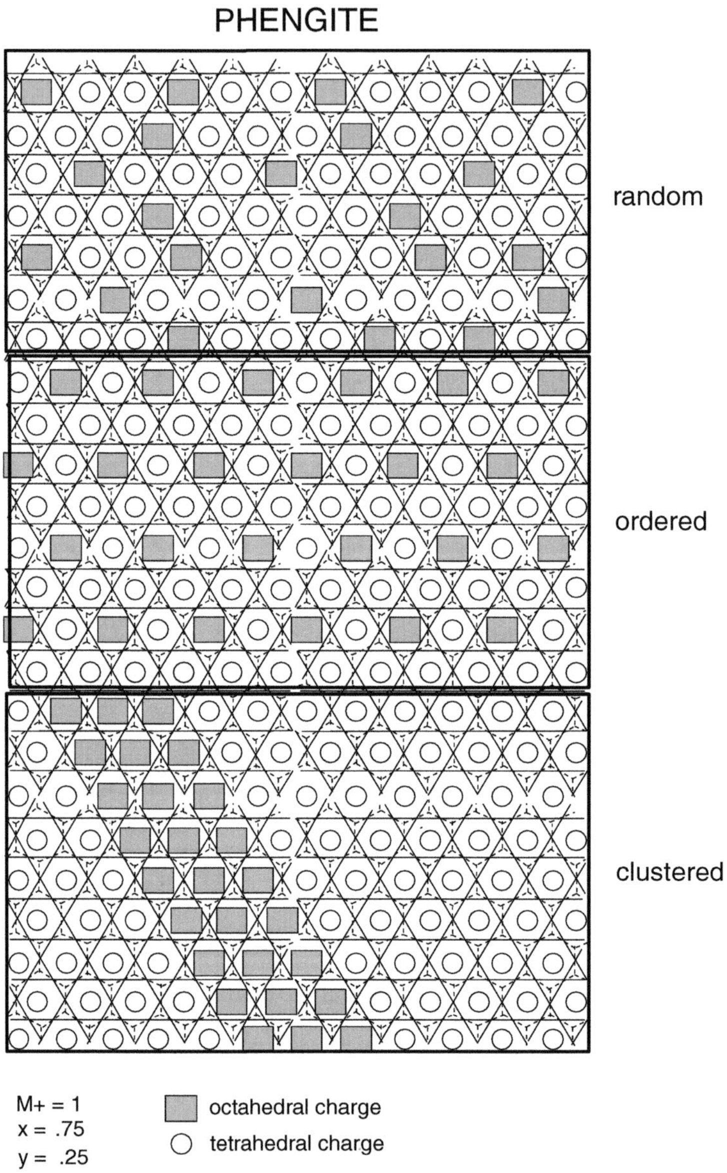

Fig. 1.15. Some theoretically possible crystal structures for phengite micas. The average layer charge is equal to 1 per $O_{10}(OH)_2$

1.2.4
The Theoretical Crystal Structure of Illite

1.2.4.1
Unit Cell Heterogeneities Inside the 2:1 Layer

Compared to micas, the amount of K^+ in illite is lower, indicating that some hexagonal cavities remain free of compensating cations. Coming back to the pioneer work of Hower and Mowatt (1966), the layer charge at the surface of a 2:1 unit is dependent on the distribution of ditrigonal cavities whose charge is due to tetrahedral or octahedral substitutions (muscovite or celadonite types respectively). Some of them (at least 10%, considering the illite end-member composition: 0.9 per $O_{10}(OH)_2$) have no charge (pyrophyllite type). The corresponding half cell unit in both facing layers must be electrically neutral: Si_4 R_2^{3+} O_{10} $(OH)_2$ (pyrophyllite-like unit). Thus, considering the theoretical crystal structures of phengitic micas (Fig. 1.15), one can deduce that of illite, considering that pyrophyllite sites must be added in a proportion of 1 upon 10. The spatial distribution of the pyrophyllite-like unit cell inside a single 2/1 layer may be randomly ordered, ordered or clustered (Fig. 1.16).

- random: pyrophyllite sites are dispersed and isolated in the 2:1 layer;
- ordered: the muscovite-celadonite-pyrophyllite-type site sequence is repeated more or less regularly in the 2:1 layer;
- clustered: pyrophyllite sites form zones distinct from that of mica-type composition (muscovite or celadonite). However, the pyrophyllite sites cannot be segregated in distinct 2:1 layers since illite d_{001} spacing is similar to that of mica and not to that of a mica-pyrophyllite mixed layer.

These models introduce the problem of the layer charge heterogeneities and the cation exchange capacity of illites.

1.2.4.2
Layer Charge and Cation Exchange Capacity

Layer Charge Heterogeneities Why are $1M_d$, 1M or $2M_1$ polytypes encountered in illites instead of the common smectite-like turbostratic stackings? Let us consider how two facing 2/1 layers are linked together. The distribution of the different ditrigonal cavities at the facing tetrahedral sheets locally determine the negative charge:

1. musco-musco: interlayer cation fixed by a full charge
2. celad-celad: interlayer cation fixed by a full charge
3. musco-celad: interlayer cation fixed by a full charge
4. musco-pyr: interlayer cation fixed by 1/2 charge
5. celad-pyr: interlayer cation fixed by 1/2 charge
6. pyr-pyr: no charge

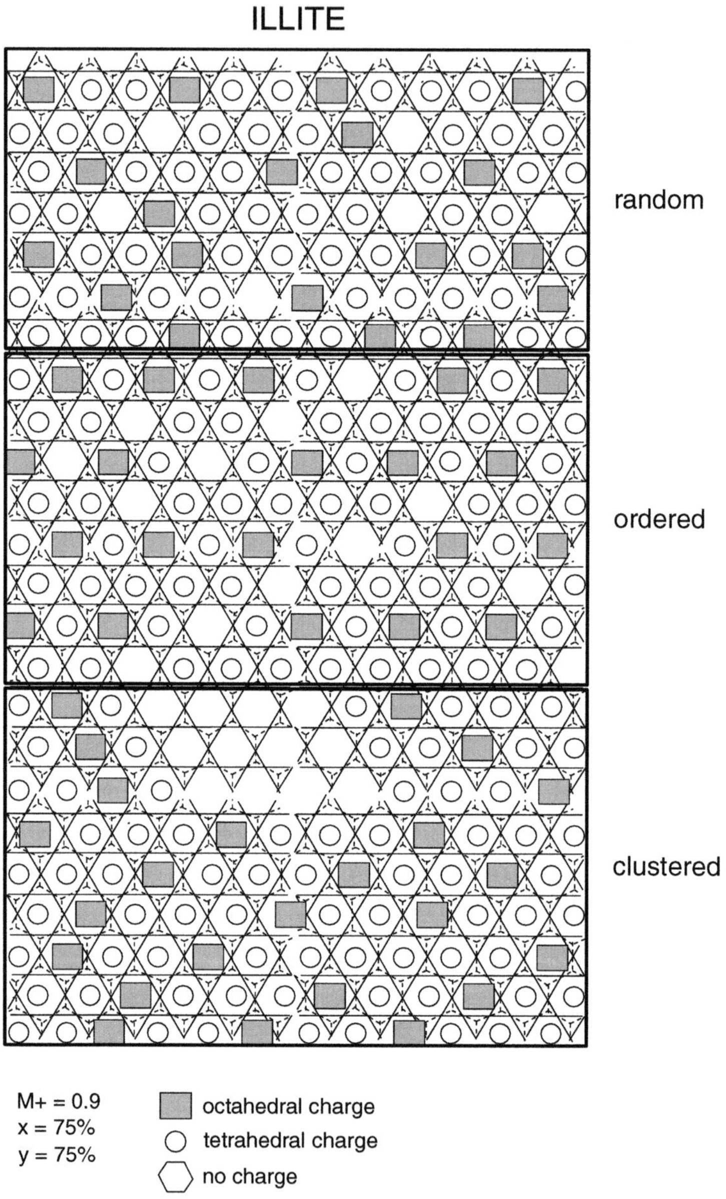

Fig. 1.16. Some theoretically possible crystal structures for illite. The average layer charge is equal to 0.9 per $O_{10}(OH)_2$. The distribution of the pyrophyllite sites may be randomly ordered, ordered or clustered

Theoretically, all the interlayer sites are of types 1, 2 or 3 in micas (phengite). Each interlayer cation (mostly K^+ and Na^+) is equally bonded to the surrounding oxygen anions which forms the facing hexagonal cavities. Thus, the lower interlayer charge of illites is explained by the presence of sites having only a half charge (types 4 and 5) or no charge (type 6) in different proportions. One can deduce that cations occupying the half-charged sites may be exchangeable and not irreversibly fixed. Indeed, the cation exchange capacity (CEC) of illite is higher than that of micas: 20 to 40 meq/100 g instead of an expected 5–10 meq. Most of exchangeable sites are located on the crystal sides. This is partially true for illite, most likely due to mineral impurities such as I/S phases. Grim (1953) showed that for kaolinite and illite, the CEC increases with decreasing particle size (Fig. 1.17a). Thus, the bulk CEC values are not sufficiently precise since they do not give any information on the energy heterogeneities of the exchange sites. On the contrary, the selective sorption may be used as a way to study these heterogeneities (Shawney 1972).

The Selective Sorption of Low Hydration Energy Cations A small fraction of the CEC of illites is known to be selective for cations having a low hydration energy such as Cs, Rb or K. The sorption of cesium ions has been intensively studied because of analogies with radioactive elements (Kim et al. 1996; Poinssot et al. 1999; Kulik et al. 2000).Three different sites have been identified: (Maes et al. 1985; Cornell 1993 and references therein):

- type a: high-selectivity sites which are supposed to be located inside the interlayer ("interlayer-edge") on stacking defects (0.2 to 0.55 percent of the CEC),
- type b: intermediate selectivity sites (2.5 to 3.3 percent of the CEC),
- type c: low-selectivity sites which are located on the outer basal surfaces (96 to 97 percent of the CEC).

The type *a* sites were considered to result from the opening of the mica interlayer zones due to partial removal of K^+ ions during weathering (so-called "frayed" sites). These sites were supposed to be located at the edges of the particle because the cesium ion is too large to penetrate inside the interlayers. Even if these interpretations are questionable, the presence of interlayer sites having different exchange energy have been undoubtedly confirmed. Therefore, another alternative for the interpretation of the crystallographic nature of the different sites may be proposed if the pyrophyllite component is taken into account. Indeed, cations may be sorbed in the half-charged sites (mus-pyr and cel-pyr) if these sites are made accessible. This implies that they are connected to the faces of the particles (Fig. 1.17b,c).

1.2.4.3
Pyrophyllite Site Segregation: Facts and Assumptions

Besson and Drits (1997a–b) showed unambiguously that pyrophyllite-like local structural environments exist in illite crystals. How are these sites distributed

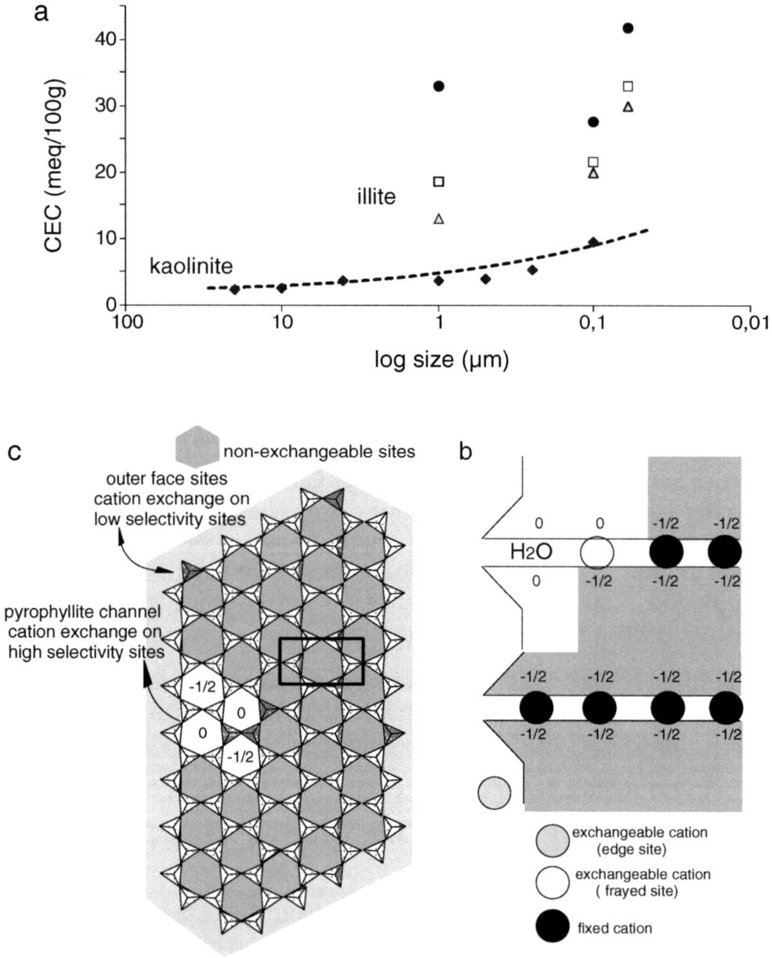

Fig. 1.17a–c. The cation exchange capacity of illites. **a** Relation of CEC to particle size for kaolinite and illite (data from Grim 1953) **b** Hypothetical representation of the highly cesium-selective exchangeable sites in pyrophyllite "channels" **c** Channel in the a-b plane formed of neighbouring pyr-pyr and mica-pyr hexagonal cavity pairs in facing layers

in the layer structure? A random distribution of the mica and pyrophyllite-type sites would lead to local charge imbalance and increase the entropy of the crystal structure. Thus, ordering would increase the stability of the illite by creating pyrophyllite clusters inside a given layer. In such a case, the strain created by differences in cell parameters would be reduced at the border of the clusters. If pyrophyllite clusters in two adjacent layers face each other, they would try to respect the pyrophyllite structure in which the tetrahedral rings are displaced along one of the pseudohexagonal symmetry axis (a/3) because of the Si-Si repulsion (Bailey 1980). On the contrary, the shift of adjacent layers is not possible in mica structures because of the superposition of ditrigonal rings around interlayer cations. Thus, an elastic energy would be created at the contact of mica-like and pyrophyllite-like regions in the crystal structure. This would increase its instability with respect to micas. However, as illite behaves as a true phase in the thermodynamic sense, the excess energy is most likely lowered by a specific distribution of pyrophyllite sites in its crystal structure. We will see further (see Sect. 1.3.2) why pyrophyllite sites are probably segregated for thermodynamic reasons.

1.2.4.4 Location of the Octahedral Vacancy in the 1M and $2M_1$ Polytype

The illite crystal structure being dioctahedral, there are 3 possible locations for the vacant octahedral sites: 2 *cis* and 1 *trans* positions. The presence of a vacant site induces deformations (angles and lengths of ion bonds) of the octahedral sheet. Particularly, it modifies the distance between the anions forming the apices of octahedra (Fig. 1.18a). However, these deformations conserve the symmetry planes of the vacant sites. Now considering the unit cell of the 2:1 layer (tetrahedral+octahedral sheets), the total symmetry depends on the position of the octahedral vacant site: a symmetry plane exists in the *trans* structure and is lacking in the *cis* one (Fig. 1.18b,c).

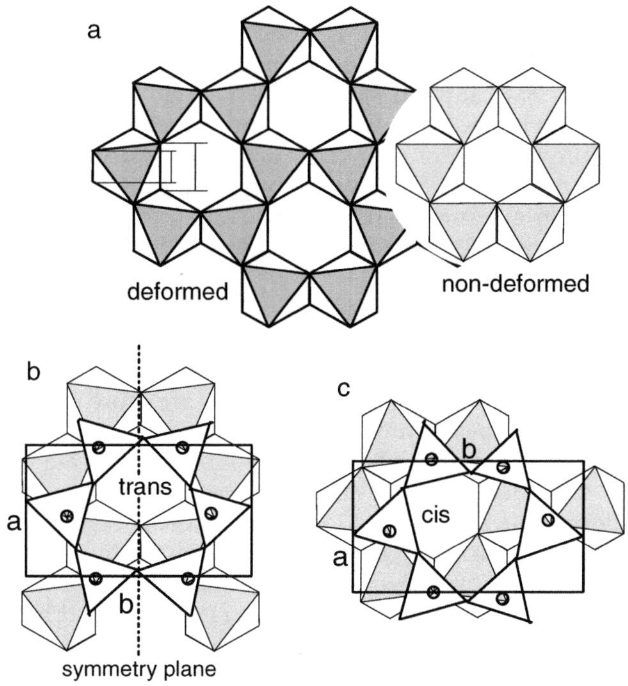

Fig. 1.18a–c. Deformation of octahedra in the vicinity of vacant sites. **a** Structure of deformed and non-deformed dioctahedral layers **b** The vacant site in the *trans* position induces the presence of a symmetry plane **c** The vacant site in the *cis* position: no symmetry plane

Table 1.3. X-ray diffraction lines characteristic of illite $1M_t$ (octahedral *trans* sites vacant), $1M_c$ (octahedral *cis* sites vacant) and $2M_1$ polytypes (from Drits et al. 1993)

$1M_t$		$1M_c$		$2M_1$		Properties
Pos. Å	*hkl*	Pos. Å	*hkl*	Pos. Å	*hkl*	
3.655	$\bar{1}12$	3.591	$\bar{1}12$			Peak shift is characteristic of $1M_t$–$1M_c$ illite polytypes
3.073	112	3.126	112			Peak shift is characteristic of $1M_t$–$1M_c$ illite polytypes
		3.885	111	3.889	$11\bar{3}$	Line common to $2M_1$ and $1M_c$ polytypes
		2.875	$\bar{1}13$	2.870	115	Line common to $2M_1$ and $1M_c$ polytypes
				3.735	023	Characteristic line for $2M_1$ polytype
				3.500	$\bar{1}14$	
				3.208	114	
				2.999	025	

The octahedron deformation in dioctahedral 2:1 layer structures has noticeable consequences on the polytype symmetry degree. Indeed, there is an additional symmetry plane in the $1M_t$ (*trans*) polytype compared to the $1M_c$ (*cis*) one (Fig. 1.19a,b). Besides, it is easy to see that a given family of atomic planes may have different d-spacing values and intensities (atomic densities) in the $1M_c$ and $1M_t$ polytypes. The atomic planes having a relatively high density in the 1M polytype are not oriented in the same directions according to the position of the vacant site. The position differences gives us a tool to identify the $1M_c$ and $1M_t$ polytypes and the ratio of their intensities can be used to measure their respective quantities (Drits et al. 1993; Reynolds and Thomson 1993). The identification method is based on the relative intensities of the $\bar{1}12 - 112$ and $111 - \bar{1}13$ peaks for $1M_t$ and $1M_c$ respectively (Table 1.3). Because of the 120° rotation of alternating layers in the $2M_1$ polytype, the position of the octahedral vacancy does not have an important contrasting effect for the relative atomic density of *hkl* recticular planes (Fig. 1.18c).

However, most of natural samples are more complicated than $1M_c + 1M_t + 2M_1$ polytype mixtures. Indeed, the presence of the partly disordered 1M polytype ($1M_d$) makes quantification more complicated. The disordered structure is difficult to identify and hence to quantify because it only has weak, non-polytype-specific reflections. Among the polytype quantification methods that have been proposed in literature, one of the most efficient for users is based on the calculation of 3-D X-ray diffraction patterns using WILDFIRE (Reynolds 1993, 1994). Grathoff and Moore (1996) showed how to quantify the relative proportions of the 1M, $1M_d$ and $2M_1$ polytypes in illite samples by measuring the intensities of some *hkl* diffraction bands.

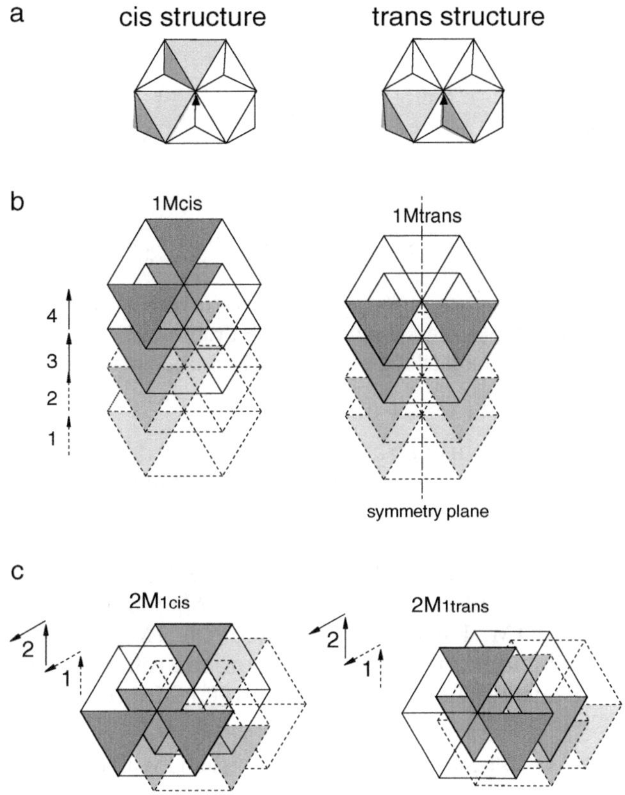

Fig. 1.19a–c. Simplified 3D atomic organisation in illites according to the location of the vacant site. **a** The position of the shift vector in *cis* and *trans* structures of the octahedral sheet **b** 1M polytypes for *cis* and *trans* structures **c** $2M_1$ polytypes for *cis* and *trans* structures

1.2.4.5
Difference Between $1M_d$ and Turbostratic Structures

Among the phyllosilicates, the turbostratic stacking is typical of the smectite group. A simple way to picture it was given by Moore and Reynolds (1980): "it is a highly disordered stacking arrangement resembling a pile of playing cards lying flat on each other but with little or no alignment of the edges" (Fig. 1.20a). Smectite crystallites may be considered as bi-dimensional lattices with no or little periodicity in the third dimension. Thus, the *hk* lines in the reciprocal space form concentric cylinders centred on the 00 line. This is shown by the electron diffraction diagrams in which portions of circles replace the hexagonal network of dots. The corresponding X-ray diffraction patterns are un-modulated bands (Fig. 1.20b). Reynolds (1992) defined a turbostratism index (Fig. 1.20c). He showed that this index regularly increases with the smectite content in I/S MLM. In 1M illite crystals, the degree of ordering increases with K^+ amounts. The higher the degree of order observed in the well crystallized illite, the greater the size of the coherent diffracting domain was. As a result, the transition to the $2M_1$ polytype is observed only for high charge illites: 0.87 to 0.9 per $O_{10}(OH)_2$ (Srodon and Eberl 1984; Meunier and Velde 1989; Lonker and Fitz Gerald 1990; Yates and Rosenberg 1997, 1998).

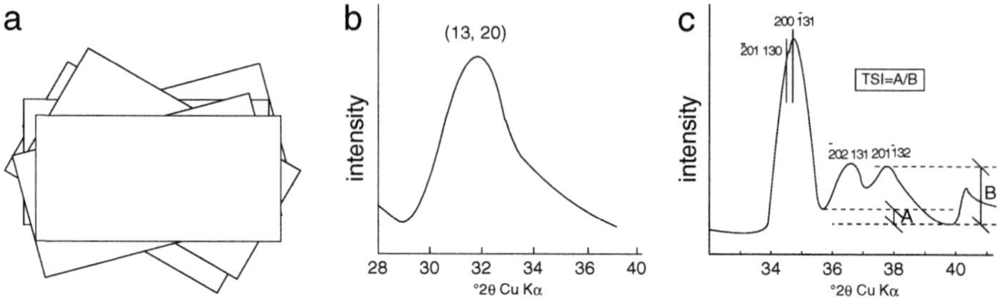

Fig. 1.20a–c. X-ray diffraction properties of turbostratic stackings. **a** The pile of playing card model (Moore and Reynolds 1980) **b** Profile of the (13, 20) *hk* diffraction band indicating disorder in the stacking sequences **c** Definition of the turbostratic index (Reynolds 1992)

Locally, turbostratism interrupts ordered stacking. The defects are frequently observed in phyllosilicates formed in low-temperature conditions. This is the case of corrensite for instance (Beaufort et al. 1997). The presence of such defects scatters the diffracted intensities. When numerous enough, they induce the disappearance of some *hkl* diffraction bands. This is the case for illite for which the diffraction peaks of the $1M_d$ polytype are much less numerous than that of 1M. Thus, the presence of the $1M_d$ polytype or turbostratic stacking may be identified by some diagnostic reflections or *hk* bands respectively as indicated in Table 1.4 (Brindley 1980; Moore and Reynolds 1989).

Table 1.4. Diagnostic reflections for $1M_d$ and 1M illite polytypes (from Brindley 1980; Moore and Reynolds 1989)

Turbostratic (montmorillonite)			$1M_d$ and 1M		
d (Å)	*I* (relative)	*hk* lines	*d* (Å)	*I* (relative)	*hkl*
4.61	10	02, 11	4.35	15	$11\bar{1}$
			4.12	10	021
			3.66	50	$11\bar{2}$
			3.07	50	112
			2.93	10	$11\bar{3}$
2.56	8	13, 20	2.69	20	023
			2.450	11	131
			2.405	4	$13\bar{2}$
2.22	3	04, 22	2.156	20	$13\bar{3}$

1.2.4.6 Fundamental Particles

The concept of fundamental particles (Nadeau et al. 1984) has been discussed in numerous papers. Fundamental particles are individual or free particles whose interfaces are capable of absorbing water and organic molecules and, as a consequence, behave as a smectite interlayer when two such particles are stacked one facing the other. If the particle is 10 Å or 20 Å thick, it is a fundamental particle of smectite or illite respectively (Nadeau et al. 1984a). Fundamental particles can be considered as subunits composing MacEwan crystallite (Eberl and Srodon 1988; Reynolds 1992). The thickness of fundamental particles is equivalent to the coherent X-ray scattering domain size (CSDS). Smectite interlayers correspond to defects in the stacking sequence (turbostratism) and an illite crystal may be composed of several fundamental particles which ordered stacking is probably of the 1M polytype. In other words, the size of the fundamental particle is equivalent to the size of the crystal in 1M particles. If lower, the polytype is $1M_d$.

However, the fundamental particle concept as it was defined by Nadeau et al. (1984) may lead to some confusion, particularly on the following points:

- the composition of the outer interfaces of the 2:1 unit is smectitic. This means that the 2:1 unit is asymmetric in charge, one tetrahedral layer having a high charge and the other a high (illite) charge. Even if such polarity is now well accepted (Cuadros and Linares 1995), it is difficult to understand how illite crystals can grow on the smectite outer interfaces of the fundamental particles.
- the size of the illite fundamental particles is not related to the polytype of the crystallites. For the $2M_1$polytype the size must be a multiple of layer pairs.
- the chemical composition of any illite particle must integrate its outer smectitic interface. This assumption has two consequences: it lowers the layer charge and imposes the pyrophyllite component to be located on these interfaces. This is not coherent with the chemical composition of individual illite particles nor with the location of the exchange sites which have a high selectivity for cesium.

Thus, for these reasons, we do not use the "fundamental particle concept" concerning illite. However, as it is another way to measure the thickness of I/S MLM or illite particles, it will be mentioned further in Sects. 2.2.4.2 and 2.3.3.2 for diagenetic and hydrothermal occurrences of illite.

1.2.4.7
Summary: Consequences of Polytype and Compositional Substitutions in Illite

Since the pioneer work of Hower and Mowatt (1966), we know that illite has a minimum layer charge of −0.75 per $O_{10}(OH)_2$ (no detectable expandable layers in XRD patterns) and that the composition of illite is can be considered as a mixture of muscovite, celadonite and pyrophyllite components. In their study, the authors did not consider the polytype of illites. Thus, the question is now: is the composition related to polytypes?

Using published illite compositions, Meunier and Velde (1989) showed that two composition domains can be distinguished in a M^+–4Si-R^{2+} plot (Fig. 1.21):

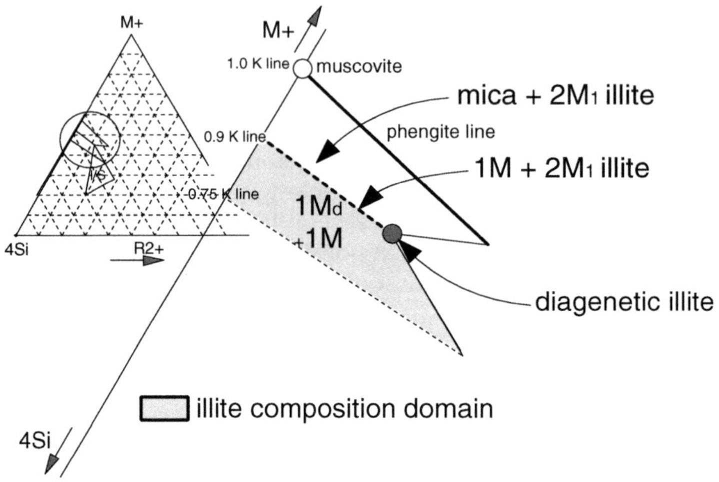

Fig. 1.21. Relations between compositional substitutions and polytypes of illite

- domain between line 0.75 K and line 0.9 K. The compositions correspond to 1M and 1Md illites;
- domain between line 0.9 K and phengite line. The compositions are given as that of $2M_1$ "illite" or sericite.

However, we must keep in mind that $2M_1$ well-crystallized illite and dioctahedral micas are not distinguishable using classical XRD patterns. Thus, this domain could correspond to $2M_1$ illite + mica mixtures. This is coherent with the assumption of an illite phase whose end-member composition is 0.9 K.

The significance of the two boundary lines at 0.75 K and 0.9 K is that there is a continuous solid solution between a "muscovite-type" end member in which the layer charge is essentially due to Al for Si substitutions in the tetrahedral sheets and a "celadonitic-type" in which the layer charge is shared between the tetrahedral and octahedral sheets. The "muscovite-type" is encountered in geothermal systems while the "celadonitic-type" is typical of diagenetic series.

1.3 Thermodynamic Stability of Illite

1.3.1 The Gibbs Free Energy of Formation of the Illite Phase

1.3.1.1 Mica, Illite and Smectite Phase Relationships Deduced from Dissolution Experiments

One method of determining the properties of illite or other minerals is to perform solubility experiments and to use the solution chemistry obtained as a basis for thermodynamic calculations concerning the mineral observed.

Kittrick (1984) studied the solubility of three "standard" illite samples using long-term experiments (more than 2 years) to avoid the effects of kinetics:

1. Beavers Bend: $[Si_{3.62}\ Al_{0.39}]\ O_{10}\ (Al_{1.66}\ Fe_{0.20}\ Mg_{0.13})\ (OH)_2\ K_{0.53}$
2. Fithian: $[Si_{3.51}\ Al_{0.49}]\ O_{10}\ (Al_{1.54}\ Fe_{0.29}\ Mg_{0.19})\ (OH)_2\ K_{0.60}$
3. Goose Lake: $[Si_{3.65}\ Al_{0.35}]\ O_{10}\ (Al_{1.58}\ Fe_{0.20}\ Mg_{0.15})\ (OH)_2\ K_{0.59}$.

As all of the potassium contents are well below those of the illite end member composition of I/S-mineral series one should suspect that the experimental materials are not pure illite (see Sect. 1.1.2).

The mineral equilibrium is approached from both understaturation and supersaturation. These experiments showed that multiple phases or components control solution equilibria in the three samples. In spite of the fact that the chemical composition of these phases could not be determined unequivocally, this confirms that illite samples from diagenetic or low metamorphic environments are not monophase as was shown by the decomposition of XRD patterns (Fig. 1.3). The presence of I/S MLM may explain the low K contents of the bulk compositions. This was confirmed later by the multiphase experimental approach of Sass et al. (1987) who showed that the Goose Lake and Beavers Bend illites behave as three-component mixtures: smectite, illite and mica having 0.24, 0.67 and 0.9 K per $O_{10}\ (OH)_2$, respectively. This can be deduced from the decomposition components of XRD diagrams shown for illite in Fig. 1.3. The "illite" and "mica" components are in fact PCI and WCI while the smectite is an ordered I/S MLM. It is remarkable that in these experiments the ordered I/S assemblage becomes stable with respect to the microcline-kaolinite at temperatures above 90 °C. This transition occurs approximately at the same temperature as that at which the random to ordered I/S MLM occurs in the diagenetic series of the Texas Gulf Coast (Perry and Hower 1970; Hower et al. 1976). The existence of a phase-controlling solubility which is of muscovite composition (1.0 K per $O_{10}\ (OH)_2$) suggested by Aja et al. (1991) is not coherent with the products of solution equilibration experiments. The end-member illite and the muscovite are distinct phases, the former being stable at temperatures below 300 °C.

Synthesis experiments from glasses (Eberl and Hower 1976) were conducted to study the phase relationships in the K_2O-Al_2O_3-SiO_2-H_2O system, neglecting the chemical components Mg^{2+}, Fe^{2+} and Fe^{3+}. Thus, the conclusions are not directly applicable to natural systems except in the case of direct precipitation from boiling fluids in active geothermal fields in which the smectite-to-illite transition is controlled by the beidellite-muscovite joint (Patrier et al. 1998), phases which do not contain significant amounts of Mg or Fe. Actually, in diagenetic or hydrothermal series, the importance of the Mg and Fe elements cannot be neglected. Güven and Huang (1991) showed the importance of Fe^{3+} ions in the synthesis of illite from gels.

If the thermodynamic parameters of the "illite" component cannot be quantified easily from dissolution experiments using "standard illite samples", how can they be determined or approached? Research on the illite end-member (see Sect. 1.3.2) would be the most appropriate way for both the experimental and theoretical studies. However, we must keep in mind that the illite end-member composition depends on two characteristics of the geological environments: 1) the chemical control of the composition of the fluids; 2) the temperature × time factor.

The fluid composition may be controlled by the local rock composition in low-permeability formations such as shales or clay-rich sandstones (low water/rock ratio) or by external input into highly permeable sandstones (high water/rock ratio). In the first case, the formation of illite end-member occurs through the illitization of a smectite precursor (smectite-to-illite transformation) in the second, illite directly precipitates on pre-existing mineral surfaces (quartz grains or kaolinite-dickite crystals). In both case, the illite end-members are characterized by a high K content ($\sim$ 0.9 K per O_{10} $(OH)_2$) but have different Al and R^{2+} contents (Fig. 1.21).

Given the polyphased characteristic of natural samples, the calculation of the illite standard free energy of formation from dissolution experiments remains indefinite. Thus, a thermodynamic calculation, even complex, may be the most useful approach.

1.3.1.2 A Theoretical Approach of the Gibbs Free Energy of Formation

A second approach to determining the thermodynamic values of a mineral is by calculating the sum of the values due to different components in the crystal structure. A first approximation of the standard free energy of formation of clay minerals is made possible by summing the free energy of any oxide making up the silicate structure. This is based on the assumption that oxides have the same standard free energy of formation in all phyllosilicates (Tardy and Garrels 1974). Several refinements have been performed by addition of empiric parameters (Tardy and Garrels 1976, 1977): ΔO^{2-} and $\Delta_{\text{hydroxide}}$ or electronegativity considerations (Vieillard 2000).

The Vieillard method is based first on the calculation of the ΔO^{2-} parameter which is defined as representing the difference between the free energy of formation of the elements in a crystal oxide and the free energy of formation of cations in aqueous solution:

$$\Delta O^{2-}\ \mathrm{M} = (\Delta G^0_{\mathrm{t}}\ \mathrm{MO}_x - \Delta G^0_{t,\mathrm{sil}}\ \mathrm{M}^{2x+})\ \mathrm{kcal\,mol^{-1}}$$

M: a given metallic cation

MO_x: the oxide of this metal

M^{2x+}: the corresponding cation of this metal in aqueous solution

x: cation valency/2

Subsequently, the accuracy of calculations of the free energy of formation of minerals is improved by the integration of crystallographic and optic data, such as the refractive index, the molar volume, the mean distances between ions, the shortest cation-oxygen bond lengths in every site of the crystal structure.

Table 1.5. Uncorrected values of the free enthalpy of standard formation of oxides in a phyllosilicate frame ($\Delta G°f$ oxides), compiled by Vieillard (2000)

Oxides	$\Delta G°f$ oxides (kJ/mole)	Oxides	$\Delta G°f$ oxides (kJ/mole)
K_2O	−322.10	Al_2O_3	−1582.30
MgO	−569.30	SiO_2	−856.30
Fe_2O_3	−744.40	H_2O	−220.00

The calculation of the enthalpy of formation of oxides ($\Delta H^0_{t,oxides}$) is improved by the consideration of five parameters (Vieillard 1994, 2000):

1. Sites are distinguished according to whether they are occupied by one or several cations;
2. A new formulation of the parameter $\Delta_H O^{2-}$ is used according to site occupancy;
3. Consideration of the fact that some polyhedra have extra-long cation-oxygen bonds;
4. Non-bridging oxygens between several neighbouring polyhedra are taken into account;
5. Introduction of the prediction of error as a function of the standard error made when measuring bond lengths.

The free enthalpy of formation of the Beavers Bend illite ΔG°_f (illite) has been recalculated in order to illustrate the performances of this method. Uncorrected values of $\Delta G°f$ oxides (Table 1.5) have been compiled by Vieillard (2000):

$$K_{0,53}(Mg_{0,13}Fe^{3+}_{0,2}Al_{1,66})(Si_{3,62}Al_{0,38})O_{10}(OH)_2 \\ \rightarrow 0.53\,K_2O + 1.02\,Al_2O_3 + 0.1\,Fe_2O_3 + 0.13\,MgO + 3.48\,SiO_2 + H_2O$$

$$\Delta G^\circ_f \text{ illite} = 0.53\Delta G^\circ_f(K_2O) + 1.02\Delta G^\circ_f(Al_2O_3) + 0.1\Delta G^\circ_f(Fe_2O_3) \\ + 0.13\Delta G^\circ_f(MgO) + 3.48\Delta G^\circ_f(SiO_2) + \Delta G^\circ_f(H_2O) + \Delta G^\circ_{ox}$$

hence the uncorrected value

$$\Delta G^\circ_f(\text{illite}) = -5170.17 + \Delta G^\circ_{ox}$$

This approximate value can be improved if the energy of formation of oxide [$\Delta_G O{=}M^{z+}$(clay)] is corrected by the effects of the electronegativity of cations in their respective crystal sites.

- Calculation of the numbers of oxygen atoms bonded to the various cations in the various sites (t: tetrahedral; o: octahedral; i: interlayer).

$$K_{0.53}(Mg_{0.13}Fe^{3+}_{0.2}Al_{1.66})(Si_{3.62}Al_{0.38})O_{10}(OH)_2 \\ \rightarrow 0.265\,K_2O(l) + 0.83\,Al_2O_3(o) + 0.1\,Fe_2O_3(o) + 0.13\,MgO(o) \\ +3.62\,SiO_2(t) + 0.19\,Al_2O_3(t) + H_2O(i)$$

Nb Oxyg.:	0.265	2.49	0.3	0.13	7.24	0.57	1 = 12
Mol. Fract.:	0.265/12	2.49/12	0.3/12	0.13/12	7.24/12	0.57/12	1/12

- Values of the parameter $\Delta_G O = M^{z+}$(clay) used in the calculation of ΔG°_{ox} (Vieillard 2000)

	K^+(l)	Mg^{2+}(o)	Al^{3+}(o)	Fe^{3+}(o)	Si^{4+}(t)	Al^{3+}(t)	H^+(arg)
$\Delta_G O = M^{z+}$(clay) (kJ/mole)	425.77	−112	−161.2	−164.1	−166.1	−197.31	−220

- Details of the calculation of ΔG°_{ox} (Table 1.6)

Table 1.6. Details of the calculation of the correction of electronegativity: $\Delta_G O = M^{z+}$(clay) according to Vieillard's procedure (2000)

[1]Nature of Interactions	12 $X_i.X_j$	$[\Delta_G O = M_i^{z+}(\text{arg}) - \Delta_G O = M_j^{z+}(\text{arg})]$ (kJ/mole)	$-12X_iX_j\{[\Delta_G O = M_i^{z+}(\text{clay}) - \Delta_G O = M_j^{z+}(\text{arg})]\}$ (kJ/mole)
Within octahedral sites			
Mg^{2+}(o)-Al^{3+}(o)	0.0270	49.23	−1.33
Mg^{2+}(o)-Fe^{3+}(o)	0.0032	52.05	−0.17
Al^{3+}(o)-Fe^{3+}(o)	0.062	2.82	−0.18
Within tetrahedral sites			
Si^{4+}(t)-Al^{3+}(t)	0.344	31.22	−10.74
Between octahedral sites and OH			
Mg^{2+}(o)-H(arg)	0.0108	108	−1.17
Al^{3+}(o)-H(arg)	0.2079	58.77	−12.22
Fe^{3+}(o)-H(arg)	0.025	55.95	−1.40
Between octahedral and tetrahedral sites			
Mg^{2+}(o)-Si^{4+}(t)	0.078	54.09	−4.24
Mg^{2+}(o)-Al^{4+}(t)	1.505	4.86	−7.32
Al^{3+}(o)-Si^{4+}(t)	0.181	2.04	−0.37
Al^{3+}(o)-Al^{3+}(t)	0.0062	85.31	−0.53
Fe^{+3}(o)-Si^{4+}(t)	0.1185	36.08	−4.28
Fe^{+3}(o)-Al^{3+}(t)	0.0143	33.26	−0.47
Between the interlayer site and octahedral sites			
K^+(l)-Si^{4+}(t)	0.1599	591.86	−94.63
K^+(l)-Al^{3+}(t)	0.0126	623.08	−7.84
		ΔG°_{ox} (kJ/mole)=	−146.88

The following value of $\Delta G^{\circ}_{f(illite)}$ is inferred = – 5170.17–146.88.

Thus the corrected value is:

$$\Delta G^{\circ}_{f(illite)} = -5317.05\ \text{kJ/mole} \quad \text{or} \quad -1270.81\ \text{kcal/mole}$$

The corrected value is 4% different from the noncorrected one. Even though small, this difference is significant.

The final, corrected value for the free energy of formation of illite developed by the calculation of thermodynamic values of the individual components compares very closely (–1271 and –1268 kcal/mole) to that found by dissolution experiments. The correspondence between experimental values and calculated values of free energy can be used as an argument for using calculations instead of long and arduous experiments on pure materials which are difficult to obtain.

1.3.2
The Stability Field of the End-Member Illite Phase

1.3.2.1
The Instability of Dioctahedral Mica in Low-Temperature Conditions

Dioctahedral micas (muscovite, phengite) are not stable in low-temperature environments. They are known to alter into expandable phases under weathering conditions (Rich 1958; Rich and Obenshain 1955; Cook and Rich 1962) and vermiculite (Meunier 1980) or into illite in hydrothermal systems (McDowell and Elders 1983; Meunier and Velde 1979). In the first case, the initial expandable phase produced by K-release is a vermiculite. It was clearly shown by experimental studies that the transformation is due to K exchange by other cations (Na^{+}, Ba^{2+}, Mg^{2+}) which enter in the interlayer region surrounded by water molecules (Scott and Reed 1966; Kitagawa and Watanabe 1970; Kodama and Ross 1972). This is accompanied by the charge-reducing exchange reaction in the 2:1 layer. The structure expands to 12.2 or 14.3 Å after glycol or glycerol saturation according to the saturating cation. The vermiculite-to-mica reaction is not fully reversible: the vermiculite does not collapse exactly to 10 Å but rather to 10.4 Å after K-saturation and heating. This is due to the presence of $Al(OH)_3$ structures in the vermiculite interlayer regions. The instability of muscovite with respect to intergrade vermiculite is explained at low-temperature conditions ($T < 25\,^{\circ}C$) by its high dissolution rate on basal surfaces (Johnsson et al. 1992) on one hand and by the catalytic properties of these surfaces on the other. They act as an activated support for heterogeneous nucleation of brucite or kaolinite (Nagy et al. 1996).

Inherited micas (muscovite) in sediments, altered under hydrothermal conditions in sediments in the Salton Sea geothermal field, were shown to be fully or partly recrystallized (McDowell and Elders 1983). New illite grains form in subparallel orientation with the original shape of the micas. In such a case, the illite formation is controlled by a dissolution-crystallization process.

The comparative dissolution experiments of Yates and Rosenberg (1996) suggest that muscovite is not stable with respect to illite (San Juan illite, 0.83 K) at temperatures $\leq 250\,^{\circ}C$. Their study confirms that the illite end-member with a layer charge of 0.9 per $O_{10}(OH)_2$ is a phase in the thermodynamic sense. This phase is considered to have no permanent interlayer cation exchange capacity, the cation adsorption occurring only at the outer surfaces of the crystals. However, this must be re-examined from a crystallochemical point of view (see Sect. 1.4.2). The instability of muscovite and phengite with respect to end member illite has been experimentally studied further by Yates and Rosenberg (1997, 1998). The alteration product of the $2M_1$ mica phases is an illite having the end member chemical composition (charge of −0.9) and conserving the $2M_1$ polytype structure but it has a typical lath morphology. In spite of the fact that the initial mica polytype is conserved, the change in particle morphology indicates that the mineral reaction was not controlled by a solid state transition process but rather by dissolution and crystallization. The potassium content of these particles increases with their thickness. This was explained by a growth process.

Thus, the illite end-member is a phase which is stable with respect to dioctahedral micas in the low-temperature range. In other words, the illite end member is able to grow either from smectite precursors or by dissolving micas in specific conditions typical of hydrothermal alterations, weathering or diagenesis. Rosenberg (2002) proposed that end-member illite is a stable phase below $360\,^{\circ}C$ (close to the upper stability of pyrophyllite). Above this temperature, muscovite coexists with andalousite and water.

1.3.2.2
The Stability of End-Member Illite with Respect to Expandable Phases

The relative stability of end-member illite and expandable phases in the K_2O-Al_2O_3-SiO_2-H_2O system have been studied through dissolution experiments (Yates and Rosenberg 1997). With increasing temperature the interlayer composition of the expandable phases changes from $K_{0.25-0.31}$ to $K_{0.50-0.54}$ per $O_{10}(OH)_2$ which correspond to low-charge smectite and high charge smectite (vermiculite), respectively. Figure 1.22 shows the presence of two invariant points: I/S-smectite-kaolinite (1) and end-member illite-I/S-kaolinite (2). Solutions in equilibrium with both three-phase assemblages are shifted to lower K^+/H^+ activity ratio and to higher aqueous silica activity with increasing temperatures. Dissolution experiments from 25 to $250\,^{\circ}C$ (Sass et al. 1987) showed that the transition from low-charge smectite ($K_{0.25}$) to ordered I/S MLM ($K_{0.67}$). A brutal change occurs in a narrow temperature range between 90 and $100\,^{\circ}C$ corresponding to the random to ordered I/S transition.

The stability of end member illite relative to dioctahedral micas in low-temperature conditions signifies that a K-deficient mica crystal structure is stable in such conditions. This means that a specific crystal structure composed of the three components muscovite, celadonite and pyrophyllite has a lower Gibbs free energy than the mechanical mixture. Several distribution patterns are possible, some of them are pictured in Fig. 1.15. Rosenberg (2002) considers that the stability of end-member illite is due to the formation of pyrophyllite domains in the 2/1 layer structure. The presence of such domains has been

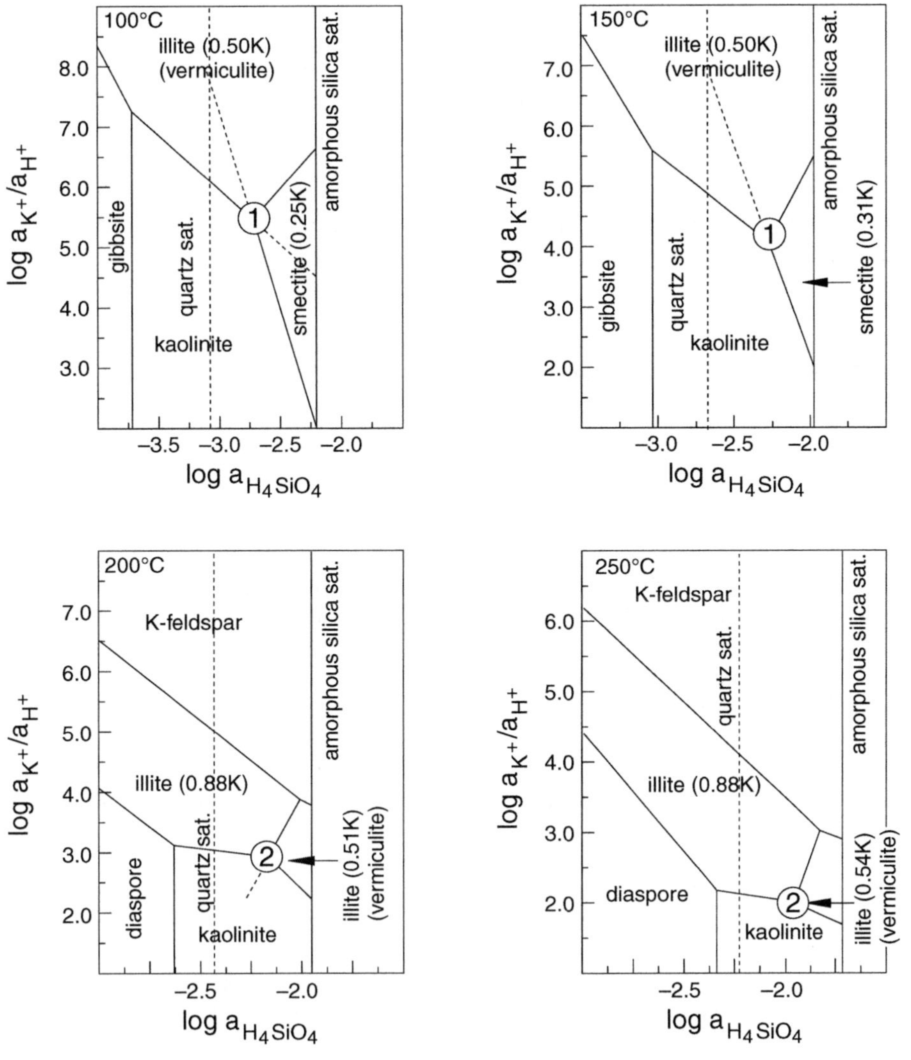

Fig. 1.22. Isothermal, isobaric phase diagram from Yates and Rosenberg (1997) showing the stability fields of end-member illite, I/S MLM and smectite. Metastable phase boundaries are indicated by stippled lines. Invariant points of different smectite composition are indicated by numbers 1 and 2. Illite 0.50 to 0.54 K is considered to be equivalent to vermiculite

identified by FTIR studies (Besson and Drits 1997a_b). However, until now we have no direct proof of the presence of pyrophyllite interlayers equivalent to the pyr-pyr sites in facing 2/1 layers. The unique indirect proof would be the "excess" water at the condition that water is in its neutral molecular state. Louks (1991) establishes that H_3O^+ hydronium ions as well as neutral water molecules are present in pyrophyllite domains of illite structure. Such "structural" water molecules have been identified by IR absorption bands at 3420 and 3260 cm^{-1} (Besson and Drits 1997a_b). As their is no evidence of any increase in the d_{001}-spacing, it is probable that these water molecules are present in the pyr-pyr sites and reduce the structural strain between the pyrophyllite and mica domains.

The thermochemical considerations lead us to the conclusion that illite is in fact a stable, individual phase, distinct from aluminous micas and distinct from I/S MLMs of high illite content.

1.4 The Growth of Illite Crystals

1.4.1 Crystal Shapes in Diagenetic Environments

One aspect of silicate mineralogy which is often overlooked is the way in which a crystal grows. Crystal growth habit is the result of preferences of development of specific crystallographic faces. Often this growth preference is guided by mineral chemistry, and hence it is useful to look into the growth habits of illite minerals (Wilkinson and Haszeldine 2002).

Illite crystals exhibit two different morphologies according to the conditions of their formation: laths and plates. The two major sedimentary environments where one finds abundant illite are shales and sandstones. The chemical systems of these two facies are different (sandstones are open systems to chemical migration and shales are closed systems) and one tends to find different crystal habits among the illites found there. The formation of lath-shaped crystals in diagenetic environments is observed either during the smectite-to-illite conversion in shales or at times from the direct precipitation or transformation of kaolinites or dickites to illite in sandstone. The initial lath-shaped crystals in shales are formed in the intermediate stages of smectite-to-illite transformation. This is particularly obvious in the Gulf Coast series as shown in Fig. 1.23. In this figure one sees tracings of crystal morphologies obtained from transmission electron micrographs of samples from a deep well in the Texas Gulf Coast. Lath and hexagonal shapes were selected from the micrographs. The large peak at approximately 15 Å is formed by large, roughly hexagonal grains which were not traced here. The peak of the new, ordered (R1) I/S MLM is shown in the figure as I/S ill. During illitization of the R1 I/S MLM, the size and shape (length/width ratio) of the particles change and have the effect of changing the FWHM of the I/S band: the sharper the diffraction peak, the bigger and less elongated the particles are (Fig. 1.23, unpublished data of the author). In the first case, the lath morphology is inherited from that of I/S MLM and the most illitic mineral in the smectite to illite transformation will be lath-shaped, and PCI. However, one can see in the lower figure that there are a small number of hexagonal crystals present. These crystals form the increasingly strong illite peaks. They are most likely the WCI illite phase.

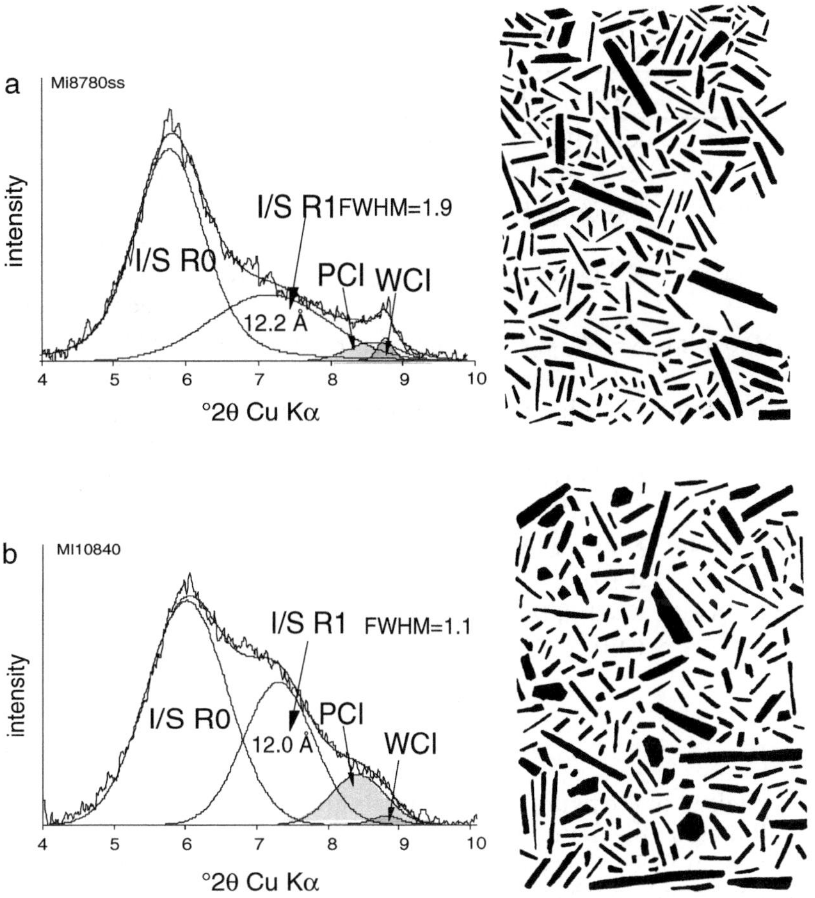

Fig. 1.23a,b. Relation between the FWHM of ordered I/S (laths in the figure) and the size and morphology of clay particles in two samples from the Gulf Coast diagenetic series. The lower sample (*b*) is found at greater depth and more evolved in the smectite-to-illite reaction series. Shapes (located at the right hand portion of the figure) were obtained as tracings from transmission electron micrographs. Only laths and hexagons were selected. The background-subtracted XRD spectra (located at the left hand portion of the figure) are decomposed into basic component curves representing the different phases present. The ordered R1 phase is I/S$_{il}$ indicated on the spectra. This curve becomes more narrow in the lower spectrum indicating the larger crystal size of the laths. Note the presence of hexagonal shapes in the more evolved sample *b*. These correspond to the illite fraction (WCI)

Using the information in Fig. 1.23 and that presented by Lanson and Champion (1991), one can reconstruct the morphological change in I/S MLM minerals forming from the smectite-to-illite transformation in burial diagenesis. This hypothesis is shown in Fig. 1.24a where the initial lath shaped ordered (R1) (1 in the figure) are ripened by dissolution-crystallization processes to form

Table 1.7. Typical chemical composition of illite crystals having different habitus and originating in diagenetic (diag) or hydrothermal (hydr) series

Habitus	Formula unit	References
Lath (Diag.)	$K_{.74}\ NH^{4+}_{.04}\ Ca_{.05}\ (Al_{1.71}\ Fe^{3+}_{.07}\ Fe^{2+\cdot}_{05}\ Mg_{.16})\ [Si_{3.36}\ Al_{.64}]\ O_{10}\ (OH)_2$	Nadeau and Bain (1986)
Hexag. (Diag.)	$K_{.91}\ (Al_{1.72}\ Fe^{3+}_{.12}\ Mg_{.16})\ [Si_{3.25}\ Al_{.75}]\ O_{10}\ (OH)_2$	Lanson and Champion (1991)
Lath (Diag.)	$K_{.79}\ (Al_{1.67}\ Fe^{3+}_{.15}\ Mg_{.17})\ [Si_{3.40}\ Al_{.60}]\ O_{10}\ (OH)_2$	Lanson and Champion (1991)
Lath (1M) (Hydr.)	$K_{.72}\ (Al_{1.63}\ Fe^{3+}_{.20}\ Mg_{.17})\ [Si_{3.45}\ Al_{.55}]\ O_{10}\ (OH)_2$	Zöller and Brockamp (1997)
Hex.(2M$_1$) (Hydr.)	$K_{.83}\ (Al_{1.65}\ Fe^{3+}_{.19}\ Mg_{.16})\ [Si_{3.34}\ Al_{.66}]\ O_{10}\ (OH)_2$	Zöller and Brockamp (1997)
(Lath)1M (Hydr.)	$K_{.83}\ Ca_{.02}\ (Al_{1.66}\ Fe^{3+}_{.15}\ Mg_{.29}Mn_{.03})\ [Si_{3.21}\ Al_{.79}]\ O_{10}\ (OH)_2$	Peacor et al. (2002)
(Hex)2M$_1$ (Hydr.)	$K_{.78}\ Na_{.16}\ (Al_{1.79}\ Fe^{3+}_{.05}\ Mg_{.15})\ [Si_{3.35}\ Al_{.65}]\ O_{10}\ (OH)_2$	Peacor et al. (2002)
Bentonite (Diag.)	$K_{.68}Na_{.01}\ (Al_{1.43}\ Fe^{3+}_{.30}\ Mg_{.27})\ [Si_{3.48}\ Al_{.52}]\ O_{10}\ (OH)_2$	Cuadros and Altaner (1998)

PCI crystals (3 in the figure) These crystals grow to large PCI crystals and very small illites (2 in the figure) grow into pseudo-hexagonal or hexagonal crystals (5 in the figure). The large PCI crystals will dissolve to form the WCI crystals by crystal growth mechanisms. This conversion sequence is not operating in porous sandstones where illite or I/S MLM precipitate directly from solutions and grow upon different mineral surfaces. Most often, the particle habitus is that of very elongated, thin laths ("hairy" illite) as shown in Fig. 1.24b.

Variations in size bring about changes in shape. Thus, under diagenetic conditions, the morphological modifications of crystals of I/S MLM series are different according to whether they were formed in sandstones or in shales (after Lanson and Champion 1991; Varajao and Meunier 1995). The overall evolution (Fig. 1.25) tends towards isometric shapes as the size increases but both series are perfectly distinct (after Lanson and Meunier 1995). The relation between polytype and crystal shape in individual particles has been investigated using the convergent beam electron diffraction (CBED) by Zöller and Brockamp (1997). They showed that illite originating from a hydrothermal vein was composed of lath- and hexagonal-shaped particle populations. Illite crystals have been observed also in porous rocks either in diagenetic or hydrothermal series. They exhibit fibrous, lath-shaped or platy habits and their composition varies from aluminous-rich to magnesium-rich end-members (Table 1.7).

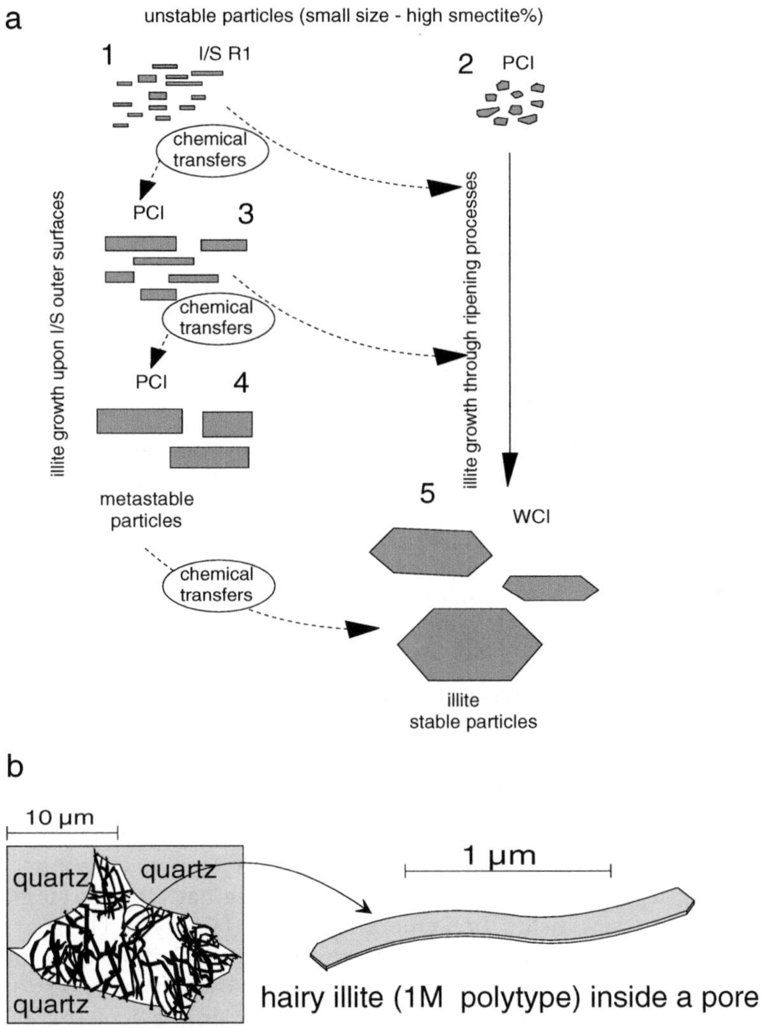

Fig. 1.24a,b. Illite crystal habits in diagenetic rocks. **a** Schematic representation of the morphological evolution (size and shape) of I/S mixed layers and illite in a given state of diagenesis. I/S, PCI and WCI refers to XRD parameters of "illitic" clays (Lanson and Champion 1991) and designate smectite-rich I/S mixed layers, poorly crystallized illite and well crystallized illite respectively. Arrows indicate the growth paths which transform the small particles into bigger ones with increasing diagenetic conditions. **b** The direct precipitation of lath-shaped crystals in sandstone pores

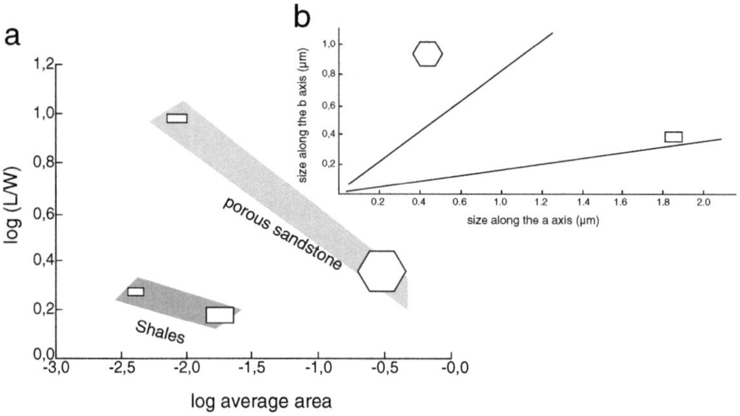

Fig. 1.25a,b. Shape-size relationships. **a** Relations for I/S MLM and illite particles in diagenetic porous sandstones and shales (from Lanson and Meunier 1995). L/W: length/width ratio for each crystal or particle. For a perfect hexagon, L/W= 1; mean surface area: length × width. **b** Relations for 1M and $2M_1$ illite crystals (from Zöller and Brockamp 1997)

1.4.2 Growth Mechanisms of Lath-Shaped Illite Crystals

1.4.2.1 Theoretical Aspects and Observations

In spite of their shape being elongated in one direction, fibrous illites exhibit spiral growth steps (Nagy 1994). Thus, illite grows like mica (Baronnet 1975), that is to say, layers stacking in the c direction originate from a single layer growing from a screw dislocation emerging on the (001) face. The step height has not been determined with sufficient accuracy to check if the growing units are composed of one or two 2:1 layers for 1M and $2M_1$ polytypes respectively. Nevertheless, it can be assumed that a single layer unit is growing, forming a spiral on the 1M lath-shaped illite crystals. Then, knowing that the ditrigonal symmetry offers a three 3-fold symmetry axis, the question is why the growth is much faster in only one direction? Güven (2001) explained the formation of the three possible crystal habits (lath, rhomb and hexagon) by the relative growth rates on the [100], $[\bar{1}10]$ and the $[\bar{1}\bar{1}0]$ directions (Fig. 1.26a). The growth fronts are respectively the (010), (110) and $(1\bar{1}0)$ planes; consequently, they do not correspond to visible crystal faces. The illite lath or fibers are formed when the growth rate is slower on the (010) front than on the (110) and $(1\bar{1}0)$ ones. Thus the (010) faces are largely developed. The rhomb-shaped crystals are formed when the growth rate on the (110) and $(1\bar{1}0)$ fronts is slower than on the (010). The hexagonal crystals are formed when the growth rate is the same for the three fronts. Güven suggested that the position of the vacant site may potentially have an influence on the growth process. The growth in

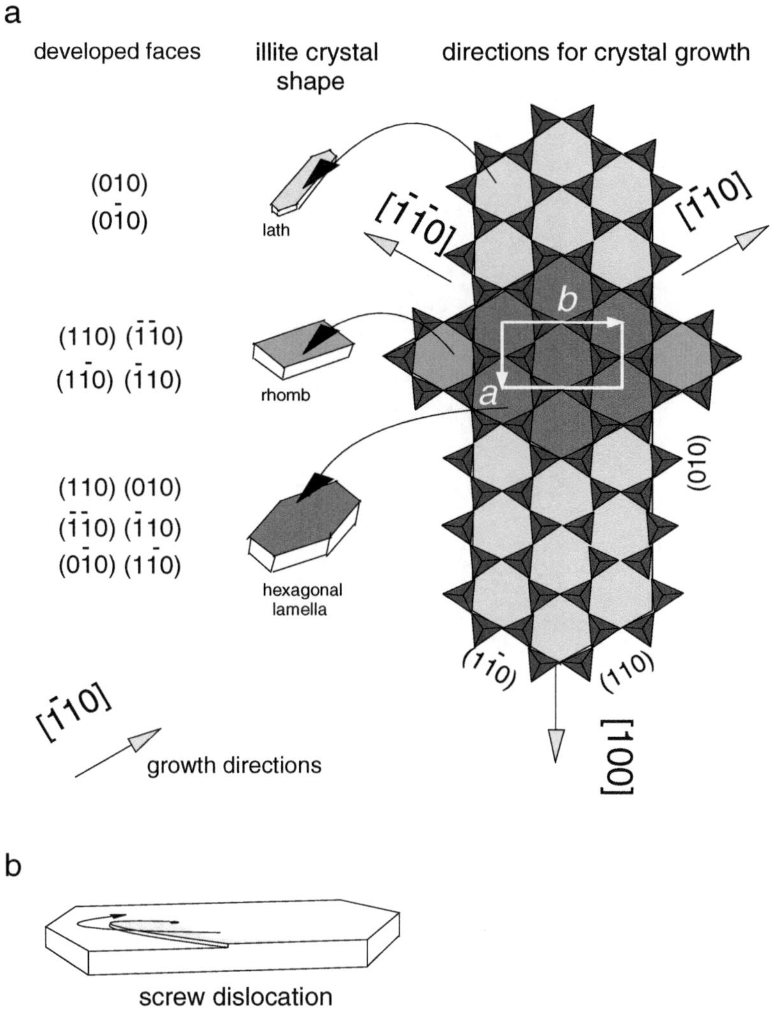

Fig. 1.26a,b. Lath-shaped illite growth processes. **a** Theoretical relationships between shape in growing faces (from Güven 2001). The white rectangle corresponds to a unit cell: the a and b axis are indicated by arrows. The light, medium and heavy grey areas correspond to crystal shapes. The developed faces are indicated for each crystal habitus. **b** Spiral growth for lath-shaped illite

the [001] direction requires either a two-dimensional nucleation or a spiral accretion from a screw dislocation (Fig. 1.26b). Even very elongated illite laths exhibit spiral growth as shown by Inoue and Kitagawa (1994) who used the Pt-Pd-shadowing technique (Fig. 1.27).

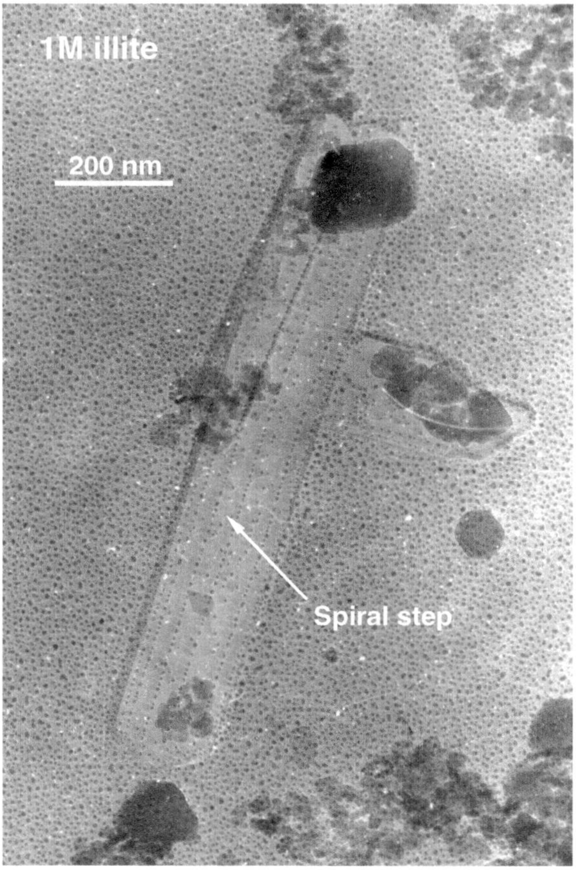

Fig. 1.27. Spiral growth of 1M illite sample from the Kamikita hydrothermal system, Japan (modified from Inoue and Kitagawa 1994). Growth steps have been decorated by the Pt-Pd-shadowing technique

1.4.2.2
The trans to cis Transition During Growth

The position of the proton for the OH groups emerging in the center of the ditrigonal cavities is not the same in trioctahedral and dioctahedral structures. Indeed, as it is equally repulsed by the three octahedral cations in a trioctahedral configuration, it is directed perpendicularly to the sheet. This is not the case for a dioctahedral structure where the proton is repulsed toward the vacant site and is oblique to the sheet. Then, it is easy to understand how the proton is directed relatively to the (110) and ($1\bar{1}0$) growing faces (Fig. 1.28a). As a symmetry plane exists in the *trans* configuration, the angle of the O-H bond is equivalent for the two faces. On the contrary, in a *cis* configuration, the O-H bond is parallel to the (110) face and makes a 60° angle with the ($1\bar{1}0$) face. This relative position should certainly make differences in the surface energy of each faces (Sainz-Diaz et al. 2001).

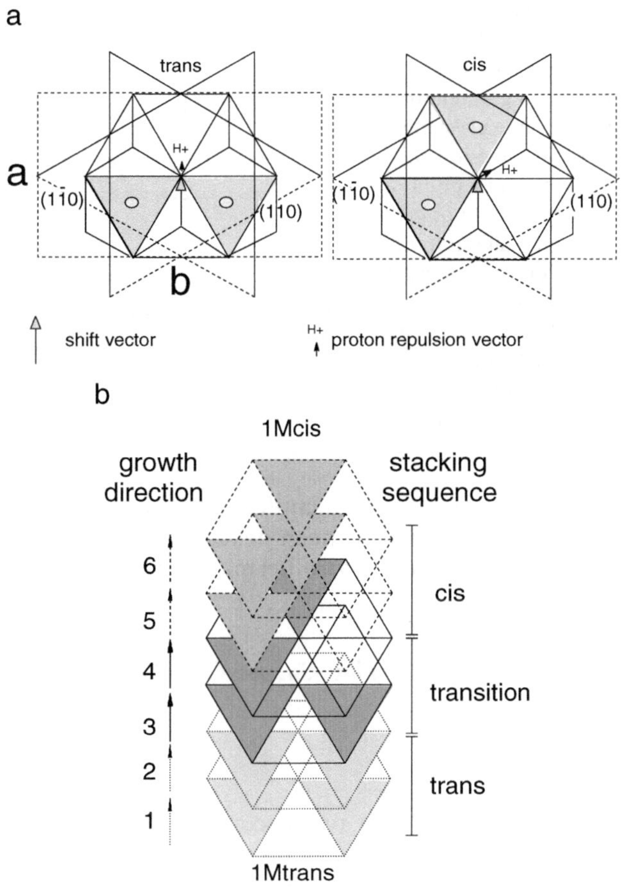

Fig. 1.28a,b. The possible influence of vacant site location on growth processes. **a** Symmetry-asymmetry of (110) and (1$\bar{1}$0) faces due to the proton orientation. Shaded area indicates the occupied octahedral sites in a single 2:1 layer. **b** Crystal structure on *cis-trans* 1M illite. The shaded areas indicate the occupied octahedral sites in pairs of 2:1 layer units

If true, the above discussion should have consequences on the crystal habits: those having a *trans* structure should grow in two directions and exhibit a more platy shape than those having a *cis* structure which should grow preferentially in one direction. Recent data (Laverret 2002) show that both are lath-shaped but the smallest ones are *trans* while the biggest ones are *cis*. In the local diagenetic conditions, the trend from *trans* at the top to *cis* at the bottom indicates that illite grows first with a *trans* structure than with a *cis* one. This leads to the formation of *cis-trans* particles whose structure is depicted in Fig. 1.28b.

1.4.2.3
Coalescence and Epitaxy

Illite is observed as single crystals or polycrystalline particles which result from epitaxial growth on different crystal supports (illite, kaolinite) or coalescence processes (Fig. 1.29a,b). The crystal faces in particles exhibit re-entrant angles. Epitaxy was frequently described in sedimentary and diagenetic rocks for illite

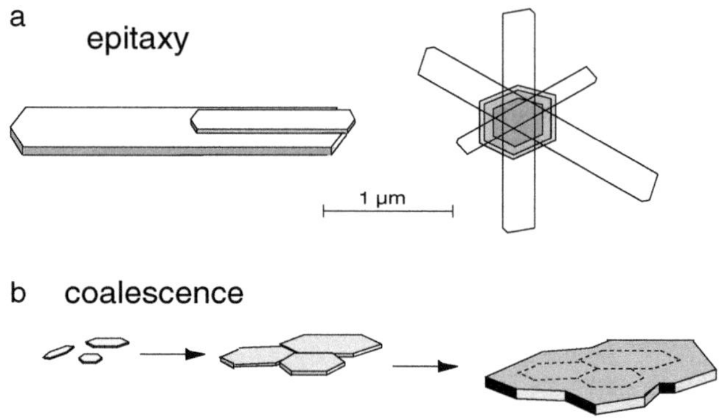

Fig. 1.29a,b. Polycrystalline particles. **a** Epitaxial growth. One crystal growing on another in the same crystallographic direction or growth from a point on the crystal but not in the same crystallographic direction of maximum extension **b** Coalescence of crystals by growth enlargement and subsequent common overgrowths

or smectite growth (Holtzapffel and Chamley 1986; Mosser et al. 1972; Güven 2002). Coalescence has been described in experimental illitization (Whitney and Velde 1993) and hydrothermal alteration of rocks (Kitagawa 1995).

1.4.3 Growth Processes for Plate-Shaped Crystals

1.4.3.1 Spiral Growth

Using an atomic force microscope (AFM), Blum (1994) showed the presence of growth step spirals on the (001) faces of isometric illite originating in hydrothermal systems. Similar observations were made using the "shadowing" technique (Kitagawa 1995): Pt or Au particles are projected on the (001) faces of illite crystals at a given angle. The dimension of the shadow allows to calculate the thickness of the particles knowing the projection angle.

Compared to the bidimensional nucleation, the spiral growth process is much less energy consuming, i.e., the needed oversaturation degree of the solution is lower (Boistelle 1982). Thus, spiral growth was observed in many illite samples from diagenetic or hydrothermal series (Inoue and Kitagawa 1994). The geometrical properties of spirals are governed by the relative growth rates of the different crystal faces. Spirals are very elongated or regular in lath-shaped or isotropic crystals, respectively (Fig. 1.30).

1.4.3.2 The 1M to $2M_1$ Transition

Observations and Problems Since the pioneer work of Yoder and Eugster (1955), the classical sequence of transformations between polytypes of Al-rich dioctahedral illite and muscovite from 1M to $2M_1$ has been established for many

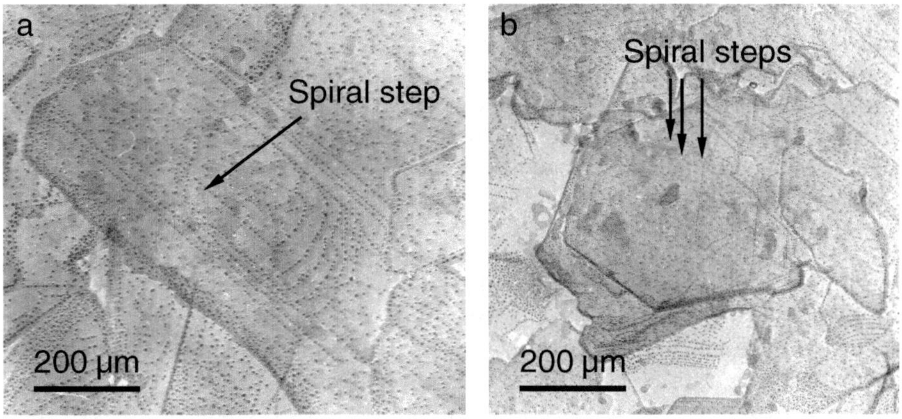

Fig. 1.30a,b. Microphotographs of spiral growth of illite crystals from the Kamikita hydrothermal system (modified from Inoue and Kitagawa 1994). **a** Illites 1M. **b** Illite 2M

years. It has been considered to be a function of increasing grade from diagenesis to low-grade metamorphism (Velde 1965; Kisch 1983; Mukhamet-Galeyev et al. 1985) or increasing temperature in hydrothermal systems (Ylagan et al. 2000). Velde (1965), using information from high temperature and pressure experiments on muscovite compositions, suggested that the stable polytype of pure muscovite is the $2M_1$, the 1M and $1M_d$ being metastable. In order to understand the importance of this point, it is necessary to come back to the elementary definition of what a polymorph transition reaction is. Whatever the reaction mechanism (overgrowth or dissolution-recrystallization), the polymorphs involved in the transition reaction conserve the same chemical composition. This is the case in the experiments on muscovite.

From reports to date, it appears that the 1M to $2M_1$ sequences for illite in most natural systems do not have the same chemical composition, and thus the 1M and $2M_1$ stacking sequences are not polytypes *sensu stricto* (Zöller and Brockamp 1997; Peacor et al. 2002). This may be confusing since the polytype definition usually describes either a stacking sequence and not a mineral reaction or a polytypism phase relation. It would be better to say that, as far as we know the illite polytype changes in diagenetic, hydrothermal or metamorphic series involved in the reaction are not an isochemical process. As a result such reactions cannot be qualified as a polymorph transition. Illite then exhibits different polytype structures under different P, T conditions which perhaps dictate different chemical compositions.

However, we must re-examine what the significance of a so-called "composition domain" for illite is. Figure 1.21 shows that, in the M^+-4Si-R^{2+} system, the composition field of illite is limited by the 0.9 K line. Except on this line, the polytypes are the 1M and $1M_d$ ones. Illite on the 0.9 K line is composed of 1M and $2M_1$ polytypes. Thus, the composition domain between the 0.9 K and the phengite lines is that of a mixture of mica and $2M_1$ illite. Consequently, the composition 1M and $2M_1$ domains of illite are only apparently distinct, the real domain of coexistence being reduced to the 0.9 K line. On that line, both polytypes have a similar composition and offer an opportunity to study the polytype transition. The coexistence of 1M and $2M_1$ polytypes has been described clearly in two natural occurrences: the Silverton Caldera, Colorado (Eberl et al. 1987) or the Broadland-Ohaaki geothermal system (Lonker and Fitz Gerald 1990). Using HRTEM, Lonker and Fitz Gerald showed that illite crystals formed in the same place are composed of either pure 1M or pure $2M_1$ or composite particles with no apparent habitus differences. In composite crystals, the textures indicate that the region with two-layer stacking may be embedded in $1M_d$ sequences even though there is a general $1M_d$ to 1M to $2M_1$ trend with increasing temperature. Even if the mechanism of the polytype transition was not determined (solid state replacement, simultaneous nucleation and coalescence or dissolution-precipitation), the authors determined the activation energy of the reaction. The value is 216 kJ mol^{-1} which is similar to that given by Mukhamet-Galeyev et al. (1985): 213 ± 42 kJ mol^{-1}.

Another problem rises from the way that polytype stacking sequences are determined for a given geological sample. For example, X-ray diffraction (XRD) and selected area electron diffraction (SAED) or convergent beam electron diffraction (CEBD) do not give the same information. XRD averages information for particle populations and thus reflects the significant, dominant sequence. SAED or CEBD give a direct characterization of polytypism in single particles. This sometimes leads to contradictory information. Using SAED, Peacor et al. (2002) claim that the 1M polytype is rare in prograde diagenetic series of pelitic rocks while the $1M_d$ is dominant. This is exactly the reverse of results obtained using XRD studies. Why is this so? The disagreement between the two procedures may be due to two reasons: 1) the individual particle analysis method implies selection of individual grains by the operator which can enhance the differences observed between single crystals while XRD averages the resulting signal of all particles in a sample whatever their size or their shape; 2) the size of the coherent scattering domain and the degree of ordering are interrelated, thus the polytype determination of a given sample may change with the size fraction under analysis. In other words, the predominance of the $1M_d$ polytype as seen by SAED is probably due to the fact that analyses were performed on thin particles while XRD patterns are representative of particles having higher diffraction energy. Thus, one should consider that the two techniques offer complementary rather than interchangeable information.

Overall the XRD method is statistically more reliable and more representative of a given geological sample.

Reaction Mechanism We saw that, from a theoretical point of view, the 1M to $2M_1$ transition may happen through solid state replacement, simultaneous nucleation and coalescence or dissolution-precipitation processes. The two first mechanisms operate under isochemical conditions in solid or solid-solution systems. The third one does not necessarily impose isochemical conditions but could occur in open systems. In such a case, the 1M and $2M_1$ crystals do not have the same composition. Unfortunately, if it exists, the composition change during the polytype transition controlled by a dissolution-precipitation process is too subtle to be detected using AEM (Lonker and Fitz Gerald 1990).

It seems that the dissolution-recrystallization process prevails in natural hydrothermal systems, e.g. Silverston Caldera (Eberl et al. 1987); Broadland-Ohaaki geothermal field (Lonker and Fitz Gerald 1990); Ponza Island (Ylagan et al. 2000); Golden Cross epithermal ore deposit (Tillick et al. 2001). Then the question is: is the 1M illite particle totally dissolved while the $2M_1$ crystals nucleate and grow in separate particles or may $2M_1$ illite overgrow 1M crystallites? For Peacor et al. (2002), illite crystals do not present a structure with both polytypes occurring as alternate units in alteration zones where the two polytypes coexist. If this is a general rule for illite no matter the geological conditions prevailing during its formation, this means that the $2M_1$ structure cannot overgrow on the 1M substrate during the particle coarsening process. The consequences of the study of Peacor et al. (2002) is that the transition reaction indicates a progression through the dissolution of 1M particles and the nucleation and growth of the $2M_1$ particles. This is contradicted by the observations of Lonker and Fitz Gerald (1990) who showed that 2M sequences are embedded into 1M or $1M_d$ stackings. Consequently, even if the dissolution-precipitation process dominates during the illite polytype transition, we must admit that overgrowths are possible (Fig. 1.31), in other words, 1M and 2M sequences either in separate or single particles during all the polytype transition period. Equivalent alternating sequences have been observed in micas (Nespolo 2001). The fact that some 1M particles dissolve while $2M_1$ illite overgrows on others leads us to consider different degrees of metastability in a particle population.

The $1M_d$ and 1M crystallites being metastable with respect to the $2M_1$, the conversion should be a part of an Ostwald-step-rule-like change from smectite to illite, each step being characterized by an activation energy (Ylagan et al. 2000). Synthesis experiments from muscovite stoichiometric composition gels

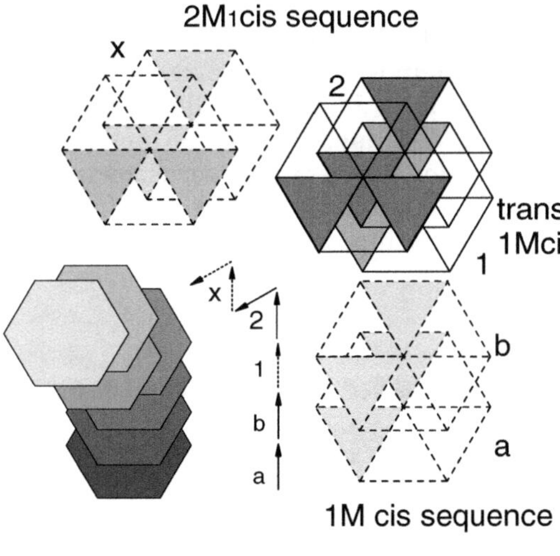

Fig. 1.31. Schematic representation of a $2M_1$ polytype overgrowth upon a 1M structure

(Mukhamet-Galeyev et al. 1985) showed that the 1M and $2M_1$ polytypes are formed at the same time, the latter being formed more slowly. The conversion of 1M to $2M_1$ begins after the complete crystallization of the gel in the 1M form. The conversion is slow and decreases with temperature. Below 200 °C, the process stops, at least on the scale of laboratory examination. The activation energy (50±10 kcal mol^{-1}) is higher than that of the formation of illite from glass: 19.6±3.5 kcal mol^{-1} (Eberl and Hower 1976). Consequently, the conversion rate should be slow. This explains why 1M polytype is observable in natural diagenetic sequences where it is a metastable state of illite. Theoretically, the least energy consuming-process would be the $2M_1$ overgrowth on 1M support compared to the direct $2M_1$ nucleation.

The conclusion to this summary of information on polytypes is that illites can have both 1M and 2M polytypes in various growth habits. The apparent transformation of one polytype to another is not isochemical and hence not a true polytype transformation but a mineral reaction. The differences in polytype preference for illites from different geological environments could very well be due to differences in composition of the illite under the influence of different pressure-temperature and chemical activities. It appears that the most stable polytype of illite (WCI) is the 2M form.

1.4.3.3
Summary of the Chemical Compositions of Illite in Individual Crystallites and Illite Layers in Mixed-Layer Minerals

The information presented above is what can be considered to be a traditional point of view concerning illite. Older, initial data has been corrected and augmented to form a more coherent and complete body of information. Given this, one can construct a first approximate definition of illite. Chap. 2 will present the published data in a more detailed form, attempting to describe the mineral occurrence of illite in nature better. This will give us a new, more refined definition of illite.

Illite is often or almost always associated with other clay minerals in nature, often with mixed layer I/S MLMs. Illite-smectite mixed layer minerals (I/S MLM) have been described in most of the geological systems at the Earth's surface: soils, sediments, diagenetic rocks, hydrothermally altered rocks. Thus in order to understand illite, or separate it from I/S MLMs it is important to distinguish the two components.

1.4.3.4
Illite Layers in Interstratified Minerals in Different Geological Environments

In spite of similar XRD patterns, the illite components of I/S MLM are not identical: both the illite and smectite components exhibit different crystallochemical properties (Table 1.8).

Table 1.8. Chemical composition of I/S MLM in different geological environments at the Earth's surface

Geological setting	Illite layer charge	Smectite layers	References
Soils	−0.47	Charge: −0.48 $Si_{3.78}\ R^{2+}_{0.21}$	Laird et al. (1991) Laird and Nater (1993)
Shale and clay-rich sandstones diagenesis	−0.9	−0.3 montmorillonite	Howard (1981)
Bentonite diagenesis	−0.9	−0.35 to −0.45 montmorillonite	Srodon et al. (1986)
Metasomatism	−0.9	−0.65 montmorillonite	Brusewitz (1986) Huff and Turkmenoglu (1981)
Hydrothermal Alteration	−0.9	−0.40 montmorillonite	Inoue et al. (1978) Meunier and Velde (1989)
Hydrothermal flow		Beidellite	Beaufort Papapanagiotou

1.5 A Working Definition of Illite

This chapter has posed several questions and given a brief resume of the different descriptions and definitions of illite in the past. Below we give a first definition of illite summarizing the available data. As a second step, in Chap. 2, we will investigate the occurrence of illite in natural systems and in synthesis studies more closely in order to give a more complete, and hopefully precise definition of illite. Below then we propose a working *definition of illite.*

We have seen that the definition of illite may be significantly different according to the geological settings (more exactly according to the cultural background of peoples studying these settings) or to the techniques used to identify it (XRD, chemical composition). It seems that combining XRD and chemical properties could help us to clarify some problems (Table 1.9).

Table 1.9. Summary of illite and illitic mineral properties

	Layer charge per O_{10}		
	0.75 to 0.88	0.88	0.88–1.00
FWHM of the (001) peak (°2θ Cu Kα)	> 0.25	≥ 0.25	< 0 25
Polytypes	$1M_d$ + 1M	1M + $2M_1$	$2M_1$
Permanent CEC (meq/100 g)	10 to 0.5	0.5	0
Mineral phases	Illite + "illitic minerals"	End-member illite	End-member illite + mica?

Given the above, one can make a first approximation for a definition of illite. Here we shall consider that illite can be identified by the following properties:

- The CEC (cation exchange capacity) of illite due to permanent low charges depends in high selectivity sites which are related to "pyrophyllite" local composition of the 2:1 layer. Most cation exchanges on illite are due to edge sites or un-completed bonds (so-called variable charge sites). These variable charge CEC values are about 5 meq/100 gm of sample. Consequently, any CEC values above variable charges indicate the presence of interstratified expandable layers, i. e., IS mixed-layer minerals. These illite-rich IS will be designated as "illitic minerals". According to Hower and Mowatt (1966), the minimum layer charge for illite is about 0.75 per $O_{10}(OH)_2$.
- The maximum layer charge is that of the end-member, i. e., 0.9 per $O_{10}(OH)_2$. Therefore, any charge in the 0.9 to 1 layer charge range is considered to be due to a mixture of two phases: end-member illite+mica. The lower charge value is not well defined.
- XRD patterns should show no sign of expansion upon glycol treatment.
- Illite will not have a disordered, turbostratic structure (1Md) but either a 1M or 2M polytype. Eventually the 2M polytype is the final, most stable phase.

We will now compare this definition with an analysis of published and new data found for minerals in different geological environments in order to better define illite in its different occurrences.

CHAPTER 2

The Geology of Illite

The geology of illite is surprisingly diverse since it forms in soil as well as in metamorphic or peri-magmatic conditions. The goal of this chapter is to give an overview of this diversity in order to outline the most important facts concerning illite formation. This information gives a basis to determine the transformation of minerals into illite according to the factors of kinetics (Chap. 3) and to apply the knowledge of illite formation to practical problems of mineral resource development and some environmental problems.

2.1 Illite in Soils and Weathered Rocks

2.1.1 Occurrence of Illite in Soils

Classically (Jackson 1964; Fanning and Keramidas 1977) and more recently (Righi and Meunier 1995; Ellis and Mellor 1995) indicate that illites have their origin in micas, usually muscovite, which is inherited in the soil sequence from the source materials from which the soil is formed. Most of the general studies describing illite genesis in this manner show large mica flakes breaking into smaller particles until they become clay-sized (Norrish 1972). However, Reichenbach and Rich (1975) indicate in a general schema that pedogenic mica occurs as a neoformed mineral.

Mineralogical X-ray diffraction analyses show that the amount of fine (1 to 0.2 µm) and very fine (< 0.2 µm) particles increase from the C horizon to the A1 horizon. This attests to the formation of clay minerals in the soil. The inherited coarse-grained minerals are fragmented and can be found in the very fine grain size fractions. At the same time, one finds a complex assemblage of mixed-layer and intergrade minerals. Organic compounds or deposits of Fe-Al-oxyhydroxides can migrate into the interlayer sheets (Fig. 2.1). Thus it is assumed that the "illitic" fraction is identified by a sharp peak at 10 Å because the mica debris conserves a large size scattering domain in the c^* direction. For the most part these descriptions are deductive, not being based upon actual observations of micas becoming illites. There is no mention of how the mica composition changes into illite except by leaching of potassium, it does not involve magnesium, iron, aluminum or other elements which distinguish illite from muscovite.

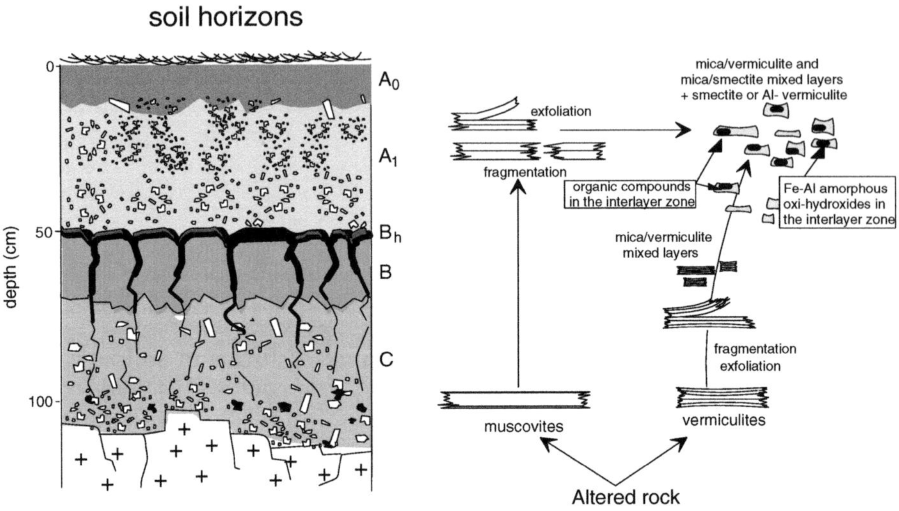

Fig. 2.1a,b. Schematic representation of the transformation of phyllosilicates inherited from the weathered granitic rock in the A, B and C horizons of a brown acid soil. Fragmentation takes place either by chemical weathering or by mechanical deformation. In the same time, the inherited phyllosilicates transform either into mica/smectite or mica/vermiculite mixed layers or into intergrade minerals in which the interlayer sheets are blocked by deposits of organic compounds or of Fe-Al-oxyhydroxides

The illites occurring in soils are often found to be unstable in studies of soil genesis, being transformed into vermiculites (Wilson et al. 1984; Wilke et al. 1984; Koch et al. 1992) or interstratified illite/smectites (Hughes et al. 1994; Dabkowska-Naskret and Dlugosz 1996; McDaniel and Nielson 1985; Bürhmann and Schoeman 1995) in the soil profile. This transformation of illite into other, less potassic minerals is observed in various types of soil-forming environments, those of high rainfall as well as conditions of low water activity. Such observations have led soil scientists to consider that illite is a transient phase in most soils. Nevertheless, illite is sufficiently abundant in some cases to warrant a place in soil classification and terminology (Harris and Zelazny 1985). All together this indicates that illite can be considered to be a soil clay mineral.

2.1.2 More Recent Studies

Two rather detailed studies of X-ray diffractograms give further insight into the apparent instability of illite and mica in soils. Velde and Peck (2002) and Velde et al. (in press) show that the illite content of a soil can be affected by the presence of plants. Continuous corn cultivation diminishes mica content (WCI and PCI) in an Illinois Mollisol after 70 years. However the illite peak (PCI) in the diffractograms of the soils, appears to change less in proportion (surface area) than does the mica (WCI) peak (Fig. 2.2).

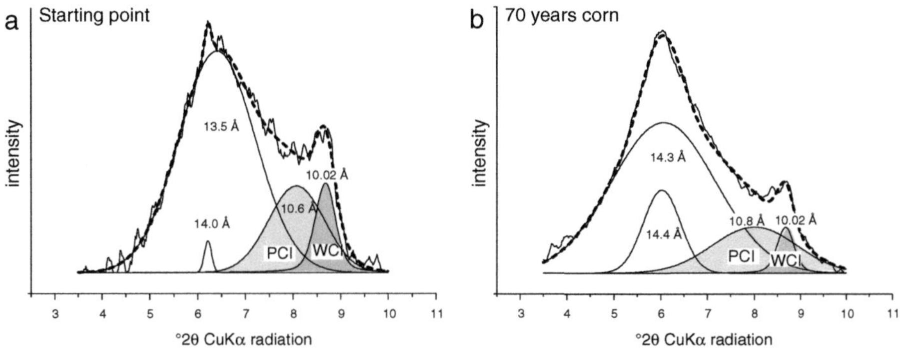

Fig. 2.2a,b. 70 years corn planting. Background-subtracted, decomposed XRD patterns. Clay mineralogy in experimental plots (Morrow plots) of the University of Illinois. A comparison is made between the initial clays in the soils and the result after 70 years of continuous corn planting. Samples in Sr-saturated, air-dried state

In these decomposed diffractogram traces, one sees four peaks, two of which are attributed to disordered mixed layered I/S minerals and two to the illite-mica group. Illite (PCI) occurs at near 10.6–10.3 Å while mica (WCI) occurs at 10 Å. The I/S minerals are dominant in the soil mineralogy here. Inspection of the diagrams shows that after 70 years of continuous corn culture the illite-mica peaks change in character and in intensity. The PCI peak widens and shifts to slightly higher spacings indicating an increase in smectite content. The intensity of the WCI peak diminishes compared to the PCI peak and the total of the illite-mica peaks decreases compared to that of the I/S minerals. Hence there is a net loss of the individualized illite-composition minerals in the soils after intense corn cultivation.

A second study investigates growth of natural grass on poldered sediments after 800 years (Velde et al. 2003) The XRD diagrams show that illite (PCI) content as well as mica (WCI) content changes depending upon the organic-plant context in the soil profile (Fig. 2.3). Initial sediments have a high illite/smectite content and the PCI peak surface area is significantly greater than that of WCI. In the organic humic zone of the soils which have developed as pasture since 1158 (the data of poldering) the PCI content decreases relative to mica. However the sum of both is approximately constant relative to the illite/smectite peak areas. Below the humic zone, the illite-mica peak areas decrease relative to the illite/smectite (I/S MLM) peak areas. There is slightly less PCI relative to WCI. In this study it appears that illite and mica tend to be replaced by I/S minerals depending upon the context of whether there is much organic or little organic matter present. In the prairie humic horizon mica-illite is stabilized relative to I/S but when this humic matter is not present, there is loss of illite-mica.

A third set of data (Huang, Institute of Soil Science, Chinese Academy of Sciences, Nanjing and Velde, unpublished) observes the effect of different fertilizer treatments on cultivated Vertisols in China. Three treatments over a period of 40 years were used; no specific amendments, chemical fertilizer

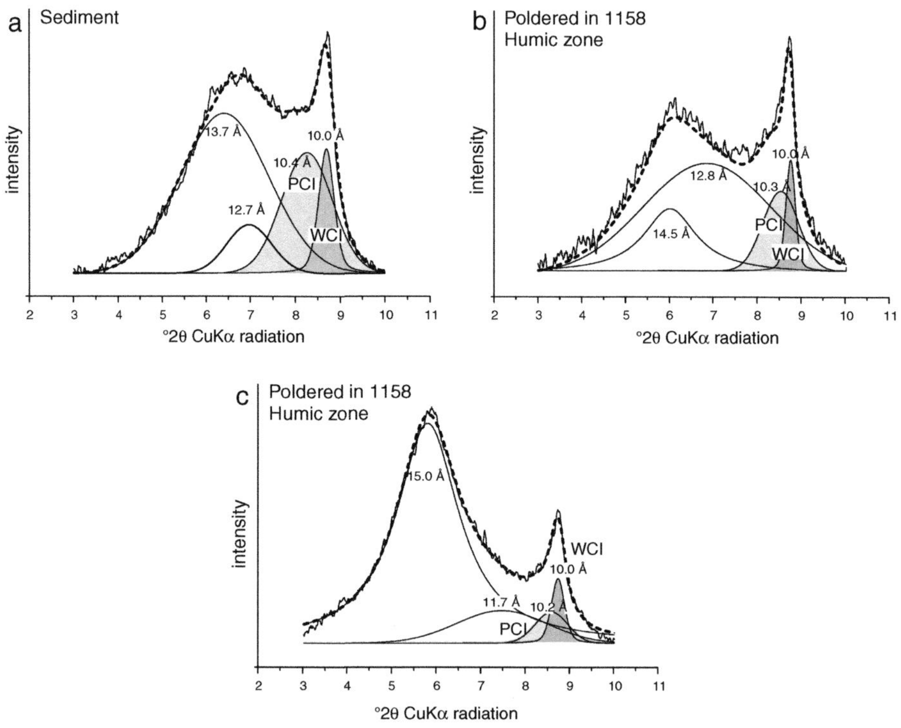

Fig. 2.3a–c. Background-subtracted, decomposed XRD patterns. Sr-saturated, air-dried state. Samples shown are from poldered sediments put into prairie for 850 years (Baie d'Authie, France). Initial sedimentary material is compared to two horizons in a soil profile developed on sedimentary material since 1158. The upper humic zone and the lower non-humic zone are compared. PCI + WCI are relatively more intense in the humic zone and less so in the lower zone compared to the sediment

(NKP) and barn yard manure. The observations are shown in Fig. 2.4. The peaks of the I/S minerals are dominant in all treatments. There is a strikingly high intensity of the mica (WCI)-illite (PCI) peak in the manure-treated soils. Here it is clear that the presence of organic matter stabilizes the micaceous minerals compared even to the standard NKP fertilizer treatments.

In a fourth study, Pernes-Debuyser et al. (in press) report on clay mineralogy in soil plots which were not cultivated but received fertilizer amendments for 70 years (Fig. 2.5). Most of the different treatments changed pH but little in the clay mineral compositions or in their relative abundance. However addition of KCl or horse manure did change the clay compositions somewhat, increasing the illite content of the illite/smectite mixed layer minerals and showed a relative increase in the mica and illite content. Increase in illite content of the I/S MLM is shown by the increase in background, and loss of definition of the 17 Å peak. This is most evident in the manure-treated sample. Calculations using the Reynolds NEWMOD program (Reynolds 1985) indicates

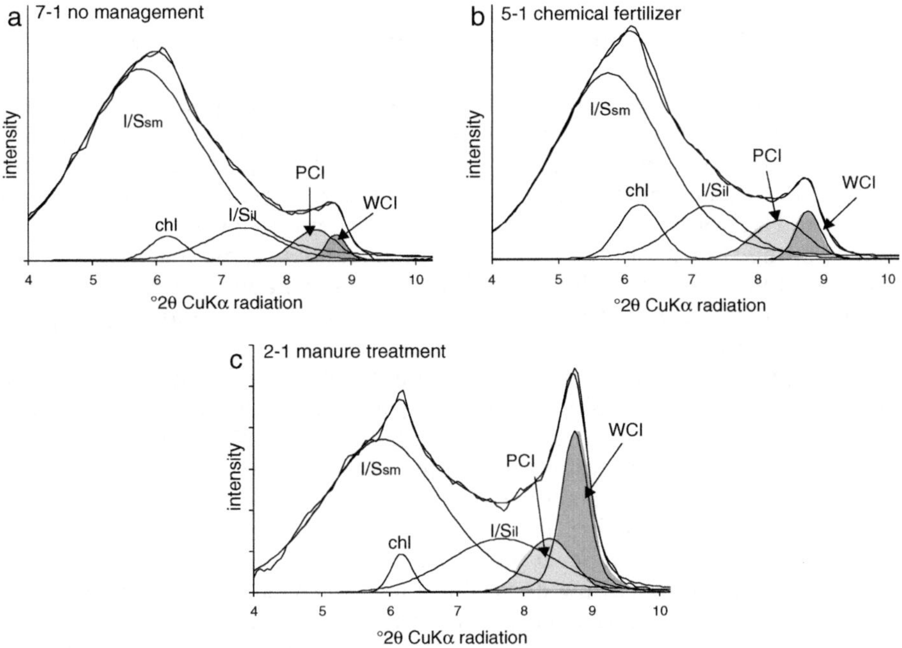

Fig. 2.4a–c. Background-subtracted, decomposed XRD patterns; Sr-saturated, air-dried state. Three soils after 40 years treatment in a Chinese Vertisol region. Manure treatment (**c**) greatly enhances the WCI portion of the soil clays, while no management (**a**) and chemical fertilizer (**b**) show much less illite components

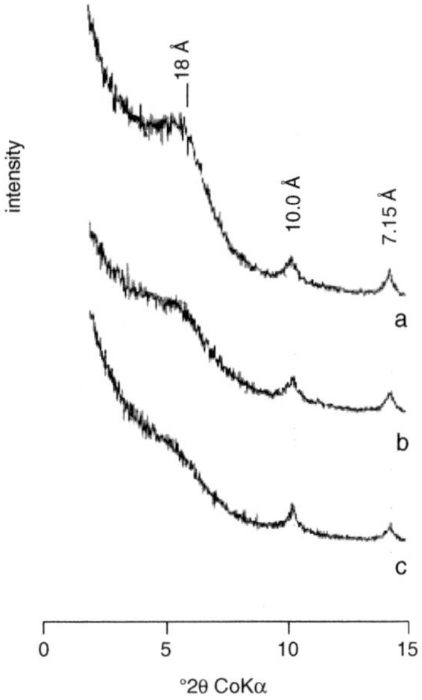

Fig. 2.5. Glycol-treated samples from experimental plots at Versailles, France: 70 years treatment or no treatment. Increase in illite content of the I/S MLM is shown by the increase in background, and loss of definition of the 17 Å peak in samples treated with KCl. The soil was not treated (**a**) potassium-treated (**b**) and (**c**) manure-treated. The untreated soils show no significant change in clay mineralogy compared to starting soils

a change of 20% illite content. Mica increases more than the illite peak surface area in the KCl treatments. This suggests that it is possible to crystallize illite in soils given a high enough activity of potassium.

Considering the results of the studies quoted above, it appears that illite can be stabilized under certain conditions in soils, when the type of organic matter present protects it from degradation (example of the pasture soils, and crop rotation). Other organic action (cultivation of corn) acts more as a destabilizing agent and lack of organic matter in certain soils tends to favour the loss of illite-mica. In the studies of the impact of fertilizers, the organic material seems to stabilize illite-mica even though KCl treatments can apparently produce illite material. It is clear that illite can be generated in soils under appropriate chemical conditions.

The question then is not really if illite forms in soil environments, but under what conditions? Singer (1989) suggests that illite can form from non-micaceous precursors in arid soil environments where the input is essentially feldspar-rich dust. Observations by Meunier (1980) indicate that illite can form from feldspar in the early stages of weathering of granites in temperate climates. The study of plant-free fertilizer amended plots shows the formation of WCI in KCl-enriched soils. Hence it seems clear that illite does form in soils.

This being the case, how does illite form and what are its mineralogical characteristics? Illite has been observed to form during the early stages of granite alteration under conditions of a temperate climate. A detailed study of the occurrence of illite formation in granite weathering is dicussed below.

2.1.3 Early Formation of Illite in Weathered Granites

Petrographic observations of granite weathering have been reported by Meunier (1980). These observations were completed by electron microprobe analyses and X-ray diffraction of micro sampled phases taken out from thin sections. The rock altered was a two-mica granite. It is massive and does not contain any visible vein or altered zone. First we will consider the stability of mica and then the formation of illite in the early stages of granite alteration.

2.1.3.1 Mica Instability in Soils: Formation of hydroxy-Al Vermiculite or Smectite (HIV-HIS)

Many previous studies have shown that muscovite transforms into soil vermiculite, i.e. a hydroxy-interlayered expandable phase in natural weathering areas (see Rich 1958). Similar mineral changes have been reproduced in laboratory experiments (Kitagawa and Watanabe 1970). The expandable phase, i.e. vermiculite, is formed when the potassium is leached out of the muscovite crystal lattice and replaced by hydrated cations. The interlayer charge is partly compensated by hydroxy-aluminum complex ions which limit the expandability. This occurs because, during the dissolution of mica layers, silica is lost from all the solids. The Si/Al ratio decreases progressively from 3 toward 1 which are the characteristic values for mica and kaolinite, respectively. The important point in these studies is that muscovite does not transform into illite, but into another or other minerals which are less siliceous.

How is muscovite altered during granite weathering? As indicated above, the basic phenomenon is the loss of potassium due to the absorption of protons (H^+). Two different processes must be considered: 1) the early alteration stage in which the shape and size of the mica is conserved in spite of K^+ ion release; 2) the final process in which mica is dissolved and replaced by a polyphased assemblage: Al-vermiculite + kaolinite.

Process 1 During the first stage, the common weathering products formed are dioctahedral vermiculites (or high-charge smectite) whose layer charge is lower than that of the parent mica. It is thought that the K^+ release is related to the sympathetic behaviour of the two OH groups in a the vacant site. The expansion of the $d_{(001)}$ spacings is not due to large changes of the layer charge in the mineral (Scott and Reed 1966). During this earliest alteration step, the *b* dimension increases during the loss of K^+ ions and return to its original values during K uptake (Kodama and Ross 1972); this is typical of a vermiculite behavior which has no hydroxi-Al in the interlayer space. The reaction: muscovite + Na^+ $\rightarrow$ Vermiculite (muscovite-Na, H_2O)+K^+ is more or less reversible and quasi-isochemical for the 2:1 layer. The $d_{(001)}$ spacings expand to 12.2 Å or 14.3 Å when K^+ ions are replaced by Ba^+ or Mg^{2+} respectively:

$$\text{Muscovite} + M^+ \rightarrow M^+ - \text{vermiculite} + K^+ \qquad (2.1)$$

Experimental alteration showed that the vermiculite inherits the crystal structure of the micas when interlayer K^+ ions are replaced by more polarizing (or hydrated) ones. Reaction 2.1 is one of the earliest mica alteration stage observed in granite weathering profiles. Indeed, the first transformations of the parent crystals occur inside microsites (a few micrometers in size) distributed along the fluid pathways (connected microfractures, intergranular joints, mineral cleavages etc...). The reactive solids, i. e. dissolving primary phases and secondary products, with the fluids in contact form "microsystems" which may be more or less closed (Meunier and Velde 1979).

Process 2 For higher degree of mica dissolution, the vermiculite $d_{(001)}$ spacings do not collapse exactly to their initial position after resaturation with potassium. The XRD peaks become broader, indicating that structural disturbances are irreversible. Advanced alteration stages of micas are easy to observe using the classical petrographic techniques for weathered granite rocks. Corroded zones appear within the mica crystals showing that intense dissolution was operating locally while secondary phases, i. e., a polyphased assemblage dioctahedral hydroxy-vermiculite (Al-vermiculite)+kaolinite, precipitate in the free spaces. The mineral reaction muscovite $\rightarrow$ Al-vermiculite+kaolinite is no longer isochemical. Indeed, if aluminum is considered as totally immobile, a simple calculation of the mass balance shows that SiO_2 must be brought into the system with H_2O and M^+ hydrated ions (Na^+, Ca^{2+}, etc.). Let us consider that the reaction produce *a* moles of kaolinite and *b* moles of vermiculite, then the reaction may be qualitatively written as follows:

$$(2a + bx + 2b + by)\underbrace{[\mathrm{KAl_3Si_3O_{10}(OH)_2}]}_{\text{muscovite}} + \underbrace{(\mathrm{H_2O + SiO_2 + M^+})}_{\text{solution}} \tag{2.2}$$
$$\rightarrow a\underbrace{[\mathrm{Al_2Si_2(OH)_4}]}_{\text{kaolite}} + b\underbrace{[\mathrm{M - Al_y(OH)_{3y} - Al_{2+x}Si_{4-x}O_{10}(OH)_2}]}_{\text{Al-vermiculate}} + \mathrm{K^+}$$

In summary, the decrease of the chemical potential of K^+ relative to that of H^+ induces first the transformation of mica to vermiculite by exchange of K^+ ions with hydrogen ions. Then the mica is dissolved and an Al-vermiculite + kaolinite mineral assemblage forms. It appears that muscovite is not replaced by illite even in the early stages of granite weathering.

2.1.3.2
K-feldspar Instability: Formation of Illite

K-feldspar alteration was observed in two different types of microsystems (Fig. 2.6a,b): 1) in the early stages of the weathering process along the muscovite or biotite – orthoclase crystal joints (contact microsystems in the saprock); 2) in the final stages of weathering where the initial structure of the parent rock is destroyed (plasmic microsystems in the saprolite: a new rock formed of parent

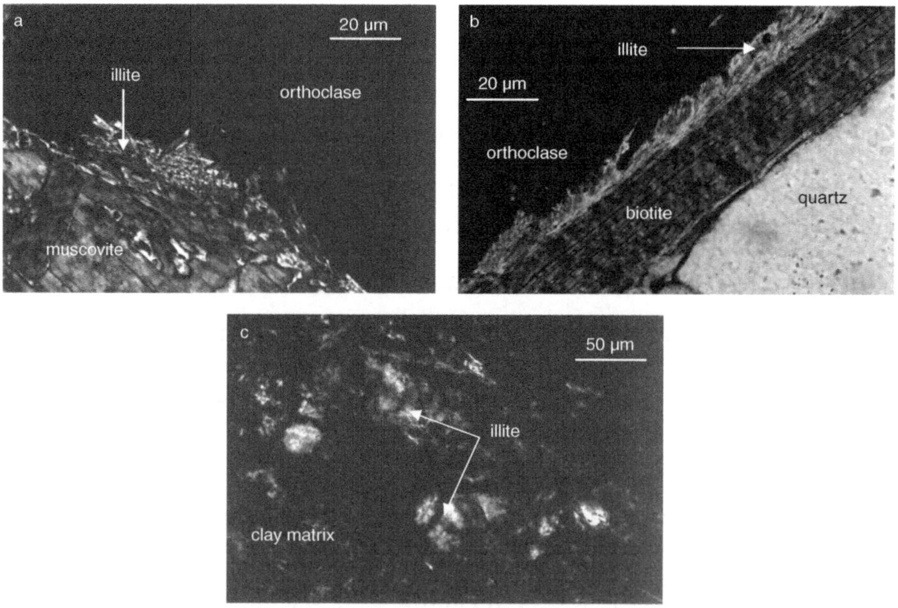

Fig. 2.6a–c. Micrographs of weathering "microsystems" in a granitic saprock where illite is formed. **a** The orthoclase-muscovite intergranular joint. **b** The orthoclase-biotite intergranular joint. **c** Illite crystallization in the clay matrix of saprolite

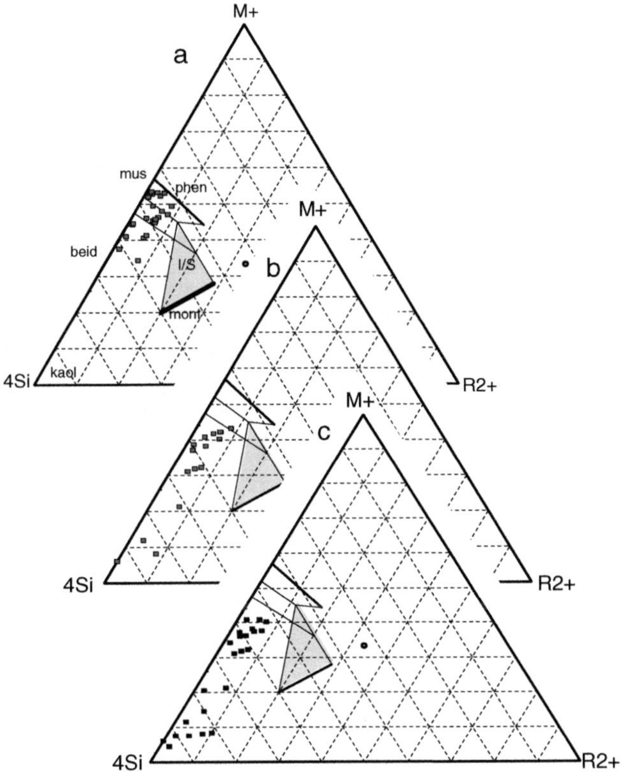

Fig. 2.7a–c. Chemical compositions of illites formed during the weathering of granite rocks (Dudoignon 1983). **a** Illites formed on the mica-orthoclase crystal joints in the early alteration stages (saprock). **b** Illites formed in the muscovite-orthoclase intergranular joint in the saprolite. **c** illite + kaolinite mixtures formed in the fine-grained matrix (mixture of parent mineral debris and clays) in the saprolite

mineral debris mixed with secondary clays and oxides, Fig. 2.6c). Illite was shown to be formed in both cases in association with other clay minerals: smectite in the early stages and kaolinite in the latest ones. Consequently, the chemical compositions of the illite formed are different: the former being richer in magnesium, the later in aluminum (Fig. 2.7, Table 2.1).

The growth of illite into orthoclase crystals from their contact surfaces with muscovites or biotites (Fig. 2.6a,b) is controlled by the dissolution of the K-feldspar:

$$K - Feldspar + H^+ + Mg \rightarrow illite + K^+ + Si^{4+} \tag{2.3}$$

Si^{4+} ions liberated in solution are "in excess" for muscovite. Such conditions should favour the formation of phases richer in silica-like smectite. However, in presence of high amounts of dissolved potassium (high potassium chemical

Table 2.1. Chemical compositions of illite from weathered granites (data from Dudoignon 1983)

Illite in the muscovite-orthoclase crystal joints (saprock)

Si	3.05	3.28	3.79	3.16	3.28	3.19	3.47	3.39	3.19	3.09	3.50	3.41	3.24	3.47	3.51	3.12	3.08	3.09	3.10	3.09	3.55	3.10	3.31	3.59	3.65
Al	2.86	2.60	1.94	2.61	2.78	2.55	2.30	2.32	2.63	2.72	2.22	2.34	2.52	2.35	2.20	2.71	2.82	2.81	2.78	2.79	2.33	2.83	2.75	2.26	2.12
Fe	0.04	0.07	0.06	0.08	0.09	0.12	0.06	0.04	0.09	0.07	0.10	0.10	0.12	0.15	0.10	0.10	0.10	0.10	0.11	0.12	0.08	0.09	0.11	0.07	0.11
Mg	0.14	0.06	0.12	0.11	0.11	0.15	0.10	0.09	0.07	0.08	0.10	0.11	0.05	0.10	0.10	0.05	0.02	0.02	0.02	0.02	0.02	0.02	0.02	0.02	0.02
Mn	–	–	–	–	–	–	0.01	–	–	–	–	–	0.01	–	0.01	–	0.01	–	–	–	–	–	–	–	–
Ti	–	–	0.01	–	–	–	–	0.01	–	0.04	–	0.01	–	0.01	–	–	0.03	0.03	0.03	0.04	–	0.01	0.01	–	0.01
Ca	0.01	0.02	0.03	0.02	0.01	0.02	0.03	0.02	0.02	0.01	0.02	–	0.03	0.01	0.02	0.02	–	–	–	–	–	–	–	–	–
Na	0.10	0.03	0.03	0.05	0.02	0.06	0.06	0.09	0.05	0.06	0.04	0.05	0.05	0.03	0.04	0.03	0.05	0.02	0.06	–	–	0.08	–	0.05	0.02
K	0.68	0.68	0.47	0.74	0.76	0.82	0.69	0.51	0.86	0.88	0.73	0.73	0.74	0.63	0.75	0.87	0.84	0.84	0.83	0.85	0.63	0.83	0.51	0.68	0.73

Illite in the biotite-orthocl, crystal joint (saprock) — Illite in the muscovite-orthoclase crystal joint (saprolite)

Si	3.26	3.33	3.18	3.11	3.14	3.23	3.09	3.15	3.17	3.74	3.29	3.22	3.28	3.64	3.10	3.93	3.22	3.21	3.17	3.18	3.52	3.22	3.22	3.06	3.08	3.10	3.12	3.09	3.24
Al	2.32	2.25	2.37	2.49	2.37	2.29	2.52	2.39	2.83	2.11	2.71	2.73	2.72	2.26	2.77	1.80	2.90	2.66	2.84	2.90	2.31	2.83	2.87	3.11	3.13	3.10	2.72	2.81	2.80
Fe	0.17	0.18	0.22	0.18	0.23	0.23	0.19	0.23	0.00	0.00	0.00	0.00	0.01	0.00	0.10	0.09	0.08	0.12	0.12	0.09	0.13	0.07	0.09	0.08	0.03	0.07	0.12	0.12	0.13
Mg	0.23	0.22	0.24	0.20	0.27	0.26	0.20	0.24	0.04	0.04	0.02	0.05	0.01	0.04	0.05	0.00	0.00	0.04	0.01	0.00	0.05	0.00	0.02	0.00	0.00	0.00	0.00	0.04	0.01
Mn	–	0.01	–	–	0.01	0.01	0.01	–	0.01	0.01	0.01	0.00	0.01	0.01	0.01	0.00	0.00	0.00	0.01	0.00	0.01	0.00	0.01	0.00	0.00	0.00	0.02	0.01	0.00
Ti	0.01	0.01	0.01	0.02	0.01	0.01	0.01	0.01	0.01	0.01	0.01	0.01	0.00	0.00	0.01	0.01	0.01	0.01	0.02	0.00	0.02	0.01	0.00	0.01	0.01	0.01	0.03	0.01	0.00
Ca	0.05	0.01	0.03	0.02	0.07	0.03	0.03	0.02	0.02	0.00	0.01	0.00	0.02	0.01	0.02	0.01	0.02	0.00	0.01	0.00	0.02	0.01	0.03	0.00	0.00	0.00	0.01	0.00	0.00
Na	0.04	0.03	0.05	0.06	0.03	–	0.04	0.06	0.05	0.01	0.02	0.05	0.00	0.00	0.00	0.01	0.00	0.00	0.00	0.00	0.00	0.00	0.05	0.00	0.01	0.00	0.00	0.00	0.00
K	0.85	0.85	0.88	0.92	0.83	0.87	0.83	0.81	0.35	0.59	0.56	0.75	0.59	0.54	0.82	0.54	0.09	0.70	0.01	0.37	0.43	0.35	0.09	0.08	0.01	0.02	0.83	0.76	0.23

illite + kaolinite, saprolite (per Si_4O_{10})

Si	3.34	3.30	3.30	3.33	3.34	3.33	3.28	3.30	3.25	3.24	3.12	3.16	3.19	3.13	3.23	3.30	3.52	3.34	3.90	3.07	3.38	3.77	3.32	3.45	3.54	3.06
Al	2.57	2.57	2.71	2.56	2.61	2.58	2.64	2.68	2.71	2.66	2.97	2.77	2.80	2.92	2.77	2.64	2.47	2.72	1.90	2.64	2.71	2.64	2.80	2.44	2.43	3.02
Fe	0.10	0.11	0.10	0.10	0.08	0.09	0.09	0.09	0.12	0.32	0.28	0.26	0.14	0.27	0.08	0.08	0.09	0.11	0.18	0.16	0.16	0.18	0.08	0.26	0.14	0.03
Mg	0.07	0.11	0.02	0.11	0.06	0.08	0.08	0.04	0.09	0.08	–	0.05	0.03	0.03	0.06	0.08	0.11	0.02	0.03	0.05	0.02	0.06	0.03	0.09	0.09	0.02
Mn	–	–	–	–	–	–	–	–	–	–	–	–	–	–	0.01	–	–	–	–	–	–	–	–	–	–	–
Ti	0.01	0.01	–	–	–	–	0.01	–	–	0.01	–	0.01	0.01	–	–	–	–	–	–	0.01	–	0.01	–	0.01	–	–
Ca	–	–	–	–	–	–	–	–	–	–	–	–	–	–	–	–	0.04	0.03	0.01	0.02	0.01	0.01	0.02	0.04	0.04	0.07
Na	–	0.01	–	0.01	–	0.01	–	–	–	0.03	–	–	0.03	0.05	0.01	–	–	–	0.02	0.04	–	0.04	0.01	0.06	–	0.10
K	0.56	0.62	0.44	0.56	0.21	0.62	0.54	0.49	0.43	0.21	0.05	0.38	0.46	0.06	0.40	0.08	0.01	0.01	0.22	0.06	0.02	0.02	0.06	0.04	0.01	0.41

potential), the local chemical conditions favour the formation of illite rather than that of an expanding mineral such as high charge smectite (vermiculite). The compositions of illite determined by microprobe are Al-rich and always contain small amounts of Mg^{2+} ions no matter the original mica type present (muscovite or biotite). Most of the K content values vary between 0.9 and 0.65 per $O_{10}(OH)_2$. Those which are lower than 0.65 are probably due to the presence of kaolinite which gives high totals of cations, above 6 for a dioctahedral mica or mica-like mineral (Table 2.1). It is important to note that the illite analyses with the highest potassium contents have a higher silica content than those of muscovite.

The earliest weathering reactions involving the dissolution of minerals by unsaturated solutions increase the rock porosity inside the microsystems. Consequently the following reactions are controlled by a higher fluid/rock ratio. The initial structure and volume (solids plus pores) of the parent rock is conserved (saprock); each primary mineral being altered separately (internal microsystems). New mineral assemblages are formed. Namely, the orthoclase crystals are transformed into expandable phases+kaolinite. The illite crystals formed in the earliest stages remain unaltered, even if the surrounding orthoclase has disappeared, being replaced by new clay minerals. The overall mineral reaction may be written as follows:

$$\text{K} - \text{Feldspar} \rightarrow (\text{illite}) + \text{smectite} + \text{kaolinite} \tag{2.4}$$

Then, the rock structure disappears in the most altered zones in which fine-grained debris of the parent minerals are embedded in a porous clay-rich matrix (saprolite). The collapse of the initial structure creates new microenvironments in which the local chemical conditions are different from those prevailing in the saprock: higher fluid/rock ratio, new mineral contacts. Under these conditions, the orthoclase debris are dissolved, creating locally high K^+ chemical gradients. Small crystals in spherolite or fan-shape appear locally in the matrix, indicating that the clay components recrystallize locally. Using the micropicking technique to obtain X-ray diffraction diagrams of alteration minerals (Beaufort 1987), it was shown that illite and kaolinite are formed concomittantly. Both microsites show that a second generation of illite is formed in equilibrium with the dominating clay species: kaolinite (Fig. 2.8) according to the following reactions:

$$\text{K} - \text{Feldspar} \rightarrow \text{illite} + \text{kaolinite} \tag{2.5}$$

$$\text{smectite} + \text{K} - \text{feldspar} \rightarrow \text{illite} + \text{kaolinite} \tag{2.6}$$

Here it is quite clear that illite, a low-potassium, high-silicon mica-like phase crystallizes under various conditions of high potassium activity at the expense of pre-existing magmatic potassic minerals such as orthoclase or expandable phases produced in early alteration stages.

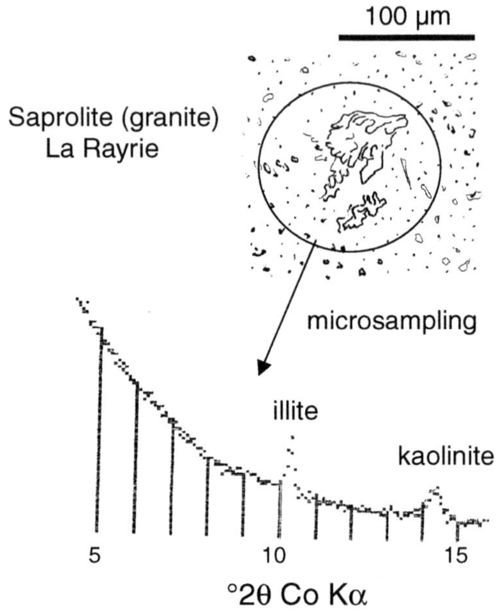

Fig. 2.8a,b. Illite formed in granitic saprolites. **a** Micrographs of saprolite microsystems in which a second generation of illite is formed. **b** XRD from microsampling

2.1.3.3
Phase Diagram

The relations described above can be placed in a chemiographic context using the two variables potassium and silica as a function of their chemical activities in aqueous solution in a phase diagram (Garrels 1964). The vermiculite (K-depleted, aluminum-hydroxy-interlayered mica) stability field must be added to the diagram calculated by Garrels. Then it is easy to represent the secondary mineral assemblages corresponding to the 6 reactions defined above (Fig. 2.9). The activity of potassium in solution (a_{K^+}) versus the activity of silica (a_{SiO_2}) pathways for mica and K-feldspar alteration processes are represented by stippled arrows. This shows that the activity of SiO_2 in solution being controlled by the solubility of quartz, illite cannot be formed during the alteration of muscovite while it is formed during K-feldspar alteration. Reactions 2.5 and 2.6 implies that a_{K^+} is able to increase even in the latest alteration stages of the granitic rocks. This is the case in the very vicinity of K-feldspar debris which are strongly attacked by meteoric fluids percolating in the porous saprolite zones.

The decrease of K^+ chemical potential induces first the destabilization of muscovite by the introduction of hydrogen ions in the place of potassium. This mineral is not stable, probably because of the strong difference in hydration energies of hydrogen ions and potassium. The potassium-leached mica then becomes unstable, dissolution occurs and finally a vermiculite+kaolinite mineral assemblage crystallizes. A similar reaction was observed in potassium-leaching laboratory experiments where Ba replaces potassium ions (Kodama and Ross 1972) Here a disequilibrated muscovite recrystallized into kaolinite.

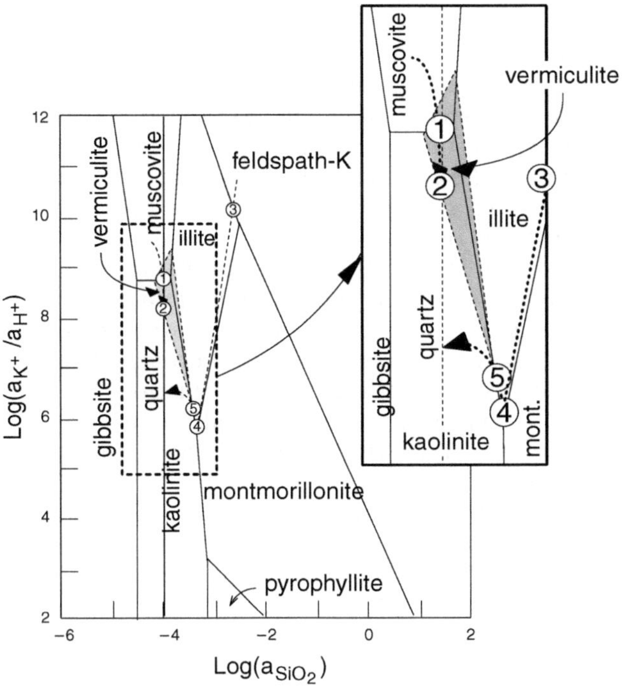

Fig. 2.9. Phase diagram for the weathering of granitic rocks in the SiO_2–Al_2O_3–K_2O–H_2O system (modified from Garrels 1984). The stability field of vermiculite is represented by the grey zone. *Numbers* correspond to the mineral reactions defined in the text

If we consider the above discussion, it is possible to establish the conditions of illite formation in granite weathering. First, in most soils, two major actions take place 1) the ion exchange in silicates by the substitution of hydrogen for a cation, 2) the eluviation of smaller minerals, essentially clays, lower in the profile where they constitute a clay-buffered system. The first action can be expressed on a simple phase diagram concerning K, H, Si and Al (Fig. 2.9). In the upper portion of the diagram potassium activity is more important than that of hydrogen. Feldspar and mica are stable. As the ratio of potassium to hydrogen ions (K to H) changes which results from the addition of slightly acidic rain water, phases such as smectite and kaolinite, containing less potassium, become stable. One can expect that illite, less potassic than muscovite, will become stable. This is described above. As the ratio of K/H continues to change in favor of more potassium-poor phases, smectites or Al-vermiculite will form. Most often an interstratified illite/smectite, or illite/high-charge smectite(vermiculite) mineral is present in illite-bearing soils.

In the studies of granite weathering it is clear that illite in a new, stable phase formed according to specific chemical constraints. Further, it is clear that muscovite (high temperature mineral) is not stable and is transformed to less potassic minerals. This is a demonstration that illite is stable under the appropriate chemical conditions in a natural setting, and that muscovite, the fully potassic mica, is not stable.

In agricultural and other soils the observed destabilization reaction of illite to illite/smectite interlayered mineral can be reversed in nature. After 40 years of treatment with KCl fertilizer or with manure amendments, illite becomes a greater portion of the I/S clay minerals. Also illite itself can be formed under certain conditions. This suggests that in fact the presence of illite in soils responds to a specific set of chemical constraints and hence that illite is in fact a soil clay mineral forming in response to specific chemical potentials. It is not just an unstable intermediate phase present in a process of degradation. In principle a metastable phase should not form after reversing the chemical potentials in a system. It can only resist change through the existence of a high activation energy necessary to effect its transformation. The fact that soils are dynamic systems of varying chemical potentials often confuses observations on mineral occurrence and stability which could have led to the conclusions of many authors in the past.

From observations of granite weathering and soil clays that have a chemical environment that has been changed by agricultural practices, one can deduce that illite is a stable clay mineral responding to chemical potentials as do other stable phases.

2.2 Illite in Diagenetic Series

The increasing abundance of illite has been largely described in diagenetic environments since the pioneer works of Burst (1959) and Weaver (1959). The fundamentals of the smectite-to-illite transformation (illitization) through illite-smectite mixed layer (I/S) series were outlined by Dunoyer de Segonzac (1969) and Hower et al. (1976). The formation of illite has been described in three different environments: 1) the recrystallization of clay minerals in shales and clay-rich sandstones 2) the direct precipitation from solutions on kaolinite or quartz grain surfaces 3) clay transformation in bentonite beds. The first occurrence is assigned classically the smectite-to-illite conversion and has been abundantly studied (see the review from Altaner and Ylagan 1997). We will consider the several types of illite occurrence in sediment and sedimentary burial conditions.

2.2.1 Illite Formed During Early Sedimentary or Eodiagenetic Processes

2.2.1.1 Ocean Sediments

In spite of what was believed until the 1960s, the formation of aluminous illite from sediments deposited in marine environments was shown to be negligible (Weaver 1967). Most of the marine illite occurrences described in DSDP reports for instance are detrital, 2M mica and not neoformed species (see the review from Srodon and Eberl 1984). Indeed, the phyllosilicates which absorb the K^+ ions in these environments are Fe-rich species: Fe-smectites (nontronites) or glauconites. This could be due to structural constraints as proposed by Velde (1985): the *b* dimension increases with Fe amount, giving room for large cations in the ditrigonal cavities. However, Fe micas such as glauconite are not formed. There is no clear evidence for the formation of illite in deep ocean sediments.

Glauconite is formed in shallow oceanic environments. It is generally not considered to be illite, despite the continuity in composition between the aluminous illite and iron-rich glauconite (see Sect. 1.2.1). We will not try to determine the differences between illite and glauconite here. The distinction between the two minerals will be made based on geologic occurrence. Glauconites are formed in pelletal form through the diffusion of elements into and out of these pellets. This form is not illite. Thus due to this mineralogical-geological definition, we will not consider the shallow oceanic occurrence of potassium mica-like minerals which are called glauconites even though their compositions apparently form a continuous series with illites (Fig. 1.9).

2.2.1.2
Sedimentary Illite (Shallow Basin)

The formation of illite in lacustrine sediments was observed in high-salinity lakes (Singer and Stoffers 1980). These authors showed the presence of three smectite-to-illite transition cycles in clay-rich sediment drill-cores (10 and 56 m) from two east African lakes. However, in spite of the shallow depth, these changes were attributed to diagenetic reactions rather than to sedimentary ones because of two reasons:

- I/S MLM are formed repeatedly during the smectite-to-illite transition,
- clay mineral and zeolite mineral reactions are concomitant.

The global reaction, smectite + K, Na zeolite $\rightarrow$ illite,+ analcime is controlled by high saline palaeo-lake water in which the K/Na ratio is buffered by the destabilization of Na, K zeolites.

An apparently rare case of massive, mono-mineral illite formation in sediments is found in the Oligocene sedimentary basin of Central France centered near Le Puy en Velay. These sediments were studied in detail by Gabis (1963) using XRD and chemical methods. The data are rather complete, even though some questions remain unanswered, as is usually the case in the investigation of geological problems. The shallow Tertiary basin, several hundreds of meters of sediments, is formed on a granitic basement. The shallow sedimentary cover indicates that the temperatures due to burial were never above 50–60 °C. Several meters of clay-rich sediment is found in the center of the basin. Sediments near the edge of the basin indicate the presence of smectite-rich I/S minerals, kaolinite and illite. In several instances the layers in the central part of the basin are massive, olive green in color and are composed of illite, some carbonate

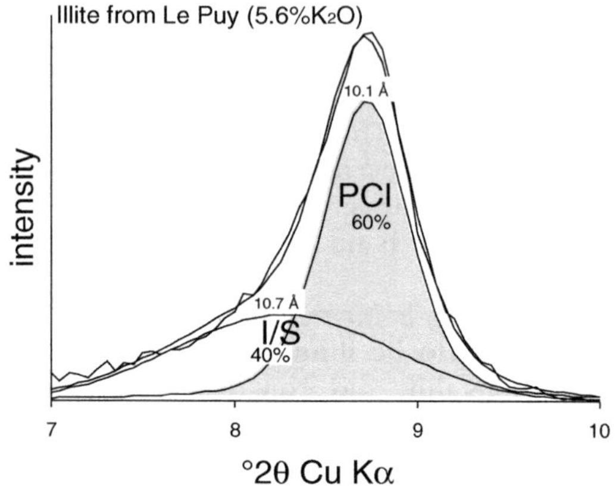

Fig. 2.10. Background-subtracted, decomposed XRD patterns. Sr-saturated, air dried sample of Illite du Puy

and quartz. The illite is of the 2M polymorph (E Nicot, University Paris VI, unpublished data). It is aluminous and contains more iron than most aluminous glauconites (Berg-Madsen 1983). This occurrence is of great interest in that it straddles the glauconite-illite compositional line and instead of being 1M as are glauconites (Hower and Mowatt 1976), this mineral it is of the 2M polymorph.

Even if we know much about the mineralogy of illite du Puy, there are some serious problems in explaining its formation despite the excellent study by Gabis. Gabis states clearly that the major volcanic activity of the Massif Central in France is posterior to the deposition of the beds containing the illite. Further the volcanic activity of this period is generally basic in nature, not that conducting to the formation of aluminous illite of I/S minerals. However, there are outcrops of acid volcanic material in the region and such sources cannot be ruled out. Further, the illite beds apparently show no chemical or mineralogical zoning, phenomena typical of transformed volcanic beds (bentonites, see following section). Thus the precise mechanism of formation and starting materials are not known for this material.

The mineralogy of this illite is generally typified in XRD studies by a strong, asymmetric peak near 10 Å. This peak is composed of two peaks in fact, one small wide one (0. 8 °2θ Cu Kα) near 10.8 Å and another, more narrow (0. 5 °2θ Cu Kα) near 10.2 Å. The first peak indicates a high illite content I/S mineral with measurable smectite content, and the second a PCI of low smectite content (Fig. 2.10). Thus the material is of the 2M polymorph and is a PCI mineral. In this rare instance one finds that 2M illite, PCI, is produced at or near the sediment interface.

2.2.2
The Origin of Illite in Shale Burial Diagenesis

2.2.2.1
Sediment Burial Diagenesis

Diagenesis reactions have been observed in great detail in three major rock type categories; shales, bentonite beds and sandstones. These categories have been chosen and investigated for their roles as indicators of mineral change in different contexts related to the production and accumulation of petroleum. Shales are the source rocks for petroleum. Thus it has been assumed that mineral change could influence the development and migration of hydrocarbons. For example the change from smectite to illite involves a loss of water from the solid phases and hence there is a possibility of water release and flushing of hydrocarbons due to mineral change in the shales.

Bentonites have been used as analogues of reactions believed to be found in shale. The reason for using bentonites is that they present very clean, almost monomineral clay mineral assemblages. It was believed that observing a simple system would give insight into reactions in systems that are more complex such as shales where detrital material results in different reactions of different importance depending upon the rock investigated.

Sandstones were investigated because these rocks are one of the major reservoirs for petroleum accumulation. As such their permeability and porosity are extremely important. The formation of clay minerals can have determinant effects on these properties. Thus the timing and importance of clay mineral formation is often critical in an assessment of petroleum prospects in a given series of sedimentary rocks.

We will consider these different sedimentary rock facies individually because they in fact represent different chemical environments. All rock types in a sedimentary basin are subjected to diagenesis which is the increase of temperature as a function of time. However, in each rock type the differences in permeability result in differences in chemical influences from outside the rock concerned. In a shale the bulk composition throughout the diagenetic process is roughly the same. This approaches a closed system in the thermodynamical sense. Bentonites are known to be permeable and to transfer of certain elements, potassium and silica being the most important. These processes occur slowly due to the relatively low permeability of the clay-rich beds. Thus bentonite beds show edges where chemical gradients are found. In sandstones permeability is often high and fluids move freely. In doing so the fluids can change the chemistry of the system, greatly provoking the dissolution of minerals or the precipitation of others. The result is that there is a range of chemical systems of varying degrees of "openness" where the overall effects of temperature and time work along with more or less rapid changes in chemistry.

We will investigate the different types of sedimentary rocks step by step.

2.2.2.2
The I/S MLM Versus Depth Sequences in Shales

Weaver (1959) brought attention to the fact that illite was a major, if not the major, mineral in old sedimentary rocks. He observed that this was less the case as the sedimentary rocks were of more recent age. This led him to postulate the formation of illite during the processes of diagenesis, which is none other than the initial stages of metamorphism, i.e. the effect of increasing temperature and pressure on sediments and other rocks. The realm of diagenesis (see Dunoyer de Segondzac 1968 for a review) is in fact that of gradual, change in temperature over different periods of time, millions to hundreds of millions of years. Normally diagenesis is restricted to the conditions of stability of clay minerals. This is not a definition, but probably a coincidence, ignored by most geologists. For Weaver and the clay mineralogists who followed him, the problem of time was not considered to be important but the temperature was. Hence diagenesis was considered to be largely an affair of thermal energy.

More recently, it became apparent that time could be a factor in the development of illite in sedimentary rocks. The idea of the smectite-to-illite reaction (Burst 1969; Hower et al. 1976) was confirmed and new sets of data led to the realization that time was a factor in burial diagenesis which has, in turn, led to different kinetic models of the reaction (see. Elliot and Masitoff 1996).

In the observed transformation of smectite to illite, there are not two independent phases which occur in varying proportions as is the classic chemical situation, but one finds that the reaction is described in its intermediate stages by minerals of mixed layering of the two components, illite and smectite. The proportion of the components in the mixed-layered minerals is taken to represent the progress of the reaction from smectite to illite. Identification methods developed by Reynolds (see Moore and Reynolds 1989) allow mineralogists to identify the proportion of mixed-layer components through the use of X-ray diffraction analysis. Systematic use of a single method of identification of the proportion of the components leads, most often, to regular

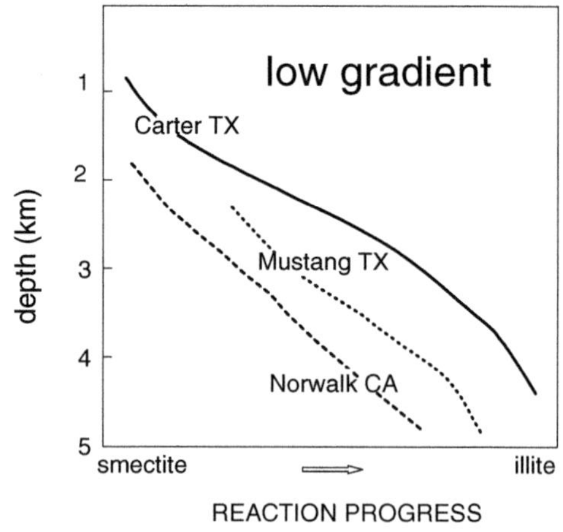

Fig. 2.11. Low thermal gradient basin depth composition (illite content of I/S MLM) curves for wells from the Gulf Coast (USA) Carter and Mustang, and California (data adapted from Velde and Vasseur 1992)

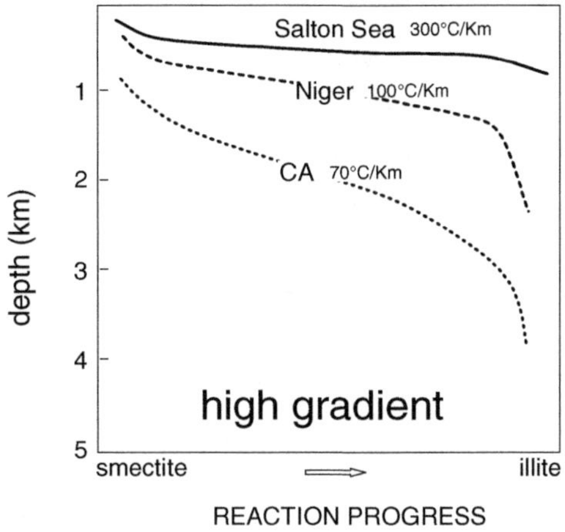

Fig. 2.12. Depth-composition plots(illite content of I/S MLM) for high thermal gradient basins Salton Sea (Lanson and Velde 1993), Niger (Velde et al. 1986), CA (adapted from Jennings and Thompson 1987)

distributions of mixed-layer illite/smectite (I/S) clay mineral composition in a burial sequence, such as those encountered in deep bore holes used to exploit petroleum resources.

Typical composition curves (i. e. proportion of illite or smectite in I/S minerals as a function of depth) of samples in sedimentary basins are shown in Fig. 2.11. These examples represent sedimentation under more or less low constant thermal gradients (20–30 °C km^{-1}) of the period of sedimentation observed.

Situations of higher thermal gradients, in geothermal areas, give similar results where a succession of compositions are observed but the shape of the composition-depth curve is somewhat different. Velde and Vasseur (1992) observed that the curve is not typical of a single reaction mechanism, but can be better described by a two-step reaction. This is valid for systems under geothermal gradients of between 20 to 40 °C km^{-1}. Higher temperatures (70–300 °C km^{-1}) lead to simulations using a single reaction (Velde and Lanson 1993) which fits the observed mineral transformation more optimally (Fig. 2.12).

2.2.2.3
I/S Phases Observed

Smectite-to-illite evolution has two sequences of continuously changing composition minerals: I/S disordered from 0 to 50% smectite and ordered I/S 50–0% smectite. Three phase changes occur:

1. smectite to disordered I/S
2. disordered I/S to ordered I/S
3. ordered I/S to illite.

One can follow these reactions by observing XRD diagrams of samples in deep bore holes.

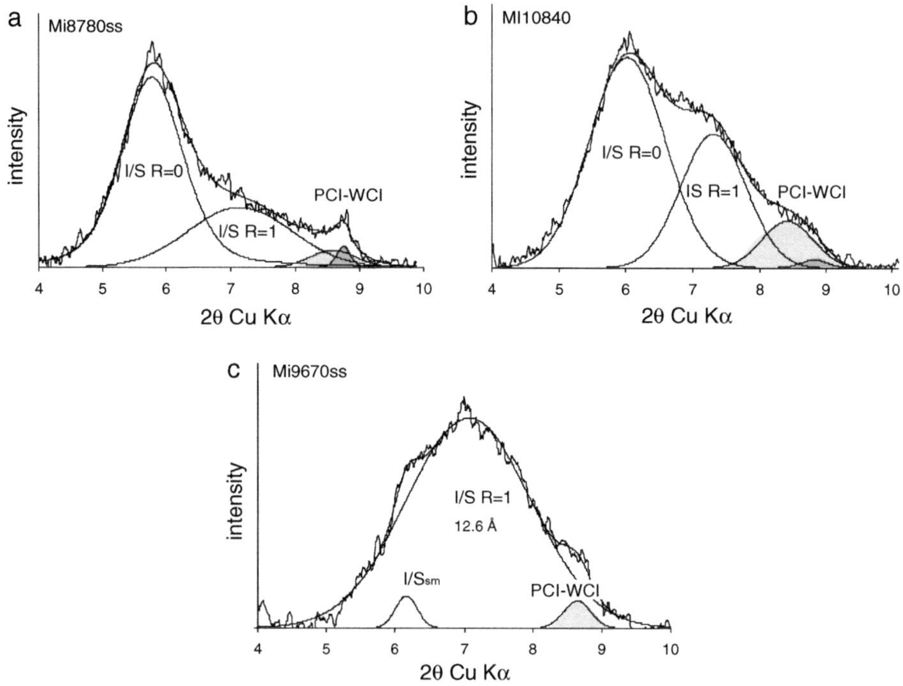

Fig. 2.13a–c. XRD decomposition of spectra for several samples from a deep well in the Gulf Coast (unpublished data). Series a to c indicates change in relative proportions of R0 compared to R1 phases. Note increasing presence of PCI

Figure 2.13 shows the transition from I/S MLM of a random structure (R0) to one with an ordered structure (R1). In the three samples one finds initially a dominant R0 phase and a small amount of ordered (R1) mineral. The second spectrum shows a stronger R0 peak, and it becomes more narrow due to crystal growth and increase in coherent diffracting domain. In the last sample of the series, one finds almost exclusively the R0 peak.

Looking at the data points in many studies, it is clear that the initiation point in the reaction (100% pure smectite) does not appear at the surface of the sedimentary sequence but at some depth, where the present temperature is at 50–80 °C between 1 and 2 km. Unfortunately, most of numerous diagenetic series studied published in available literature never investigate the upper part of the stratigraphical sequence in which the formation of pure smectite may occur because of lacking economic interest. However, a few studies showed that montmorillonite is formed during that early diagenetic stage, as for instance in the Colorado river deltaic sedimentary formations (Jennings and Thompson 1987) and Neogene Japanese sedimentary basins (Velde and Iijima 1988). Indeed, the mineralogical composition of the clay fraction (< 2 µm) of the sediments is variable in the upper part of the sequence; at a given depth which depends on the geothermal gradient, i. e. temperature of about 50 or

80 °C in the Neogene Japan and Colorado series, respectively, it becomes remarkably homogeneous. Thus it appears that the reaction of smectite to illite is dependent upon a transformation of sedimentary materials into the smectite of diagenetic reaction. This is the first diagenetic reaction step:

1. Inherited sedimentary clay mineral assemblage → smectite (montmorillonite)

The second step would be:

2. smectite (montmorillonite) → disordered mixed layer (I/S)

In the description of clay mineralogy given in Hower et al. (1976), one finds that there are two types of I/S minerals present in sedimentary sequences, one is a disordered I/S (designated as R=0) and the other is based upon an ordered, two-layer mineral (designated as R=1). For the most part the sequence seems to be more or less continuous in smectite/illite compositions, the R=0 obtains up to about 50% smectite and the R=1 begins, by definition, at 50% smectite. The third step of the smectite to illite transformation is then:

3. I/S (R=0) 50% smectite → I/S (R=1) 50% smectite

Lanson and Champion (1991) showed the evolution in crystal size and composition of the R=1 form as a function of diagenetic progress. The fourth step is:

4. I/S 50% → I/S low% smectite (PCI)

The ultimate step in the reaction sequence is the production of illite. This occurs at the expense of the R=1, ordered I/S mineral. The compositions of the illite formed by diagenetic reaction are different from those of the coexisting ordered I/S minerals (Lanson and Champion 1991). WCI begins to form during the stages of development of illitic, R1 I/S mineral development. The last step of the reaction is then:

5. low% smectite I/S (PCI) → illite (WCI)

Much work has been done on the shape, thickness and composition of the R=1, ordered I/S (Nadeau et al. 1984a,b; Inoue et al. 1987 among others) which has been clearly shown to be different from the predecessor disordered I/S mineral. The high smectite-content mineral has a rather irregular shape based upon a hexagonal form while the ordered I/S mineral has a lath shape. These differences in shape and internal structure led Velde and Vasseur (1992) to postulate a reaction between the two I/S minerals, one forming from the other hence the two-step kinetic formulation of the illite-to-smectite transformation in diagenesis.

However, it should be noted that in most instances illite (PCI and WCI) is present along with the I/S minerals in sedimentary rocks and its presence is usually greater in older sequences, as noted by Lanson et al. (1998). This relation is illustrated in Fig. 2.14 on XRD traces. These observations led us

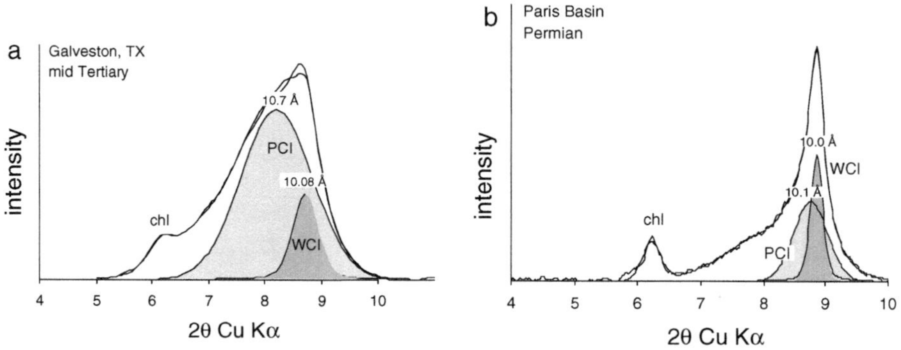

Fig. 2.14. Decomposed, background-subtracted XRD patterns of Sr-saturated specimens in the air-dried state. Both samples are near the end of the I/S evolution and show very much illite. The PCI is much more important in the younger than the older materials

to postulate yet another independent mineral formation which is the precipitation of illite during the I/S disordered to ordered I/S mineral reaction and during the evolution of these phases. Illite (WCI) probably forms from the materials which dissolve to form the I/S phases. The proportion of PCI-WCI in clay XRD patterns increases in older basins compared to that in younger basins. This is shown in Fig. 2.14 for a sample from the mid-Tertiary Texas Gulf Coast (Velde, unpublished) which experienced low thermal gradients, and a sample of similar clay evolution from the Permian of the Paris Basin. The proportion of PCI and WCI is strikingly different in the two examples which are near the end of the I/S MLM smectite-to-illite transformation series. The older sediment has a much greater proportion of WCI than the younger one.

Hence PCI, distinguished by a peak at near 10.3 Å with a width greater than 0.5 °2θ Cu Kα is the product of I/S evolution towards illite by reduction of the proportion of smectite layers in the crystallite (see Sect. 1.4.1). The more narrow peak at 10.0 Å (width less than 0.4 °2θ Cu Kα) represents the pure illite material which forms along the course of the illitization process during diagenesis. The compositions of the two types of illite are probably very much the same but they form by different reaction paths.

As far as one can tell, the ordered I/S minerals have a lath shape while the illites have a hexagonal shape (Lanson and Champion 1991; Bauer et al. 2000). Both of these mineral shapes grow (become larger in their *a–b* dimension) during diagenesis, showing typical patterns of growth morphology. Hence the end point of the I/S ordered reaction is not really illite, but a very illite-rich mineral. According to Gharrabi et al. (1998) the poorly crystallized illite (PCI of Lanson) would be a highly illitic phase with some (< 5%) smectite layers, while the well-crystallized illite (WCI) would be pure illite. This pure illite recrystallizes to fine-grained mica in the transition towards metamorphic rocks.

Lanson et al. (1992) show that the evolution of the crystals, as observed by XRD methods, is not precisely the same in all sedimentary basins. The crystal-diffracting domains, a crude indication of crystal thickness and hence size, are affected by geothermal gradient and thus growth processes.

2.2.3 Illite Crystallinity

In the 1970s, methods were developed to estimate the relative maturity or state of diagenesis of illite mixed-layer clay minerals in sedimentary rocks (Kübler 1968). The method was called the crystallinity index of illite. Due to the recent generalization of X-ray diffractometers and their systematic use in the study of clays, a simple, comparative method could be established. The idea was to estimate the overall effect of temperature and age of the sedimentary rocks as they affected the evolution of the illitic clays. The method is quite simple, it consists of the measurement of the illite peak width at half height. The measurement is made in millimeters where 10 mm = 1 °2θ Cu Kα in peak width. Generally speaking, values of less than 2.5 mm indicate metamorphic (epigenesis) conditions, values of 2.5–5 mm indicate low-grade metamorphism (anchizone) and values greater than 5 mm indicate conditions of diagenesis (see Jaboyedoff et al. 2000; Gharrabi et al. 1998).

As we have seen in the preceding sections (2.2.1 and 2.2.2), the illite peak is most often a combination of overlapping bands of different phases; I/S, PCI and WCI. Hence the width at half height of an illite peak takes into account the presence and relative intensity of the different phases. This can be seen for the examples of Fig. 2.14, shale samples from the Paris Basin where the total band complex has an overall illite crystallinity index of 17 (1.7 °2θ Cu Kα) in the upper Cretaceous indicating diagenesis whereas that of the Triassic has an index value of 7 near the anchi-epizone boundary. Figure 1.3 shows several illite samples with various proportions of WCI, PCI as well as second order illitic I/S phase peaks. Peaks at less than 10 Å are due to second-order reflections of I/S phases which contribute to the overall width of the illite peak. The first-order peaks of this phase are of very low intensity and not easily discernable. In these samples it is clear that there is a great variety of combinations of phases which contribute to the overall band width near 10 Å which is considered to be the illite peak in the illite crystallinity determinations.

Despite the problems of multi-peak overlap, the overall narrowness of the illite peak does in fact give insight into the maturity of the shales in the reaction conversion of smectite to illite and in the crystallization of the PCI and WCI mineral phases. The illite index is most useful in establishing the final stages of diagenesis and the change into metamorphic mineral assemblages. Illite crystallinity indexes are usually in the range of 6–3, or 0.6–0.3 °2θ Cu Kα FWHM. This is best seen using the third order 003 peaks as suggested by Gharrabi et al. (1998). In this range, it is easier to distinguish the different components and effects of artificial peak broadening are to a large extent eliminated. However, interference from other phases, such as quartz, often make these determinations difficult.

However, the bulk of the studies in the literature use the 001 basal spacing, and for reasons of comparison and continuity, we will continue the discussion

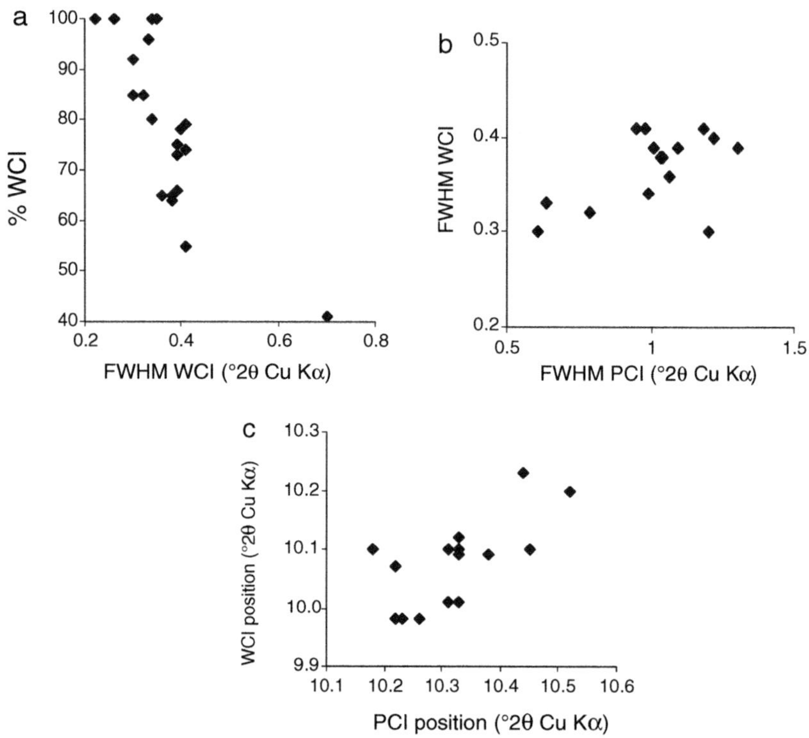

Fig. 2.15a–c. Samples taken from Paleozoic rocks in the Ardennes area (France and Belgium). These rocks have experienced high-grade diagenesis or low-grade metamorphism. Part **a** shows the covariations of % PCI vs. FWHM, **b** the covariance of WCI and PCI FWHM, **c** WCI and PCI peak positions

using this peak. One can recall the calculations of width and peak position given in Chap. 1 (1.1.2.3) which describe the PCI and WCI phases observed in sedimentary rock series by Lanson et al. (1998). A summation of the two values would give the raw illite crystallinity index, without the overlap of I/S mineral bands which contribute a certain amount of width to the illite overall peak when they are present.

Figure 2.15 shows some data (unpublished, B Velde and J-P Sagon, Univ Paris VI) for lower Paleozoic shaley rocks along the French-Belgian border between Rocroi and Givet along the Meuse river. These samples contain no I/S phases. Two relations are evident in XRD traces of these samples: the width of PCI and WCI both diminish together, indicating that there is a simultaneous growth of these crystals (thickness affecting the coherent diffracting domains) and the relation between WCI width and peak surface area of the PCI-WCI pair. As WCI intensity increases, relative to PCI, the peak width decreases, indicating larger diffracting domains and conversion of PCI into WCI even though PCI crystallites are increasing in size, and showing more narrow peaks. This suggests that the PCI grows metastably in the last stages of its transformation into WCI. As has been pointed out earlier, the formation of illite (WCI) depends on time more than temperature. In old rocks, one finds significant amounts of WCI but in rocks having experienced thermal pulses (Salton Sea or Niger Delta) the PCI is the major component. Therefore the illite crystallinity index will depend on the absolute age of the samples as well as the temperature at which they evolved.

One of the initial hopes of using the illite crystallinity index was to determine a correlation between organic-matter maturation and that of I/S and illite minerals. As pointed out by Vasseur and Velde (1992), the kinetics of illite or I/S silicate maturation are largely dependent on time and temperature while those of organic matter are largely dependent upon temperature (see Velde and Espitalié 1989). Given this difference in dynamics, it is unlikely that there would be a strong correlation between organic matter maturation and that of clay silicate minerals in spite of the efforts of several workers (see Kisch 1990; Héroux et al. 1979 for example). Recent work has tended to focus on metamorphic conditions and curve decomposition methods (Stern and Mullis 1991; Wang and Stern 1995; Warr and Rice 1994; Warr et al. 1991). Several examples of non-correspondence have been shown, notably by Lanson and Velde (1993). This problem will be discussed in more detail in Chap. 4.

In general, it seems that the crystallinity index is only for very general use, not being subtle enough to follow the changes in proportions of the different components I/S; PCI and WCI in evolving pelitic rocks.

2.2.3.1
A Possible Model for Smectite-to-Illite Conversion in Diagenetic Shales

Given the observations discussed above on mineral types discovered by XRD methods, one can begin to sketch an evolution of clays in the change from smectite to illite in sedimentary pelitic rocks.

Formation of Montmorillonite A 100% expandable new mineral of montmorillonitic chemical composition is formed, previous to illitization during the early diagenetic stages of sediments:

$$\text{Inherited sedimentary clay mineral assemblage} \\ \rightarrow \text{low} - \text{charge montmorillonite} \qquad (2.7)$$

The layer charge varies according to the alteration process (dissolution-crystallization of phyllosilicates). Whatever the layer charge differences, in all the cases, the smectite which experiences the illitization process does not have any charge in the tetrahedral sheets.

The Random I/S MLM Stage: Vermiculitization – Illitization of Montmorillonite The chemical compositions of the diagenetically altered shales, sandstone or bentonite series studied by Srodon et al. (1986, 1992), Ramseyer and Boles (1986), Nadeau and Bain (1986), Awwiller (1993) or Cuadros and Altaner (1998) indicate the formation of high-charge expandable layers together with the illitization process (Fig. 2.16). The fixed-cation contents K^+ of I/S scatter significantly

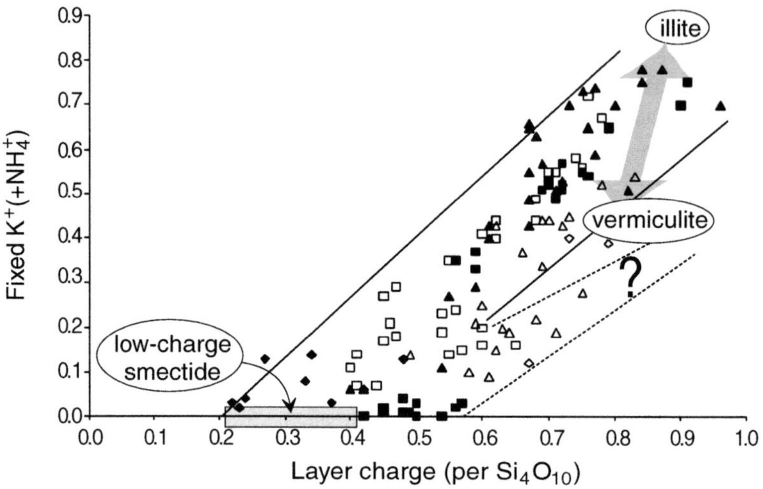

Fig. 2.16. Relation between the total layer charge and the fixed K^+ cations in natural I/S series from shales, sandstones or bentonites. Some data include NH_4^+ cations. *Solid diamonds*: Ramseyer and Boles (1986); *open diamonds*: Srodon et al. (1992); *solid squares*: Nadeau and Bain (1986); *open squares*: Srodon et al. (1986); *solid triangles*: Cuadros and Altaner (1998); *open triangles*: Awwiler (1993). The *large double arrow* outlines the K content of the illite-vermiculite domain

(large double arrow) for a given layer charge, e. g. 0.8 per $O_{10}(OH)_2$. The lower value clearly indicates the presence of high-charge layers not saturated with K^+, and which are thus still expandable. Similarly, the experimental illitization of montmorillonite has shown that the proportion of high-charge layers increases during the first step of a dual reaction (Whitney and Northrop 1988). These high-charge, expandable layers are reminiscent of the vermiculitic component described by Shutov et al. (1969), Foscolos and Kodama (1974), and Drits et al. (1997a).

A question we can ask is, how are R=0 I/S MLM formed from a montmorillonitic smectite? From the above, the R=0 I/S MLM are known to be composed of three types of layers (Fig. 2.17a): smectite (fully expandable), vermiculite (partially expandable) and "illite" (non-expandable). The first step in montmorillonite-to-illite transformation then must involve vermiculitization of the smectite layers through the formation of the high-charge, tetrahedrally substituted layers from the original low-charge montmorillonite (Drits 1985): This can be described by:

$$\text{Low charge montmorillonite (100\%\% exp)} \quad (2.8)$$
$$\rightarrow \text{vermiculite (high charge tetra-octa smectite, 100\% exp + quartz)}$$

The second step is the transformation of vermiculite layers into non-expandable ones, i. e. illite layers. Reactions 2.8 and 2.9 increase the tetrahedral charge by replacement of Si^{4+} by Al^{3+}. This reaction contributes to SiO_2 exportation from shales to neighboring sandstones where quartz overgrowths are observed:

$$\text{Vermiculite} + K^+ \rightarrow \text{illite} + \text{quartz} \quad (2.9)$$

The formation of non-expandable layers (0% expNa) occurs when two conditions are respected: 1) the layer charge is high enougth; 2) K^+ ions are available in the (dissolution of detrital mica or K-feldspars).

How are smectite particles transformed during their transition to R=0 I/S MLM? The vermiculitization of initial montmorillonite layers account for the decrease of the turbostratic index described by Reynolds (1992) as illitization proceeds. This certainly changes their morphological aspect but does not change the shape and size of individual layers. Two processes may control reactions 2.8 and 2.9: solid state transformation (SST) or dissolution crystallization (DC)? Baronnet (1997) gave simple criteria to identify each (Table 2.2).

Vermiculitization (formation of high-charge smectite) is presumably a solid-state transformation because the crystal habit does not change (flakes) and the smectite polytypism is at least partly conserved (turbostratism). The initial chemical and structural characteristics of the smectite (montmorillonite) are thought to change by ion diffusion, particularly by replacement of Si^{4+} by Al^{3+} in the tetrahedral sheets. If true, a tetrahedral charge is added to the inherited octahedral one leading to the formation of high-charge layers (Fig. 2.17b). This process can be related to the laboratory experiments of Howard and Roy (1985) which show that the reaction has a low activation energy of approximately 3 kcal/mol.

The illitization of R=0 I/S MLM may be controlled by two processes:

- chemical diffusion: The high-charge layers are known to retain preferentially K^+ ions. These ions are irreversibly fixed; thus the CEC is reduced and the interlayer becomes non-expandable.
- crystal growth: illite layers are added on crystals whose outer interfaces were "previously prepared" by the vermiculitization process.

How does vermiculitization and then illitization proceed? 1) solid state transformation (see Altaner and Ylagan 1997); 2) crystal growth (addition of illite

Table 2.2. Criteria for discriminating between the two main intermineral transformation mechanisms from Baronnet (1997). SST: solid state transformation; DC: dissolution-crystallisation

Criteria	SST	DC
Crystal habit	Retained	Lost
Topotaxy	Yes	Yes (epitaxy) or no
Polytype inheritance	Yes	No
Chemical and structural homogeneity	No	Yes
Chemical inheritance	Yes	No
Inheritance of crystal defects	Yes	No

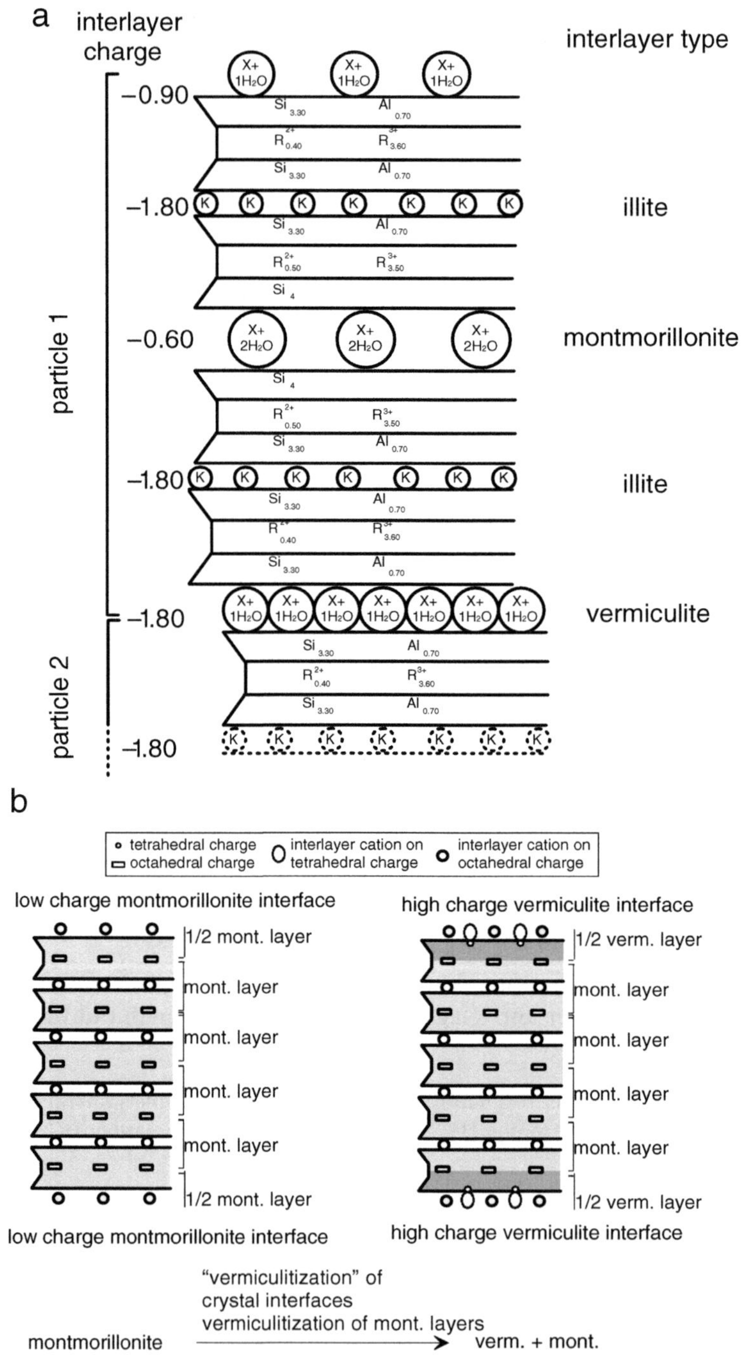

Fig. 2.17a,b. The vermiculitization process. **a** The three types of layers forming R=0 I/S MLM **b** A possible diffusion-controlled mechanism of interface vermiculitization

layers) The evolution of the size and shape of particles is different for the two processes (Table 2.2). Unfortunately, such an evolution has not been confirmed for the R=0 I/S MLMs when illite content varies from 0–50% in natural environments. However, if crystal growth is active, it should not change the crystal shape much, which remains flaky. This is confirmed by the experiments from Whitney and Velde (1993) who showed that particle size and aspect ratio did not systematically change as a function of reaction extent during R=0 I/S MLM illitization. Particles exhibited rounded edges, suggesting a dissolution of the starting material. Thus, the first stage of the smectite illitization, producing the randomly ordered I/S MLM series from 100 to 50% expandable layers is probably dominated by a solid state transformation process. The chemical composition of the illitized layers changes by ion diffusion as depicted in Fig. 2.18a.

In diagenetic systems (shales), the relative stability of single particles in given temperature-chemical conditions varies with their composition. Thus, the composition range induces an increasing unstability with smectite amount in R=0 I/S MLM. The smectite-rich particles are dissolved, while the illite-rich ones are conserved or may grow.

Using Markovian probabilities (Drits and Tchoubar 1990) for a given average composition (60% smectite–40% illite), the particle composition domain varies from nearly pure smectite to nearly pure illite end-members (Fig. 2.18b). Assuming that I/S MLM are composed of only two-layer types (illite and smectite), this wide particle composition range was used to reproduce the large width of the typical XRD bands of randomly ordered I/S MLM. It is now admitted that the width of XRD bands is related to the heterogeneity of the smectite component which may contain 1 or 2 water or ethylene glycol layers (Drits et al. 1997). Thus, it is highly probable that the particle composition range of R=0 I/S MLM is not as wide as imposed by the Markovian statistics. Moreover, other distribution functions (Gauss, log-normal, Pearson) can be used (Fig. 2.18c). Whatever the function considered, the composition limits should be at shorter intervals.

The Random to Ordered I/S MLM (50% Smectite–50% Illite) Transition The random to ordered I/S MLM transition is observed for illite amounts close to 50%. This implies that the ordered mixed layer is regular (rectorite-type). It was referred to in the past under the name allevardite:

$$\begin{aligned}&\text{R=0 I/S MLM(50\% illite} - \text{50\% smectite)}\\&\quad\rightarrow \text{regular I/S MLM(allevardite)}\end{aligned} \tag{2.10}$$

The perfect regularity is detectable in XRD patterns by rational harmonic series of diffraction peaks: 27 Å, 13.5 Å, 9 Å, ... in the ethylene-glycol-solvated state (see Fig. 2.13). Their intensities depend on the coherent scattering domain size (CSDS). Unfortunately, when detected in shale diagenetic series, these regular mixed MLM have not been studied in detail. In particular, we ignore if the CSDS of "allevardite" particles are limited to few 27 Å super-layers in ethylene-glycol-solvated state or not. This is important to understand how allevardite is transformed in illite-rich MLM because the XRD patterns of these I/S cannot

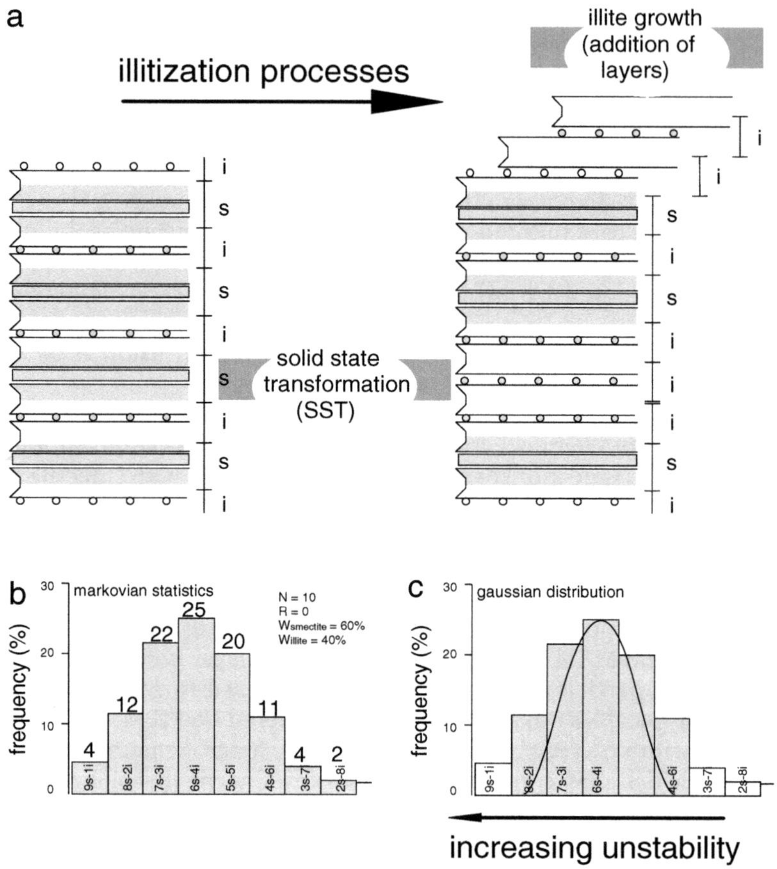

Fig. 2.18a–c. The illitization of R=0 I/S MLM. **a** Schematic representation of the illitization of a smectite layer through a solid state transformation (SST) process. **b** The Markovian distribution leads to a wide range of particle composition for a R=0 I/S MLM 60% smectite-40% illite (from Drits and Tchoubar 1990). **c** The composition range is certainly less extended in nature whatever the distribution function considered (a Gaussian distribution example is given here)

be explained by a simple growth illite layers on allevardite particles. This point will be discussed further.

However, the "rearrangement" of the random stacking into a regular sequence is similar to a polymorph transition because it apparently does not change the composition of the mineral. It could be compared to the kaolinite-dickite transition, for instance. If so, it results necessarily from a dissolution-recrystallization process. This is coherent with the observed morphology transition from flakes (R=0 I/S MLM) to laths (R=1 I/S MLM).

The crystal structure of the regular I/S MLM is not fully understood at present. Indeed, two patterns are theoretically possible: the alternation of smectite and illite layers (McEwan crystallite model) or stacking of polar layers exhibiting illite and smectite sides. The presence of polar layers has been shown by Cuadros and Linares (1995) through illitization experiments of a bentonitic clay. They showed that the symmetric-layer concept (McEwan crystallite) is inconsistent with the quantification of the amounts of illite in I/S MLM based on XRD and DTA data. The general structure of isolated and interstratified polarized layers are shown in Fig. 2.19a. As the charges are not equivalent in the two tetrahedral sheets of the 2:1 layer, an illite "unit" is composed of the two half 2:1 layers exhibiting a tetrahedral charge centered symmetrically to the interlayer zone.

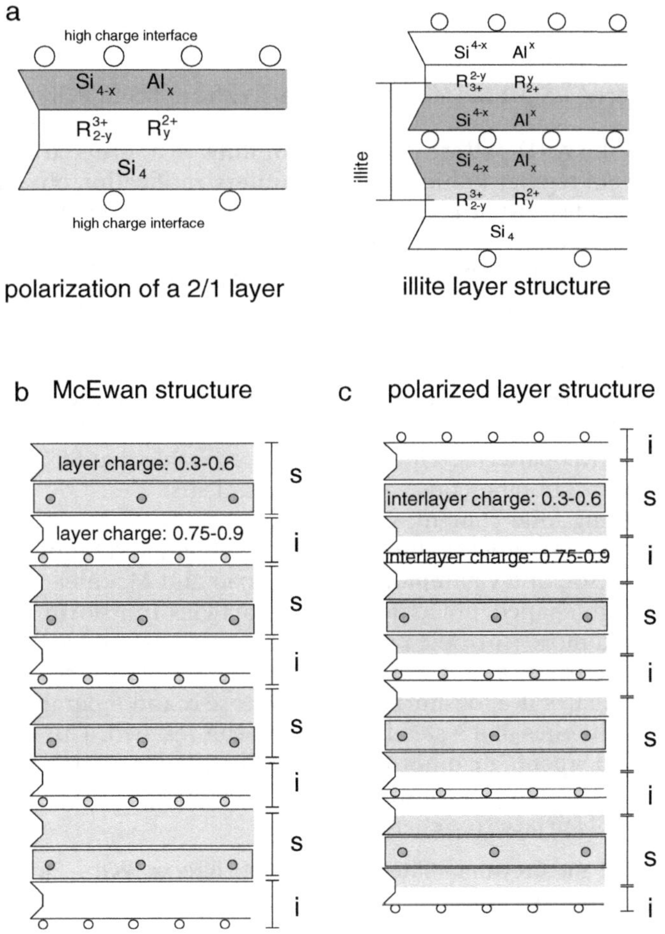

Fig. 2.19a–c. The possible crystal structures for a regular I/S MLM (allevardite=rectorite-type). **a** Structure of a polarized layer isolated or interlayered with smectite. **b** Allevardite structure according to the McEwan crystallite model. **c** Allevardite structure based on polarized layers

Thus, the possible "allevardite" structures are schematically represented in Fig. 2.19b–c. Both would diffract X-rays in the same way, but, from the standpoint of crystal growth, only one is acceptable. The structure, based on a regular alternation of illite and smectite McEwan crystallites, is hardly conceivable in diagenetic series because during illitization, the fluids are supersaturated with respect to only one phase: illite. However, the model for the growth process which controls the illitization of ordered I/S MLM, depicted by Meunier et al. (2000) cannot satisfactorily explain the development of a regularly ordered structure in which each 2/1 layer is polarized. The formation of an "allevardite-type" regular I/S MLM remains an open question at present.

The Illitization of Ordered I/S MLM (from 50 to 95% Illite) The reaction under consideration is:

Regular I/S MLM
→ ordered I/S MLM (50 to 95% illite, i. e. from I/S to PCI) (2.11)

According to the XRD pattern properties of illite-rich, ordered I/S MLM, the conversion from regular to highly ordered interstratification observed in natural environments must be conform with two conditions: 1) illite layer growth on the outer surfaces. The interfaces must be illitic; 2) SST transformation of some of the expandable sublayers inside the regular stacking sequences into nonexpandable ones (Al^{3+} and K^+ diffusion). Indeed, if condition 1 is the only one respected, the I/S particles should be composed of an "allevardite" nucleus on which illite layers are stacked during crystal growth. Thus, the XRD patterns should be that of a mineral mixture. This being not the case, it appears necessary to "break" the regularity of the stacking in order to be coherent with the observed XRD patterns. Thus it is highly probable that SST and addition of illite layers are simultaneously active mechanisms.

The increasing illite content, up to 95%, is related to crystal growth controlled by a ripening process (Inoue et al. 1988; Eberl and Srodon 1988; Eberl et al. 1990; Lanson and Champion 1991; Varajao and Meunier 1994). Indeed, small-sized, lath-shaped, smectite-rich I/S particles transform progressively into large-sized more isotropic particles. Observations carried out on several diagenetic series show that, no matter the depth, the size distribution in normalised coordinates is a log-normal curve whose α and β parameters remain constant. This implies that a steady state has been reached. This suggests that, even if Ostwald ripening cannot be applied directly (see Baronnet 1991), the driving force of crystal growth during the illitization process is related to the minimization of surface free energy.

The Precipitation and Growth of Illite Crystals (100% Illite or WCI) Finally, one may note that the proposed conversion/growth model (Fig. 2.20) may explain why the completion of the illitization reaction is so slow. In other words, why it seems so difficult to reach the 100% illite limit. The inner interfaces, initially with a montmorillonite character, which are not involved in the vermiculite stage, become increasingly less accessible to K^+ as I/S crystals are growing in the *ab* plane. As a consequence, it is likely that these inner interlayers will

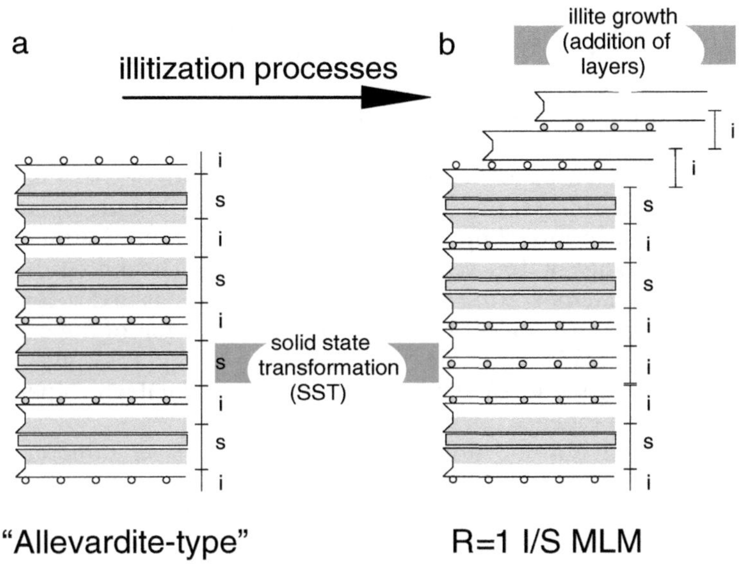

Fig. 2.20a,b. Schematic representation of the illitization process of the regular I/S MLM (**a**). The progressive addition of illite layers and the illitization of vermiculite layers in the stacking produces a series of I/S MLM in which the initial montmorillonite "nucleus" is progressively illitized via a vermiculite intermediate (**b**)

remain expandable. They can be considered as crystal defects. This makes these particles less stable thermodynamically than pure illitic ones (having the end-member composition: 0.9 K per $O_{10}(OH)_2$). How are such stable particles formed?

The illitization of I/S MLM is made possible because fluids are oversaturated with respect to illite. Even if we ignore what could be their oversaturation degree during the different diagenetic stages described above, it is highly probable that it reached values allowing the nucleation of illite particles. Then, growth was maintained by the dissolution of the unstable phases such as the smectite-rich I/S MLM and K-feldspars. The stage at which the oversaturation level necessary for nucleation was reached remains unknown. It could be very early in the diagenetic history, for instance as soon as the illite layers develop on the R=0 MLM outer surfaces.

$$\text{over-saturated solutions} \rightarrow \text{illite (WCI) nucleation and growth} \qquad (2.12)$$

This reaction is observed to begin before the last stages of illitic I/S formation.

2.2.3.2 Illitization Controls

Kinetics Several models have been proposed to account for the change from smectite to illite minerals in pelitic sedimentary rocks (see Elliot and Matisoff

1996). These models depend upon the reaction proposed (one-step, two-steps or more) and the methods of obtaining the constants of the kinetic equation. Two constants are generally used, one which affects the calculated thermal influence and the other which is related to the chemical relations of the reaction, perhaps the starting minerals, or the solution chemistry. Different methods have been used to obtain these values: laboratory experimentation or observations on natural series of sediments fitted to the kinetic equations. The importance in such modeling is to develop a description of mineral change applicable to a range of time scales and temperatures. The more restricted the range of time and temperature applications, the easier it is to develop a model. The important point is to be able to predict the evolution of the clays and to deduce the thermal history of the sequence. Since stratigraphy gives us a time frame, the missing parameter will be the thermal gradient which pertained to the sedimentary sequence during its burial history.

Thus the relative proportion of illite in a mixed-layer mineral will give information on the thermal history of the sample observed. It should be remembered, however, that a pure illite phase forms independently of the I/S reaction. There is a convergence on the illite (PCI) phase at the end of the reaction sequence but there are still two minerals present, one pure illite (WCI) and the other with a small percent of smectite layers present (PCI).

Whole Rock Chemistry Normally, shales are considered to be highly aluminous, impermeable and pelitic sediments. This translates, in terms of phase chemistry, into closed systems, i. e. where chemical transfer in and out of the rock is very limited. For the most part this is probably true. However, some indications of systematic bulk chemistry of pelitic, illite-forming sedimentary rocks (Berger et al. 1999 among others) suggest that the shales of middle to late diagenesis are less than impermeable. Altaner (1986) has suggested that the rates of feldspar dissolution would not limit the rate of change in the illitization process. Berger et al. (1997) found that the initial stages of illitization in shales was probably dependent on potassium feldspar as a major source of material for the formation of illite layers in I/S minerals. However, the low concentrations of potassium released by feldspar result in a very low rate of formation of independent illite crystals. In the latter part of the illite enrichment of I/S minerals (ordered, R=1 forms) there seems to be an indication that sources outside the shale layers come into play. Hence, the first portion of the smectite-to-illite reaction occurs in an essentially closed system, whereas in the latter stages outside sources can become important. The reliability of the smectite-to-illite transformation stage as an indicator of temperature relies on the shales being closed systems or at least that all systems investigated show similar characteristics of permeability to chemical transport.

Summary The formation of true illite (WCI) during diagenesis comes through the direct precipitation and growth of the mineral. This occurs while the reaction of smectite to illite proceeds, taking a portion of the necessary elements from the solutions which nourish the I/S crystals in their growth-illitization progress. The I/S minerals always have a small portion of smectite present (less than 5%) while the illite crystallites appear to have no smectite present.

This pure illite is normally well-crystalized (WCI) and has a crystal shape reasonably close to a hexagon.

It is important to note that the illite phases (PCI and WCI) both recrystallize to become mica under of conditions of metamorphism. This indicates that illite has a clear stability field and that it is not a mica.

2.2.4 Bentonite

2.2.4.1 I/S MLM Sequences

A special case of illite formation in sedimentary rocks under conditions of diagenesis is that of potassium bentonite or K-bentonites. Normally bentonites are almost monomineral rocks on the scale of a hand specimen with an I/S mineralogy, where the smectite content is initially very high. The origin of these beds is in the sedimentation of glassy pyroclastic material. Volcanic eruptions covering regions with fine layers of ash are the starting point of this smectite-to illite-reaction. The glass is first transformed into a smectitic I/S mineral in early stages of sedimentation and burial. Such materials are not homogeneous in their XRD characteristics, indicating different types of layer charge even in the initial stages of smectite formation in the ash layer. This initial smectite can then be transformed into a strongly illitic I/S by diffusion of elements into and out of the bed. Older bentonites often contain very illitic I/S minerals (Elliot and Aronson 1987) and in some cases very old (Proterozoic) beds can have almost pure illites present (Moe et al. 1996). It is possible to calculate the rate of illitization, considered essentially as a function of potassium diffusion (Altaner 1989; Elliot et al. 1991). The concentration of illitic layers in the I/S minerals is not a linear function of distance from the edge of the bed (Brusewitz 1986; Inoue et al. 1990) as would be expected in a diffusion process (Altaner 1989). Hence the outer edge of a bentonite bed is much more illitic than the successive layers toward the center of the bed. Looking at Fig. 2.21 (Altaner et al. 1984), it is obvious that I/S K–Ar dating does not give an age for a geological event. Indeed, the 56 to 50 Ma range is at least 25 Ma younger than the depositional age. Besides, for a given bed, the edges are about 6 Ma older than the center. This will be discussed in Sect. 2.4.2.

In such a series of mineral compositions, the smectite-to-illite transformation is clearly seen by chemical and mineralogical (XRD) observations (see Byström 1954). In fact, the bentonites have been and still are used as a model for the transformation of detrital smectite into illite during diagenesis. The reasons for this analogy are essentially ones of commodity: a single-phase material is easier to study than a multiphase material. However, Sucha et al. (1993) present data for the I/S reaction (illite content of I/S phases) for samples of pyroclastic and shaley rocks in the same sedimentary basin which have experienced the same diagenetic effects (East Slovac Basin, present geothermal gradient, 50 °C). The reaction progress is not the same for the two types of material, the bentonites transform more slowly than do the shales (Fig. 2.22).

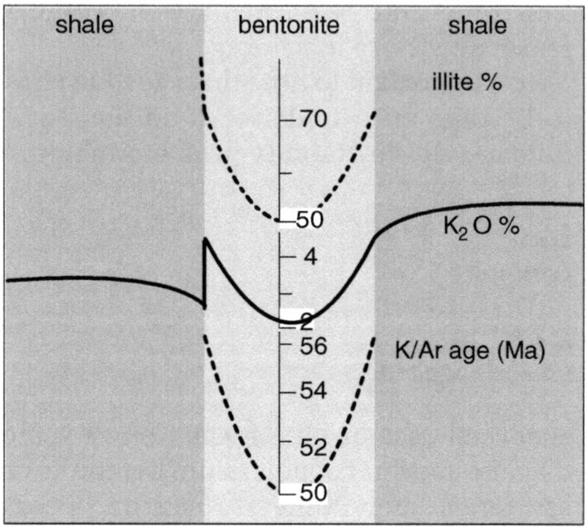

Fig. 2.21. Mineralogical, chemical and K–Ar variations in a 2.5 m thick K-bentonite deposit enclosed in shales from the disturbed belt, Montana (modified from Altaner et al. 1984)

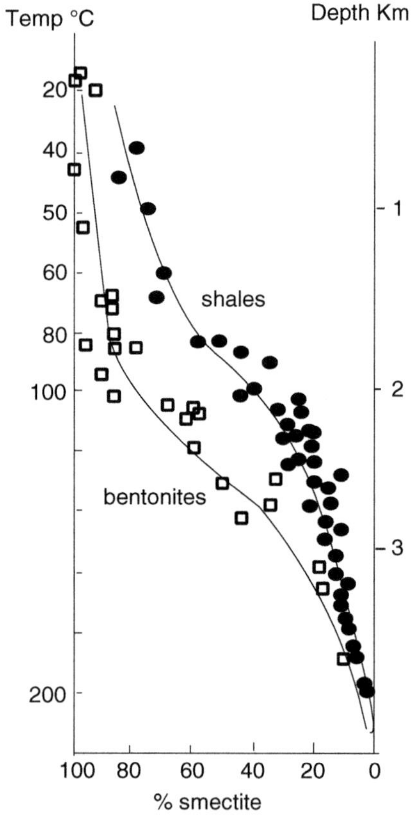

Fig. 2.22. Comparison of shale and bentonite reaction progress as a function of depth in an East Slovak basin drill hole. The bentonite smectites initially show a slower reaction rate than that of shales. Data from Sucha et al. (1993)

This is reasonable in that the shales or pelitic sediments outside of the bentonites will furnish the necessary potassium for bentonite transformation. As such the fluid solutions should be in equilibrium with an I/S at least as illitic as the mineral which will form in the outer layer of the bentonite. As would be expected then, the two sequences of smectite-to-illite mineral transformations (bentonite and pelite) arrive at 100% transformation at the same depth (time, temperature point) in the sediments. However the reaction trends are not parallel which suggests that the kinetics are not quite the same. If one uses the Velde-Vasseur model of a two step-reaction (Velde and Vasseur 1992), the deduced kinetic values of activation energy and pre-exponential function are quite different for the first half of the sequence (disordered I/S from 100 to 50% smectite). The values approach one another for the second half of the reaction.

One striking observation which can be made in studying the decomposition results of these XRD patterns samples from the Tertiary basin is the apparent lack of WCI or illite (see Fig. 2.23) when the I/S transformation is largely accomplished.

The low smectite content mineral assemblages show a rather wide band at near 10.3 Å with no apparent sharp WCI component. This would suggest that the smectite-to-illite reaction produces only a PCI end-member and not an independent illite phase as is the case in older pelitic rocks. Investigation of the older Ordovician Swedish bentonites (Brusewitz 1990) show sharp peaks at less than 10 Å (9.82, 9.26 and 9.21) which appear to represent the third order of a three-layer ordered interstratified mineral (IISm) again, indicating an absence of a true illite at 10 Å. Here again it appears that there is no true illite produced during the reaction smectite to illite.

Moe et al. (1996) report data for a bentonite from Proterozoic rocks in Montana. The major peak is at 10.16 Å. It is rather narrow (0.5 °2θ Cu Kα) which indicates a large average diffracting domain (greater than 10 units). In order to model (NEWMOD) the spectrum given by Moe et al. it is necessary to

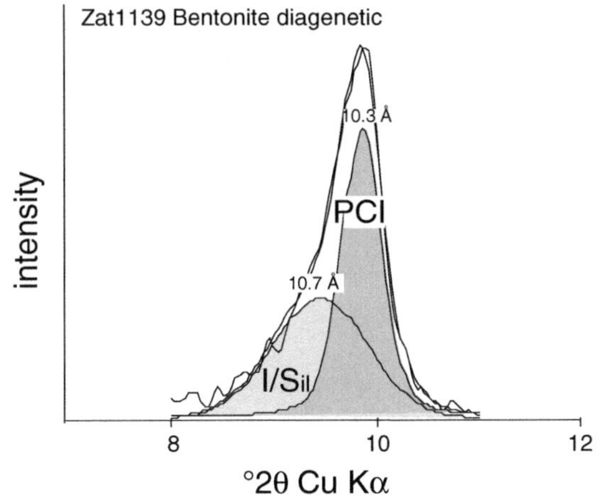

Fig. 2.23. Decomposed XRD pattern of a diagenetic bentonite (Zat1139, sample courtesy of V Sucha) showing a lack of the WCI band

assume that the glycol-saturated smectite layers present are less than 5% and contain only one glycol layer which shows a vermiculitic behavior. This gives a peak at 10.16 Å which is narrow. This mineral has a 2M polytype as shown by the authors.

It appears that in the K-bentonite transformation smectite to illite is effected by diffusion of potassium into the beds. There is in fact a mineral transformation but no parallel formation of illite, even in old beds. The end-product is PCI or a highly illitic mixed-layer mineral with less than 5% smectite content. The smectite is probably high-charge (vermiculite).

Thus, bentonites do not produce true illite, i. e. a WCI mineral seen by XRD. The smectite-to-illite transformation leaves a small amount of smectite in the illite rich, PCI structure.

2.2.4.2 Thickness of Fundamental Particles in Shales

Few papers are devoted to the statistics of fundamental particles and crystal thicknesses versus illite content using direct measurement techniques (Srodon et al. 1992; Inoue and Kitagawa 1994; Dong and Peacor 1996). All these authors consistently show a steady thickening of I/S crystals/fundamental particles with increasing illite content. The natural I/S series analyzed in bentonitic formations of Carboniferous and Silurian ages (Srodon et al. 1992) are roughly consistent with an illite-layer addition model on starting 1 layer thick particles (Fig. 2.24a). The two series indicate a transition from smectite fundamental particle (1 nm) to illite particles (2 to 5 nm). Additionally, the number of fundamental particles in crystals (S) decreases with increasing illite percent (Table 2.3, Fig. 2.24b).

Table 2.3. Particle thickness versus % illite of I/S mixed layers. Carboniferous and Silurian bentonites; Silverton hydrothermal series (Srodon et al. 1992). *N* is measured by layer counting from HRTEM lattice fringe images

Carboniferous		Silurian	
% illite	Thickness (nm)	% illite	Thickness (nm)
12	2.9	34	3.6
55	2.8	62	4.7
59	2.9	75	5
61	5	85	3.7
71	2.7	94	7.2
80	4.7		

2.2.5 Sandstones

Illite or I/S may be formed without any transformation of a pre-existing smectite phase. It forms by direct precipitation from fluids or by the transformation

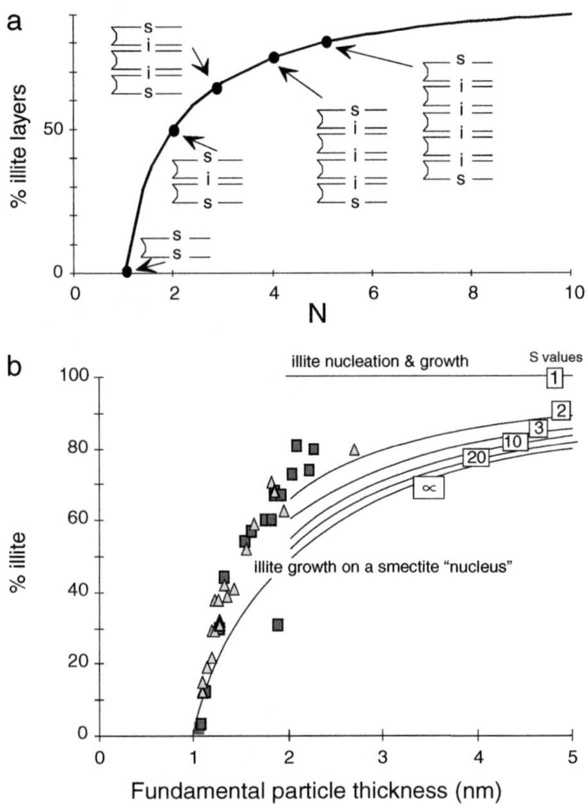

Fig. 2.24a,b. Particle thickness versus % illite of I/S mixed layers. **a** Theoretical variation of illite% versus the number of layers in the fundamental particle (N). **b** I/S from Carboniferous (*solid squares*) and Silurian (*empty triangles*) bentonites (Srodon et al. 1992); T_{fp} is measured by layer counting from HRTEM lattice fringe images

of kaolinite. This is the case of the kaolinite-rich sandstones or hydrothermally altered volcanic rocks which are largely studied as oil or geothermal reservoirs respectively. The chemical composition of illite is significantly different from that formed through a smectite illitization. In shales, illite is formed in a system chemically buffered by the dissolution of the smectite component, whereas in sandstones, illite precipitates from fluids of which the composition is controlled by external factors.

2.2.5.1
Illite Overgrowths

Rarely described, but most likely a common occurrence, white micas (detrital) in shaley sandstones are often overgrown by illite material. Nicot (1987) has

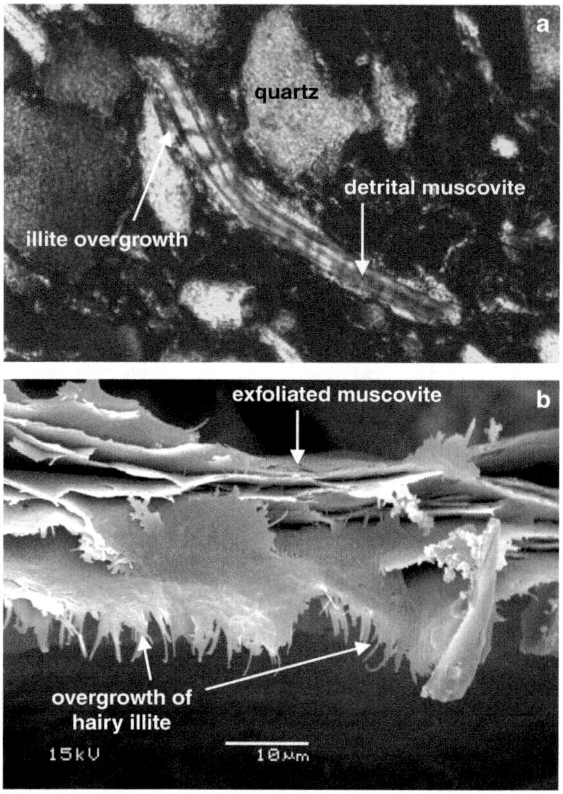

Fig. 2.25a,b. Overgrowth of illite on the basal surfaces of detrital muscovite. **a** Micrograph from a thin section of a Kimmeridgian North Sea sandstone (from Nicot 1987). **b** SEM micrograph of a Proterozoic sandstone from the Athabasca basin (Laverret 2003)

made electron microprobe analyses of these overgrowths. They appear to be of an illitic composition. It is not possible to demonstrate the mineral structure by XRD, but the occurrence merits attention. It is clear that this material does not form from the conversion of smectite to illite, since the compositions are close to illite and associated with micas. Figure 2.25a shows such an occurrence in a North Sea sandstone of Kimmerigian age. In most instances the mica serves as a substrate for the new illite growth. There is no clear relationship of replacement or pseudomorphic replacement. The edge of overgrowth shows a sharp contact with the detrital mica. The edges of the mica are at times overgrown also, but do not show a frayed or fan-shaped structure commonly associated with kaolinite growth within the old mica grain. Therefore, it appears that this occurrence of illite is one of precipitation of material from solution to form a new mineral with a smectite precursor.

Illite overgrowth on detrital mica is frequently observed in the Proterozoic sandstones in the vicinity of uranium ore deposits. In the Athabasca basin, muscovite crystals are exfoliated and their edges are fringed by hairy illite crystals (Fig. 2.25b). Illite is of the $1M_t$ polytype. It has been formed in the "hydrothermal plume" which rises from the faults crosscuting the Archean basement (Laverret 2003).

2.2.5.2 The Illitization of Kaolinite-Bearing Sandstones

The minerals of the kaolin group have been described as precursors of illite in deeply buried sandstones (Bjorlykke et al. 1986; Erhenberg and Nadeau 1989; Bjorlykke and Aagaard 1992; Lanson et al. 1996; Berger et al. 1997). Several morphologies were described for illite occurring in sandstones: hairy illite,

Fig. 2.26a–c. Illite crystal habits in diagenetic sandstones from Proterozoic formations of the Athabasca basin (Canada). **a** Elongated, filamentous illite crystals growing on platy illite crystals previously formed during the diagenesis stage. **b** Lath-shaped illite crystals formed on diagenetic platy illite. **c** Isometric pseudo-hexagonal crystals in a diagenetic sandstone. Microphotographs from Laverret (2003)

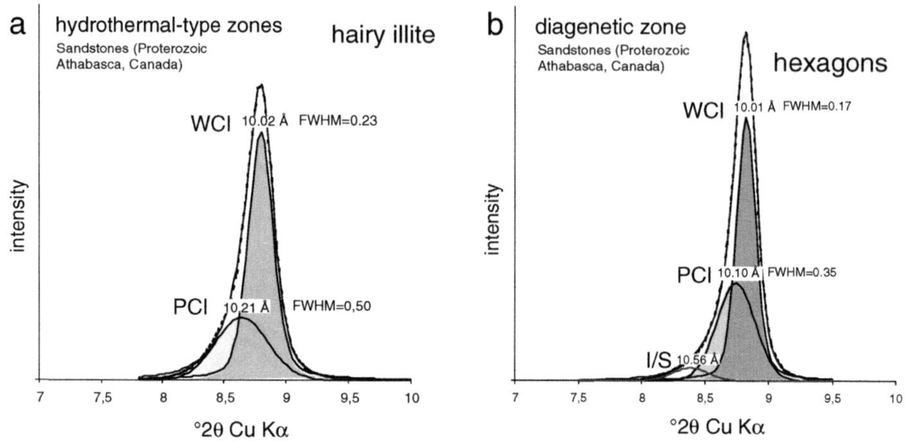

Fig. 2.27a,b. Decomposition of XRD patterns of the illitic phases in the Proterozoic sandstones of the Athabasca basin (Canada). The samples correspond to a and c pictures of Fig. 2.26. XRD patterns result from the mixture of illitic phases originating from a hydrothermal-like event and diagenetic processes. **a** Hydrothermal dominating. **b** Diagenetic dominating

lath-shaped or isometric to hexagonal plates (Fig. 2.26). The size and shape of crystals depends on temperature (Lanson et al. 1996) rather than on chemical constraints (Small et al. 1992). Indeed, with increasing paleo-burial depth, the morphology changes from elongated one-dimensional fibres (hairy illite) to more rigid laths which width increases progressively. In the North Sea reservoirs, the laths are the dominant population for illite crystals even when they coexist with isometric, pseudo-hexagonal-shaped ones. The morphological evolution occurs simultaneously with the increase of "crystallinity". In such a case, crystallinity depends on the relative proportions of PCI and WCI, as indicated by XRD decomposition (Fig. 2.27).

The precipitation of illite or illitic I/S MLM in the North Sea area was thought to occur at 120 °C (about 3500 m depth). These minerals were considered classically to be formed at the expense of vermicular kaolinite at shallower depth and to coexist with dickite at deeper ones. The dickite formation was frequently attributed to a late diagenetic stage which postdate the illite precipitation (Hurst 1985; Haszeldine et al. 1992). Recent studies showed that both kaolinite and dickite exhibit dissolution features (Beaufort et al. 1998; Lanson et al. 2002). The fact that these dissolution features are more developed in kaolinite crystals suggest that kaolinite dissolves faster than dickite during the illitization process. Indeed, dickite being more stable than kaolinite in the temperature range of burial diagenesis, the oversaturation of solutions with respect to illite must be higher.

Numerous studies devoted to the North Sea reservoirs showed that feldspar dissolution controls the formation of kaolinite (reaction 2.14) or illite (reaction 2.15) if temperature conditions are less or up to 120 °C respectively:

$$\underbrace{2KAlSi_3O_8}_{\text{K-Feldspar}} + 2H^+ + 9H_2O \rightarrow \underbrace{Al_2Si_2O_5(OH)_4}_{\text{kaolinite}} + 4H_4SiO_4 + 2K^+ \quad (< 120\,^\circ C) \tag{2.13}$$

$$\underbrace{2KAlSi_3O_8}_{\text{K-Feldspar}} + \underbrace{Al_2Si_2O_5(OH)_4}_{\text{kaolinite}} \rightarrow \underbrace{KAl_3Si_3O_{10}(OH)_2}_{\text{illite}} + \underbrace{2SiO_{2(aq)} + H_2O}_{\text{quartz}} \quad (120\,^\circ C) \tag{2.14}$$

Reaction 2.14 is maintained by protons which are provided by meteoric fluids and organic matter. The source of the K^+ ions necessary for the illite formation in reaction 2.14 is provided by the dissolution of K-feldspar. However, this reaction is not perfectly coherent with petrographical observations which show that significant amounts of kaolin minerals coexist with K-feldspars in partly illitized sandstones (Erhenberg 1991). If reaction 2.14 is truely controlling the formation of illite, it should take place until exhaustion of one of the left members in reaction 2.14: K-feldspar or kaolinite. Further, illites have been observed to be formed at lower temperature conditions: 90–95 °C (Erhenberg et al. 1993; Cassagnabère et al. 1999). In fact, the K^+/H^+ activity ratio controlled by the K-feldspar solubility at 120 °C is not high enough to overcome the energy barrier for the formation of illite (Berger et al. 1993). To reach the critical K^+/H^+ ratio, the fluid must be highly oversaturated with respect to illite. In that case, the fluids are also oversaturated with respect to K-feldspar which can no longer be dissolved. Thus, an external source of K^+ ions or a pH increase is needed to reach the critical K^+/H^+ activity ratio value (reaction 2.15):

$$\underbrace{3Al_2Si_2O_5(OH)_4}_{\text{kaolinite}} + 2K^+ \rightarrow \underbrace{2KAl_3Si_3O_{10}(OH)_2}_{\text{illite}} + 2H^+ + 3H_2O \tag{2.15}$$

It is likely that the energy barrier to illite crystallization decreases with increasing temperature. In other words, reaction 2.14 may be predominant at high temperatures while reaction 2.15 prevails at lower temperatures.

The illitization of kaolin minerals in the Broad Fourteen Basin (North Sea) was shown to be related to a rapid "hydrothermal-type" event during Kimmerian tectonics at about 150–160 Ma. Therefore, the formation of illite or illitic I/S mixed layer mineral reflects a prevailing temperature during the illitization process rather than the progress of a smectite-to-illite transformation. Whatever the depth of the kaolinite-dickite bearing sandstones, I/S MLM or illite, no matter their composition (I/S from 20 to 5% smectite) or their crystal habit (wiskers, lathes, pseudo-hexagonal plates), were shown to be formed during the intense fracturing period of the Kimmerian tectonics at 155 Ma (Lanson et al. 1996). Composition and crystal habits are dependent on local temperature conditions and fluid oversaturation degree.

The illitization of kaolinite or dickite requires highly oversaturated solutions with respect to illite. The dissolution of K-feldspars cannot produce such highly oversaturated solutions. Thus, the source of K^+ ions was assumed to come from laterally adjacent Zeschtein evaporite formations in the vicinity of faults. Because of high oversaturation, the fluids trigger the nucleation and growth of I/S MLM or illite crystals during short periods of time (see Sect. 2.4.3). They can be considered as instantaneous phenomena compared to the progressive illitization of shales.

2.2.5.3
Crystal Structure of Illitic Minerals in Diagenetic Sandstones

The $1M_t$ to $1M_c$ Polytype Transition The morphological evolution from, fibers, laths to isometric pseudo-hexagonal-shaped crystals occurs simultaneously with changes in the 3-D crystal structure. The first step is the transition from the 1M polytype with octahedral *trans* sites vacant ($1M_t$) to 1M with the octahedral *cis* sites vacant ($1M_c$). The quantitative distinction between the two structures is based on the relative intensities of $\bar{1}12$-112 and 111-$\bar{1}13$ peaks for $1M_t$ and $1M_c$ respectively as shown in Fig. 2.28a (Drits et al. 1993; Reynolds and Thomson 1993). The morphological features observed in particles that are fibers or laths for the $1M_t$ polytype may be due to the fact that all the vacant sites are lined up along the unique crystallographic *a* axis. On the contrary, the more equant and platy shape for $1M_c$ particles may be due to the fact that vacant sites are lined up along two different crystallographic directions.

Illitization in the Broad Fourteen basin is characterized by increased proportions of $1M_c$ polytype with temperature rather than by a progressive smectite-to-illite conversion (Lanson et al. 1996). At a given depth, the proportions of the $1M_c$ polytype increases with the size fraction (Fig. 2.28b). This indicates that the phase forming on I/S or $1M_t$ illite particles is $1M_c$ illite.

The 1M to $2M_1$ polytype transition (Patrier et al. 2003) The Paleoproterozoic to Mesoproterozoic McArthur sedimentary basins (Northern Territories, Australia) have been investigated for uranium. The oldest preserved formations (Kombolgie subgroup from 1822 to 1730 Ma) are composed of conglomerates and quartz sandstones covering the early Proterozoic metamorphic basement. Sediments exhibit high compression features: e.g. stylolites, interlocked microstructures. The sandstone porosity is strongly reduced by compression and the quartz overgrowths. Very few detrital white micas have been observed. The pores are filled by a well crystallized dickite which is partially or totally replaced by platy illite crystals of the $2M_1$ polytype (Fig. 2.29). Their average diameter is uncommonly high: 2 to 10 µm. The typical chemical composition of illite and detrital mica are significantly different:

- mica: $Na_{0.05}K_{0.96}(Al_{1.53}Fe^{2+}_{0.28}Mg_{0.18}Mn_{0.01})[Si_{3.34}Al_{0.66}]O_{10}(OH)_2$;
- illite: $Ca_{0.005}Na_{0.01}K_{0.86}(Al_{1.91}Fe^{3+}_{0.065}Mg_{0.025}Mn_{0.005})[Si_{3.135}Al_{0.865}]O_{10}(OH)_2$

Illite is also formed in a "hydrothermal-like" environment in the vicinity of uranium deposits at the basement-sedimentary formation unconformity.

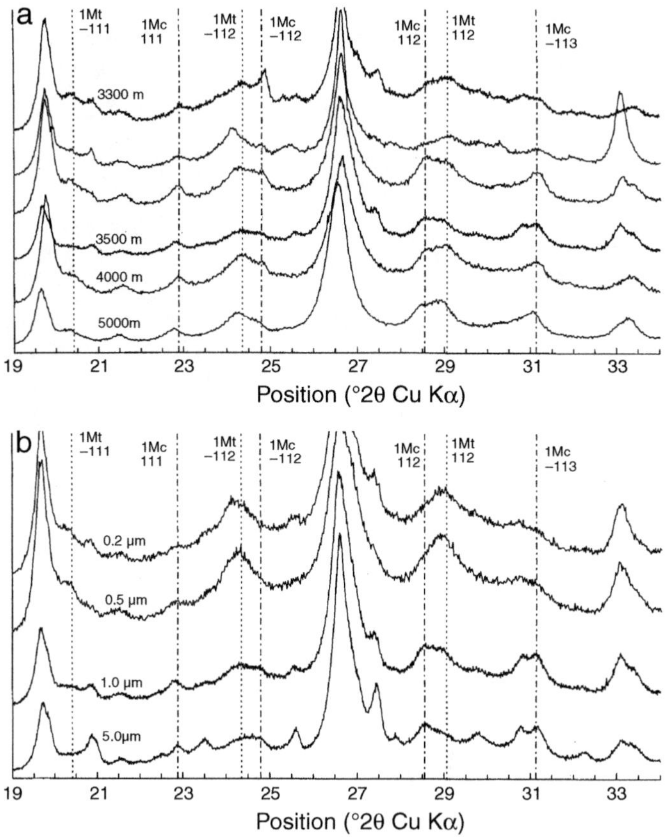

Fig. 2.28a,b. Transition from the $1M_t$ to $1M_c$ crystal structure of illite minerals in the Rotliegend sandstones in offshore Netherlands (modified from Lanson et al. 1996). **a** Proportion variation with depth. **b** Proportion variation with particle size

Crystals are elongated, small, lath-shaped and of the $1M_t$ polytype. The dickite illitization process has been estimated to take place at 1550–1650 Ma while the uranium deposit has been dated at about 1640 Ma (Kyser et al. 2000). This shows that age is not related to polytype formation. It has been shown that it is the maintained high temperature conditions for long periods of time in deeply buried sediments (more than 5000 m deep) which induce the formation of $2M_1$ large crystals.

The chemical composition domain of the $2M_1$ illite populations (with very small amounts of $1M_t$ polytype) are centered on the $0.9\,M^+$ line in the M^+-4Si-R^{2+} chemographic projection (Fig. 2.30a,b) the negative charges are low; −0.02 to −0.05. In spite of an identical layer charge, illites formed in Proterozoic sandstones have a chemical composition different from those formed during the conversion of smectite.

Fig. 2.29a,b. SEM microphotograph of platy $2M_1$ illite crystals replacing dickite at different depths in the Kombolgie sandstones from the Proterozoic McArthur basin, Australia (from Patrier et al. 2003). **a** Lath-shaped to hexagonal particles. **b** Perfect hexagonal particles

2.2.5.4
Summary

Illites, PCI and WCI, are formed by the rapid transformation of kaolinite or feldspars or precipitated from solutions. This formation short-circuits the normal, smectite-to-illite transformation seen in shales. The open-system nature of the sandstones probably favors these rapid reactions. Very often the rapid growth of illite crystals is expressed as a fiber shape giving the name of hairy illite.

Various crystal structures are found, 1M *cis* and *trans* and 2M forms. These illites do not form by a stepwise transformation of an I/S mineral.

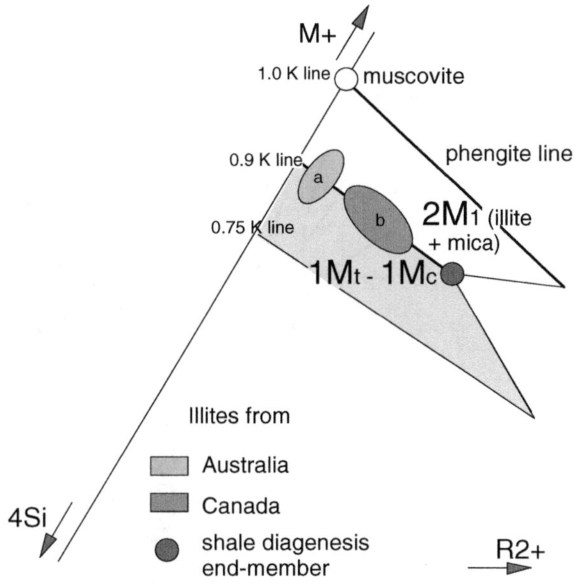

Fig. 2.30a,b. Chemiographic projection of compositions of diagenetic illite formed in sandstones from Proterozoic sedimentary basins. **a** 1M elongated lath-shaped illite from Athabasca Basin, Canada (Laverret 2002). **b** $2M_1$ platy illite from Kombolgie formation in the McArthur basin, Australia (Patrier et al. 2003)

2.3 Illite in Fossil and Active Geothermal Fields and Hydrothermal Alteration Zones

2.3.1 Sericite and Illite in Fossil Hydrothermal Systems

2.3.1.1 Hydrothermal Alteration: Mineral Zonations

Hydrothermal alteration is due to the interaction of aqueous fluids with crystalline rocks at high temperatures. High temperature usually means temperatures above 100 °C. According to Utada (1980) hydrothermal alteration occurs under conditions of local thermal gradients higher than those of the general geothermal gradient of the area. In fact the definition involves interaction of fluids with rocks, where the fluid can transport material in solution which interacts with crystalline rocks. Usually these rocks are of meteoritic origin. Hydrothermal alteration represents an open chemical system where fluids enter and leave the rock bringing dissolved material in and out of the system, which is the contrary to systems of burial diagenesis, where the rock and water mass are constant, even though the rock materials and constituent minerals can have experienced largely the same temperatures in hydrothermal and burial diagenetic conditions.

In general, the rock alteration sequences of this hydrothermal alteration show an enrichment of potassium relative to calcium, magnesium and, to a lesser extent, sodium. This is compared to the composition of the altered

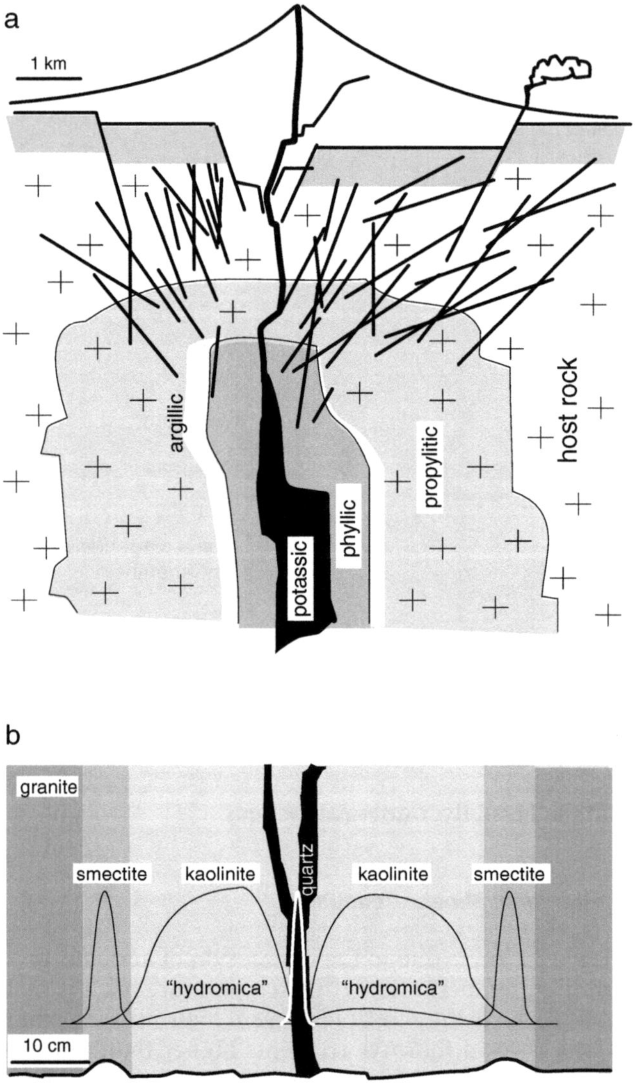

Fig. 2.31a,b. Schematic representation of the mineral zonation in hydrothermal system. **a** Porphyry copper-type zonation (from Lowell and Guilbert 1970). **b** Typical vein zonation from the Caribou mine, Idaho (from Bonorino 1959). The now obsolete term "hydromica" refers to illite and illite/smectite mixed layers

rock itself. The question of which elements increase or are introduced into the rocks and which leave the system is not always easy to determine or explain (see Berger and Velde 1992). However the alterations associated with the occurrence of sericite show low contents of Fe, Mg, and Ca with high contents of K. In general one finds that the rocks affected contain more hydrous, clay minerals which has led to the description of the chemical constraints as those of H, K and silica (see Hemley 1959; Hemley and Jones 1964). Lowell and Guilbert (1970) have codified the mineral facies due to these alterations into potassic: phyllic: argillic: propylitic where the content of clays increases as the series progresses with an accumulation of Mg and Ca in the last facies (Fig. 2.31a). The propylitic facies is not directly associated in the sequence in that it is probably of higher temperature origin than the argillic mineral facies.

Two types of hydrothermal alteration can be described as extreme cases, vein alteration and pervasive alteration. The difference between the two is essentially one of scale. The change in mineralogy around a vein (a more or less linear surface which is open to fluid flow) is similar to that of pervasive alteration which affects large volumes of rock, from hundreds to thousands of meters. In the case of vein alteration, usually a precipitation of material is seen to fill the vein itself and outward from this deposition one sees alteration of the encasing "wall" rock mineralogy. The clay-mineral assemblages found occurring outward from the vein usually show increasing characteristics of clays formed at lower temperature, and those with a higher smectite content or lower alkali content (Inoue et al. 1992). These relations have led mineralogists to assume a lower temperature of formation. On a larger scale, pervasive alteration shows the same sequence of clay mineral zonation as that seen in veins (Ylagan et al. 2000). These relations are shown schematically in Fig. 2.31b.

2.3.1.2
The Origin of Sericite and Illite in Hydrothermal Systems

Illite represents the highest-temperature alteration conditions, according to most authors, and high potassium content solutions. Yet higher temperatures and potassium activity produce the potassic alteration phases characterized by potassium feldspar crytallization. The illite found in phyllic alteration is often nearly monomineral, being largely WCI in XRD characteristics. It is often called sericite.

Sericite is an old name given to fine-grained mica which is not quite a mica. The potassium content is low and the silica content high (Grim 1953). In most circumstances this would be called illite, but the geological occurrence of sericite gives it its name. This is a hydrothermal mineral which has been largely observed under the optical microscope. Hydrothermal alteration is most often observed in granular rocks which are amenable to microscopic study. Eberl et al. (1987) have gone the furthest to identify and describe this mineral in a series of occurrences in Colorado, USA. Sericite is of course illite under another name. This mineral seems to have a pure mica XRD diagram (spacing less than 10.3 Å) and a rather narrow peak (FWHM less than 0.6 °2θ Cu Kα, as observed from scans of published diagrams in Srodon et al. 1987; Parry et al. 2002) which would identify it as a WCI mineral. Chemical compositions indicate potassium content of from 0.80 to 0.95 atoms per unit cell of 22 negative charges (Bishop and Bird 1987; Parry et al. 2002; Srodon et al. 1987).

The origin of the minerals is still somewhat in doubt, whether they occur via a smectite-to-illite transformation or from solution precipitation (Ylagan et al. 2000). Page and Wenk (1979) found a sequence of change in minerals indicating a smectite-to-illite transformation in a vein halo indicating the classical transformation sequence. If the clays found in hydrothermal alteration form through the transformation of smectite to illite, as is the case of minerals found in burial diagenesis, one would expect to find multimineral associations of different I/S mineral types as they transform one into the other, such as R0 and R1 phases or R1 and illite. However, the XRD patterns of Ylagan et al. (2000) show minerals due to pervasive alteration which appear to show single-phase material. The same is true for unpublished data of Inoue and Lanson for hydrothermal vein material from Japan. Hydrothermal sericite vein material composed of highly illitic phases given by Eberl et al. (1987) also indicate almost monomineral phase material. Thus, it appears that, in general, hydrothermal alteration produces phase assemblages of almost single-phase mineralogy. This assessment may be exaggerated as a general case, but it seems clear from the data of Inoue and Lanson (unpublished), that veins are composed of pure, monomineral material. Parry et al. (2002) deduce illite or low smectite content I/S to have been formed at temperatures near 200–350 °C, (phyllic alteration) and intermediate I/S composition minerals below 200 °C which are values often cited in older papers.

The characteristics of potassic hydrothermal alteration mineralogy are a range of smectite to illite compositions, most often in nearly monomineralogic form. The rock which contains these minerals shows a decrease in alkali content as the alteration facies is more smectitic. In general, it is agreed that higher temperatures form illites and lower ones smectites (see Velde 1985 for a summary of older work). Thus both temperature and bulk composition or chemical potential of altering fluids which affects bulk composition are agents of potassic alteration of eruptive rocks into clay mineral-bearing assemblages. Precipitation of minerals in veins results in nearly monomineralogic deposits.

These observations are very important for our understanding of the origin and stability of the I/S clay minerals. If in fact a nearly monomineralogic I/S mineral of a given percent smectite content is present in a vein due to mineral precipitation it means that specific conditions of fluid composition (chemical activity) and temperature produce this mineral. If this mineral is produced without a precursor (ex. smectite or another I/S phase of higher smectite content) this means that the precipitation of this I/S mineral responds to specific temperature and chemical constraints. This being the case, the I/S mineral is not always a product of mineral transformation (in the sense that a precursor is needed for it to form) but can be a phase or mineral form of its own.

One can compare I/S minerals to plagioclase, for example, which is a solid solution mineral formed of units of varying proportions of albite an anorthite. Specific chemical and thermal conditions produce different, recognizable plagioclase composition minerals during metamorphic or magmatic events of crystallization. Change in thermal conditions in a system of constant composition are generally assumed to change the composition of the plagioclase. The same could be true, under certain circumstances, for I/S minerals. As in the case of plagioclase, I/S minerals change composition when not within their chemcial or thermal stability range.

In the case of hydrothermal alteration, both chemical activity and thermal energy change as the fluids interact with the host rocks. This will tend to fix the I/S mineral composition or freeze it in the rock as thermal energy decreases, making mineral change more difficult. If the system introduces and maintains a temperature gradient over a period of time in a given volume (as is the case around a vein) the unstable magmatic minerals will respond to the warming phase of the alteration in a successive manner, producing a sequence of alteration minerals from low temperature and low chemical activity of alkali for instance, to ones of higher temperature and chemical activity nearer the source of fluid-rock interaction. This will give a zoning of minerals around the vein. Precipitation of a new phase in the vein will show the final phase of alteration, closing the system at a given set of temperature and chemical conditions.

2.3.2 Instability of Muscovite Relative to Illite

2.3.2.1 Fossil Hydrothermal Systems

Meunier and Velde (1982) have given chemical analyses of mica and illite minerals which occur in a phyllic and argillic alteration of a two-mica granite. There is a difference in mica compositions of the original magmatic muscovite and a new, muscovite mineral observed to form a mosaic in the older mica mineral grains. These new minerals are more magnesium-rich than the magmatic precursors. Sericite is observed to form on the edges of the muscovite micas (Fig. 2.32a). It is still more magnesium-rich than the mosaic secondary micas but it contains less potassium than the micas (Fig. 2.32b). All of these minerals can be recognized as micaceous materials by their birefringence under a cross-polarized light microscope in a petrographic thin section.

Potassic beidellite is observed in feldspar alteration zones in the rock (Fig. 2.32c). This mineral represents the lowest temperature and lowest potassium activity of the fluids affecting the rock.

The temperature of formation of sericites in the phyllic facies of alteration can be generally assumed to occur near 200 °C (Velde 1985). Thus the sericite mineralogy is due to high temperatures and high potassium activity at the same time. The beidellitic mineralogy can probably be attributed to higher temperature conditions than that of montmorillonites (Velde 1985, p 143) and hence their occurrence in hydrothermal alteration is not unexpected. The key minerals in the major facies of potassic hydrothermal alteration are potassium feldspar – sericite – illite/beidellite and beidellite minerals. The potassic alteration facies occurs in high temperature assemblages, above 250 °C or so, the phyllic (sericite) facies near 200 °C and the argillic mineralogy below 200 °C. However, one can observe these facies occurring on either side of a vein or veinlet where the alteration zone is of centimetric dimensions. Initially it would seem difficult to imagine a thermal gradient of a hundred degrees over distances of centimeters in rocks at a kilometer depth. However,

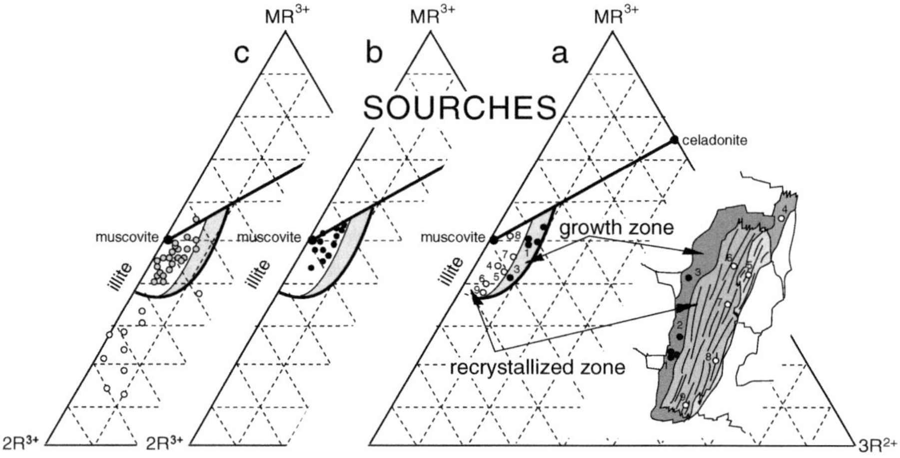

Fig. 2.32a–c. Chemical compositions of illite and K-beidellite in a hydrothermally altered granite at Sourches (Deux-Sèvres, France) plotted in the MR^3-$2R^3$-$3R^2$ coordinates. **a** Microprobe analyses of illite in the growth zone and the recrystallized magmatic white mica. **b** Microprobe analyses of other recrystallized primary white micas. **c** Microprobe analyses of sericites (*dots*) and grey multiphase zones (*circles*) in feldspars (kaolinite + K-beidellte ± feldspar remains)

it is possible to establish a thermal gradient by pulsing high temperature fluids which allow thermal differentials of up to 75 °C (Turpault et al. 1992b). It seems possible to envision the formation of illite (sericite) under conditions of strong chemical variation (potassium activity) and strong thermal gradients.

These conditions can be reasonably compared to laboratory experiments where different amorphous materials have been directly transformed into illite (Velde 1965; Velde and Weir 1979; Small 1993; Huang et al. 1993). In such experiments there is no smectite precursor and the minerals formed must be responding to overall chemical forces (bulk composition) and temperature changes. Given that the initial conditions of experimentation involve a certain thermal inertia (temperature increases gradually, at least for several hours) the starting phases of the reaction of unstable, amorphous materials to high temperatures will undoubtedly create an initial unstable mineral assemblage. This assemblage should react to form the ultimate reaction product, which represents the mineral stable at the temperature conditions of the experiment. Whatever the problems of reaction rate of metastable minerals to the ultimate, assumed stable phases are, they do not appear to involve the conversion of smectite to illite.

2.3.2.2
Active Geothermal Fields

Muscovite is unstable in hydrothermal conditions as was shown in the Salton Sea geothermal field (California, USA). The inherited muscovite crystals protected by the low permeability calcite-cemented sandstone remain unaltered (McDowell and Elders 1983). However, when in contact with hydrothermal fluids, they are replaced by partial or complete rims of illite grains in subparallel orientation. The overall reaction is a two step-process: 1) recrystallization producing illite (Mg gain and Al loss) and 2) transformation of illite into I/S MLM by K for Mg exchange in the interlayers. Allogenic chlorite and biotite crystals react also producing secondary chlorite.

Illite is formed during the first step in equilibrium with chlorite + alkali feldspar + albite + aluminous titanite. The mineral reaction induces the dissolution of the parent muscovite; the chemical composition of illite is controlled by the fluids and buffered by the other members of the paragenesis:

$$\underbrace{Na_{0.13}K_{0.84}Ca_{0.01}(Al_{1.84}Fe_{0.09}Mg_{0.02}Ti_{0.04})[Si_{3.07}Al_{0.93}]O_{10}(OH)_2}_{\text{Muscovite}} + \underbrace{(Mg, Fe, Si)}_{\text{solution}}$$
$$\rightarrow \underbrace{Na_{0.09}K_{0.69}Ca_{0.01}(Al_{1.42}Fe_{0.31}Mg_{0.27}Ti_{0.01})[Si_{3.31}Al_{0.69}]O_{10}(OH)_2}_{\text{illite}} + \underbrace{(K, Al)}_{\text{solution}} \quad (2.16)$$

The reaction above is similar to that observed in the hydrothermally altered granite from Sourches. This shows that muscovite is unstable in hydrothermal conditions and recrystallizes in another phase: illite (sericite).

2.3.3
Crystallochemical Characteristics of High-Temperature Illites (Sericite)

2.3.3.1
Chemical Composition

Sericite (fine-grained micaceous material) are often described in fossil and active geothermal fields (Parry et al. 1984; Eberl et al. 1987; Lonker and Fitz Gerald 1990; Bishop and Bird 1987; Bove et al. 2002). They crystallize in veins or permeable rocks forming different mineral assemblages according to temperature conditions within a 175–350 °C range (Table 2.4):

- intermediate argillic alteration zone: sericite + clays ± carbonates
- propylitic alteration zone: sericite + chlorite ± epidote ± pyrite
- phyllic alteration zone: sericite + biotite

The tetrahedral charge of sericites from Salton Sea and Coso Hot Springs was shown to increase with temperature (McDowell and Elders 1983; Bishop and Bird 1987). However, the illite (sericite) solid solution is not uniquely

Table 2.4. Age, temperature of formation and cristallographical properties of hydrothermal "sericites"

Location	Polytype	% smectite	age Ma	Temp. °C	References
Silverton Caldera	1M	< 10	20	178	Eberl et al. (1987)
Imperial Valley	–	< 10	Pliocene-Pleistocene	175	Jennings and Thompson (1986)
Salton Sea	–	< 10	–	185	McDowell and Elders (1980)
Amethyst vein	–	–	–	195±15	Horton (1985)
Silverton Caldera	$2M_1$	< 5	13,8	270–322	Eberl et al. (1987)
Bingham	$1M–2M_1$	6–7	–	200–220	Parry et al. (2002)
Golden Cross	1M	0	Pliocene-Pleistocene	140–220	Tillick et al. (2001)
Coso Hot Springs	–	0	–	215–250	Bishop and Bird (1987)
Roosevelt Hot Springs	–	0	Peistocene	225–235	Parry et al. (1984)
Southwest Tintic	–	0	Tertiary	280–360	
Santa Rita	–	0	Tertiary	250–390	
Lake City	1M	0–17	$330\ 10^3$ a	–	Bove et al. (2002)
Broadland-Ohaaki	$1M–2M_1$	0	Pleistocene	200	Lonker and Fitz Gerald (1990)

due to the variation of the tetrahedral charge. Plotted in the M^+–4Si–R^{2+} system, the chemical compositions of sericite are situated in a large solid solution area limited by the mica line ($M^+ = 1$) and the "illite" line ($M^+ = 0.75$) corresponding to the composition of the 0% smectite defined by Hower and Mowatt (1966). This large domain is divided into two parts by a line corresponding to the end-member illite ($M^+ \sim 0.9$) where the 1M and $2M_1$ polytypes coexist (see Fig. 1.21).

The sericite compositions in fossil geothermal systems depend on the temperature conditions prevailing during the alteration processes (Fig. 2.33a):

- argillic alteration-Lake City (Bove et al. 2002): about 0.75 K per $O_{10}(OH)_2$, 1M polytype,
- phyllic alteration-Silverton (Eberl et al. 1987): 0.9 K per $O_{10}(OH)_2$, 1M+$2M_1$ polytypes,
- phyllic-potassic alterations-Roosevelt (Parry et al. 1984): 0.9 to 1 K per $O_{10}(OH)_2$ according to associated minerals in the alteration parageneses.

The $2M_1$ polytype dominates in high temperature zones (propylitic and phyllic zones) in association with chlorite or biotite (Parry et al. 1984) while the 1M type is encountered in the intermediate argillic alteration zone in association with clays and carbonates. This is much less clear in active geothermal fields where the 1M and 2M polymorphs are frequently associated. The compositions vary from the "illite diagenetic pole" to an Al-rich one close to muscovite (Fig. 2.33b). This indicates that kinetics controls the polytype transition.

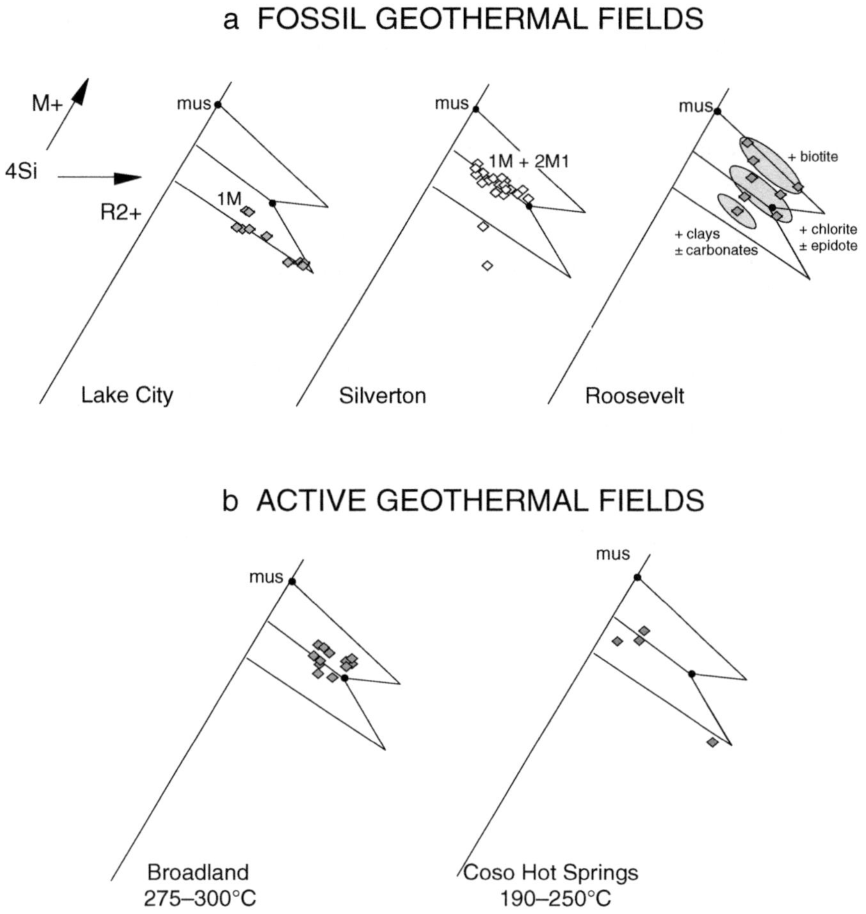

Fig. 2.33a,b. Chemical compositions of "sericites" in fossil and active geothermal fields (references in Table 2.4) plotted in the M^+–4Si–R^{2+} coordinates

Summary It seems that the compositon of illites or sericites formed under hydrothermal conditions depends on the bulk chemistry of the rocks in which they form, or perhaps the composition of the altering fluids and temperature. In any event both 1M polytypes and 2M polytypes are found: the boundary between their respective composition domains being the 0.9 K per $O_{10}(OH)_2$ line. Given the apparent instability of muscovite which is recrystallized into illite with a 0.9 potassium content during hydrothermal alteration, it seems that two independent phases occur, one muscovite-phengite and the other illite (WCI). However, in prograde sequences in which conditions of increasing temperature are present, one would expect to find illite becoming unstable and transforming into muscovite. Under these circumstances, electron microprobe analyses would show a bulk composition of the sericite minerals to be between the muscovite and illite compositions, as seen in Fig. 2.33 for the Roosevelt field minerals of phyllic-potassic facies alteration.

2.3.3.2
Thickness of Fundamental Particles and Crystals – % Illite Relation in Natural I/S Series

The illite-rich I/S mixed layer mineral series from hydrothermal systems (Srodon et al. 1992; Inoue and Kitagawa 1994) are given in Table 2.5. The relation between illite% and particle thickness is shown in Fig. 2.34. Data

Table 2.5. Particle thickness (*N*) versus % illite of I/S mixed layers. 1) Silverton hydrothermal series (Srodon et al. 1992). *N* is measured by layer counting from HRTEM lattice fringe images. 2) Kamikita hydrothermal system (Inoue and Kitagawa 1996). *N* is measured by Pt-shadowing

Silverton		Kamikita	
% illite	thickness (nm)	% illite	thickness (nm)
91	10	88	5.9
93	10	97	8.9
94	23	96	8.6
97	32	97	9.2
–	–	100	15.8
–	–	100	16.6
–	–	100	17.1
–	–	100	22.7

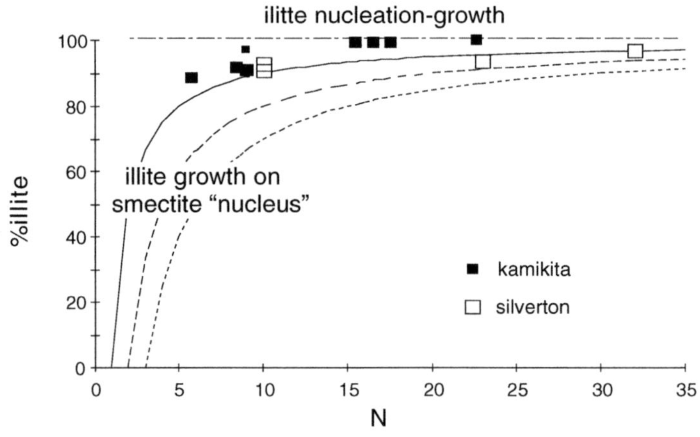

Fig. 2.34a,b. Illite and illite-rich I/S mixed-layer series from hydrothermal areas. **a** Silverton caldera, Co (*triangles*; Srodon et al. 1992), **b** Kamikita, Japan (*squares*; Inoue and Kitagawa 1994). The thickness of fundamental particles (*N*) was measured by layer counting from HRTEM lattice fringe images and Pt-shadowing, respectively

points are located in the domain limited by the growth of free illite particles and the addition of illite layers on initial 1-layer-thick smectite fundamental particles. This could indicate that the analyzed particles result from the two processes, as direct precipitation of illite from hot solutions has been established in hydrothermal systems (Eberl et al. 1987). Nevertheless, this remains hypothetical since I/S with less than 85% illite are lacking in the two systems.

2.3.4 The Smectite-to-Illite Conversion in Geothermal Fields

2.3.4.1 Montmorillonite-to-Illite Conversion

According to Utada (1980), hydrothermal alteration is due to chemical interactions of rocks with fluids of elevated temperature emanating from regional geothermal gradients. This is the case for different scale geological situations: 1) geothermal fields in which temperature anomalies cover large volumes of rocks (several km^3); 2) local thermal anomalies around fractures in which hot fluid flows (few cm^3 or mm^3). In both cases, newly formed minerals (clays and zeolites) replace primary ones. Numerous studies show that the smectite-to-illite formation at the scale of the geothermal field is similar to the high-temperature diagenesis sequence observed in XRD studies. The sequence smectite, random I/S MLM, ordered I/S MLM, illite has been systematically observed (Steiner 1968; Harvey and Browne 1991; Inoue et al. 1992; Inoue 1995; Ylagan et al. 2000, among others). Figure 2.35a shows the classical shape of the % smectite vs depth curve in a geothermal field.

The chemical characteristics of the 0 to 100% illite I/S MLM series are a continuous increase of Al for Si tetrahedral substitutions and a concomitant increase of amounts of potassium. The Fe and Mg contents are nearly constant and the 100% smectite end-member is typically a montmorillonite in composition (Inoue 1995; Ylagan et al. 2000). In spite of local petrographical and chemical properties of the altered rocks, the composition of the 100% illite end-member are rather similar: Mg and K contents are respectively about 0.25 and 0.75 per $O_{10}(OH)_2$ (Table 2.6). A continuous transition between a montmorillonitic clay and illite is observed at Ponza Island (Fig. 2.35b).

Table 2.6. Chemical composition of the 100% illite end-member of I/S MLM series from hydrothermally altered rocks in Japan (Inoue et al. 1987) and Ponza Island, Italy (from Ylagan et al. 2000)

Formula unit per $O_{10}(OH)_2$	Shape	Polytype	Reference
$K_{0.76}Ca_{0.06}(Al_{1.73}Fe_{0.07}Mg_{0.24})[Si_{3.22}Al_{0.78}]O_{10}(OH)_2$	Lath	1M+	Inoue et al. (1987)
$K_{0.73}Ca_{0.07}(Al_{1.69}Fe_{0.06}Mg_{0.24})[Si_{3.37}Al_{0.63}]O_{10}(OH)_2$	Plate	$2M_1$	Inoue et al. (1987)
$K_{0.74}(Al_{1.68}Fe_{0.04}Mg_{0.27}Ti_{0.02})[Si_{3.51}Al_{0.49}]O_{10}(OH)_2$	L+P	$1M+2M_1$	Ylagan et al. (2000)

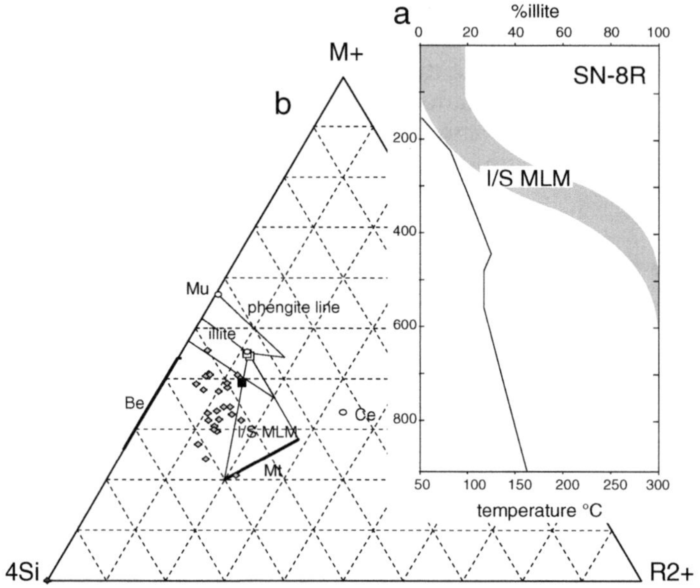

Fig. 2.35a,b. Illitization in active geothermal fields. **a** The classical shape of the % smectite vs depth curve in a geothermal field. **b** Chemical compositions of I/S MLM from Ponza Island plotted in the M^{+}-4Si-R^{2+} coordinates. Illite end-members from Table 2.6: *squares* (Inoue 1987); *filled squares* (Ylagan et al. 2000)

During conversion, the octahedral structure changes progressively from *cis*- to *trans*-vacant with increasing illite content. Concomitantly, the order in stacking sequences from turbostratic through $1M_d$ to a mixture of 1M and $2M_1$ polytypes in the 100%illite samples.

The sequence of mineral phases, smectite to illite, that is observed either in fossil and active geothermal fields produces an I/S MLM similar to that of the burial diagenesis of pelitic rocks. Indeed, the smectite composition is close to that of the low-charge montmorillonite end-member. This is the case for the sequence described in the Ponza Island (Ylagan et al. 2000): the composition of the pure smectite: $M^{+}_{0.35}(Al_{1.44}Fe_{0.17}Mg_{0.38}Ti_{0.02})[Si_{3.97}Al_{0.03}]O_{10}(OH)_2$ is more or less identical to that of the smectite layers interstratified in MLMs (Fig. 2.34b). Illite produced in the highest temperature zones has a layer charge of 0.74 per O_{10} $(OH)_2$ which originates both in the tetrahedral (2/3) and octahedral sheets (1/3).

This diagenetic-type I/S sequence is typical of mineral reactions controlled by the regional geothermal gradient which is stabilized during a long period of time by the heat diffusion from the hot intrusive rocks. Ylagan et al. (2000) showed that the transition from smectite to illite involves a temperature dependent dissolution-crystallization process.

2.3.4.2 *Beidellite to Illite (Chipilapa)*

The progressive sequence from smectite to illite through random and ordered I/S MLM described above is sometimes interrupted by "anomalies" in active geothermal fields. The high-temperature zones where hot fluids flow (porous reservoirs or fractured rocks) are characterized by the presence of aluminous-rich clay minerals: smectite, ordered I/S MLM or illite (Table 2.7, Fig. 2.36). The I/S MLM which appear at unusually high temperature conditions never exhibit random interstratification (Papapanagiotou 1992; Patrier et al. 1996; Patrier et al. 2003). The crystallochemical characteristics of the smectite formed in boiling zones are totally different from those of the montmorillonite involved in the heat diffusion controlled I/S MLM sequence. They exhibit beidellitic properties. This was particularly well demonstrated in the Chipilapa (Salvador) and Bouillante geothermal fields (Guadeloupe Island) where the aluminous-rich clays are paradoxically formed in magnesian volcanic rocks (andesite, dacite).

Montmorillonite does not crystallize in spite of the high amounts of Mg in the parent rocks. Because of high energy input in the hot fluid flowing zones, beidellite is formed as the Al-bearing dioctahedral phase while Mg is incorporated in a trioctahedral one. This is coherent with alteration experiments showing that montmorillonite is replaced by a beidellite+saponite assemblage at high temperature (Yamada and Nakasawa 1993; Sato et al. 1996). Because of the tetrahedral charges, the composition of beidellite is closer to that of Al-rich illite than montmorillonite (Table 2.7). Thus, the smectite-to-illite conversion is made easier and more direct: no random beidellite/illite MLM has been observed. The composition of low-temperature illite in geothermal fields depends on the geological settings:

1. at the scale of the field, the dominating thermal regime is controlled by heat diffusion. The smectite-to-illite conversion process is similar to that prevailing in diagenesis. The depth for a total conversion is shortened because of the higher geothermal gradient.
2. In the boiling zones, the fluids become over-saturated with respect to a dioctahedral Al-rich smectite which is an illite precursor.

Table 2.7. Average formula units of smectite and illite samples from Chipilapa geothermal field (from Papapanagiotou 1992)

Nb Part.	Average formula unit per $O_{10}(OH)_2$	X-ray Pattern	Temp. °C	Drill-hole
89	$Na_{0.01}K_{0.11}Ca_{0.19}(Al_{1.56}Fe_{037}Mg_{0.09}Mn_{0.01})$ $[Si_{3.50}Al_{0.50}]O_{10}(OH)_2$	Beidellite	185	CH-7 580–606 m
14	$K_{0.85}Ca_{0.01}(Al_{1.69}Fe_{0.19}Mg_{0.10})$ $[Si_{3.25}Al_{0.75}]O_{10}(OH)_2$	Illite	210	CH-9 1751 m
16	$K_{0.80}Ca_{0.01}(Al_{1.70}Fe_{015}Mg_{0.14})$ $[Si_{3.34}Al_{0.66}]O_{10}(OH)_2$	Illite	180	CH-7 1500 m

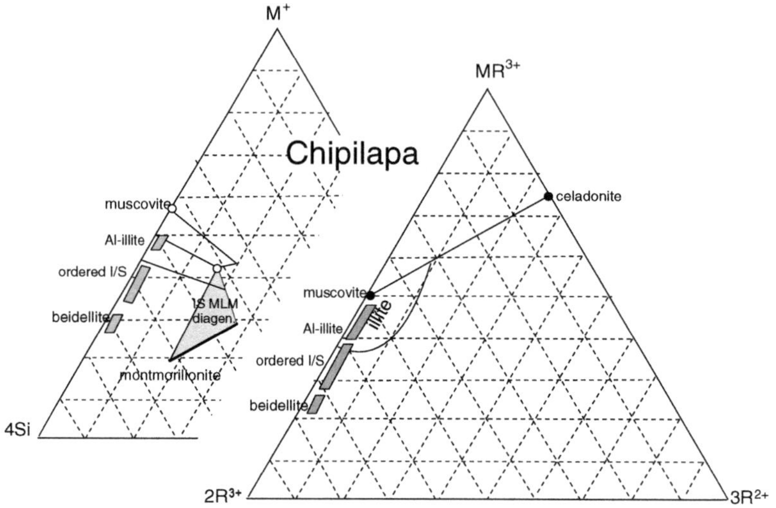

Fig. 2.36. Active geothermal field of Chipilapa (El Salvador). Chemiographic representation of the smectite, ordered I/S MLM and illite composition fields in the MR^{3+}-$2R^{3+}$-$3R^{2+}$ and M^{+}-4Si-R^{2+} coordinates (data from Papapanagiotou 1992)

Summary In this section we see that the effects of high temperatures transmitted by fluids can produce illite (PCI and WCI) as well as I/S minerals of 2M and 1M polytype. Make up of the phases can be influenced by rock composition concerning the phengite substitutions (Mg, Fe) but the potassium content seems to be more related to the thermal conditions and perhaps the potassium potential in the altering solutions. Most likely the high potassium-content sericites (0.9 to 1.0 K ions) are due to mixtures of illite (WCI) and muscovite formed during prograde alteration to higher temperature and potassium content facies.

The geothermal and hydrothermal minerals, often called sericite, come the close to a pure WCI mineral content in many instances. These are the closest that natural deposits come to containing illite (WCI).

2.4 The Illite Age Measurement

2.4.1 Fundamental Concepts

2.4.1.1 The K–Ar Dating Problem

Illite, as we know, contains potassium, this is a part of its definition. This potassium can be used to obtain an age, or period of existence of potassium in an illite crystal. The estimation of time is obtained through the identification and measurement of radioactive decay and decay products. A radiometric age is obtained by determining the amount of potassium in a mineral and the amount of radioactive decay argon (see the review from Clauer and Chaudhuri 1995). However, this is not as easy as it sounds for the following reasons:

- mixing with other K-bearing phases such as detrital mica or K-feldspar (contamination effect) in the samples analyzed,
- Ar loss or capture during crystallization,
- duration of K accumulation in the illitic phases, i.e. illite and illite-smectite mixed layer particles (I/S MLM) compared to the dating experimental error (2σ).

The contamination effect due to authigenic-detrital phase mixing has been discussed in numerous papers (Pevear 1992; Srodon 1999; Ylagan et al. 2000; Srodon 2000; Srodon et al. 2002). Whatever the mixing law used by the different authors, detrital as well as authigenic phases are considered to have a fixed K–Ar age. These assumptions are only acceptable if the source of the detrital K-bearing phases is the same in all the studied samples on one hand, and if the authigenic phases have a given constant age on the other hand. This last point is debatable since it refers to the duration of K accumulation compared to the experimental dating error (2σ). In other words the crystal growth processes of illite and I/S MLM phases should be taken into account before any interpretation of K–Ar ratio in terms of geological dating.

How can one study the effect of the growth processes related to illitization in diagenetic environments on K–Ar dating? A theoretical approach is difficult because the growth processes are still not well-known at present (see Srodon et al. 2002). However, some published K–Ar data concerning different size fractions of clay samples can be used to estimate the effect of crystal growth.

2.4.1.2
The K–Ar Dating Principles

In illite ^{40}K is transmuted either into $^{40}Ca + \beta$ particle (89.9%) or into ^{40}Ar. The values of the transmutation constants are $0.581 \times 10^{-10}\,y^{-1}$ and $4.962 \times 10^{-10}\,y^{-1}$ for $\lambda_{K\text{-}Ar}$ and $\lambda_{K\text{-}Ca}$ respectively. Therefore, the transmutation constant λ of ^{40}K is equal to $(0.581 + 4.962)10^{-10}\,y^{-1}$. The following K–Ar age equation is inferred: $t = 1/\lambda(^{40}Ar^{*}\lambda/^{40}K\lambda_{K\text{-}Ar})$. The calculation of the standard deviation (σ) of the dating has been established by Cox and Dalrymple (1967). The ^{40}Ar, which is an uncharged, neutral atom, can be readily displaced from the site it occupies by an appropriate cation. The only true obstacle to its displacement is its diameter, about 1.9 Å. Such loosely held atoms are subject to thermal agitation and migration. Theoretically, the isotopic system is closed when the ambient temperature decreases below a threshold, i.e. a temperature below which little or no diffusive loss occurs. For fine-grained minerals (clay), the closure temperature of the radiogenic argon is of the order of $260 \pm 30\,°C$

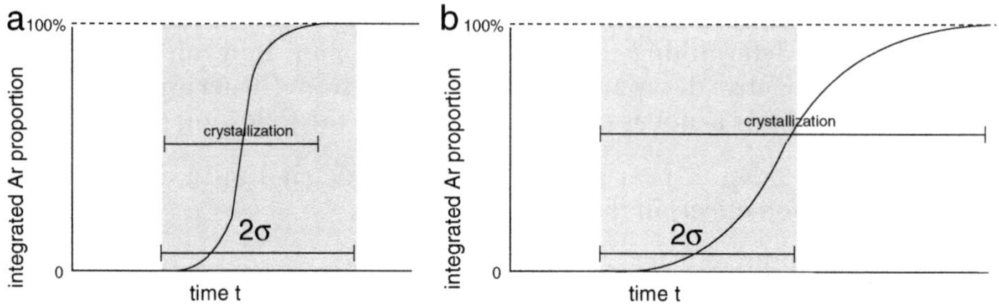

Fig. 2.37. Theoretical representation of the relation between the crystallization and the ^{40}Ar accumulation durations. **a** For an instantaneous event the accumulation takes place during a period shorter than 2σ (shaded area). An actual age of formation is measured. **b** If the duration of the crystallization period exceeds 2σ, the measured K–Ar ratio corresponds to an "integrated" age

(Purdy and Jäger 1976). Closure temperature here indicates the end of significant diffusion of argon. Hence at temperatures of diagenesis below 250 °C, one can expect that little radioactive decay material is lost through diffusion in mica-type layers.

Closure can be considered instantaneous (Fig. 2.37a) when the time required for mineral formation is lower than the experimental error of the dating ($< 2\sigma$). If, in the system, the apparent closure time is observed to be very long ($> 2\sigma$), the age measured for the sample based upon decay products will not be the actual age of the mineral formation but, rather, "an integrated age", whose value depends on the duration of the growth process (Fig. 2.37b). The age will be lower than that expected.

The existence of short and long Ar accumulation periods was shown in sandstones from UK South Central Graben. Two generations of illite have been observed to coexist in certain rocks, one is related to the burial stage, the other to a subsequent fluid invasion (Darby et al. 1997). In this example, the overall age will represent two periods of illite growth and fixation of potassium which generates radioactive decay products. The more recent age observed will not be the result of diffusion or leakage of argon from the illite crystallites or lack of closure but will be due to a new generation of illite.

2.4.1.3 Disturbances Due to the Opening of the System

The loss of argon from crystal lattices is controlled by two different processes: 1) destruction of the lattice (dissolution or fusion), and 2) diffusion within lattices. The diffusion effects are negligible at diagenetic temperature conditions. Conversely, the adsorption of argon bears witness to the conditions at the instant *t* of the trapping:

- the isotopic composition of this argon is equal to that of the atmospheric argon ($^{40}Ar/^{36}Ar = 295.5$);

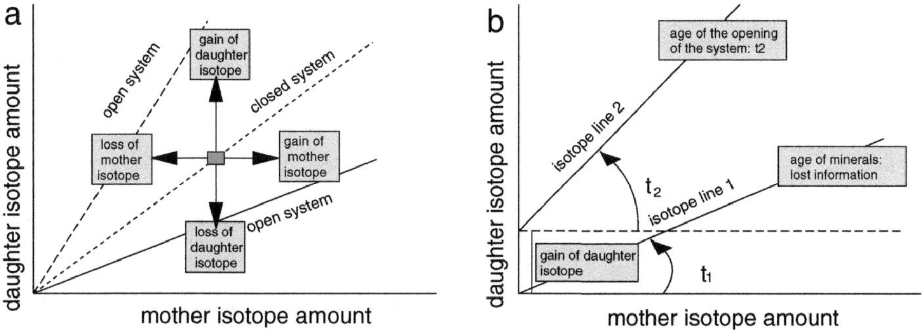

Fig. 2.38a,b. Representation in Harper's diagram (1970) of the influence of the opening of the system on the isotopic composition of minerals (modified from Bonhomme et al. 1995). **a** The opening of the system can increase the slope of the isotopic line either by gain of daughter isotopes or by loss of mother isotopes. It can decrease the slope either by loss of daughter isotopes or by gain of mother isotopes. **b** In the case where the opening of the system at an instant t_2 leads to a gain of daughter isotopes, the isotopic line permits measurement of the age of the opening of the system, and not of the age of the formation of minerals

- the isotopic composition of argon is more or less enriched in ^{40}Ar. Any impoverishment indicates that the sample measured is a mixture of mineral phases.

The opening of the system disturbs the relationship between the amounts of mother and daughter isotopes, and consequently distorts the calculation of the age of formation of minerals as given by the equation: $t = 1/\lambda(^{40}Ar^{*}\lambda/^{40}K\lambda_{K-Ar})$. Figure 2.38a shows the four possibilities of modification of mother and daughter isotope quantitities, and their effect on the slope of the isotopic line (Bonhomme et al. 1995). In reality, this slope measures the age of the opening of the system and erases any information about the age of formation of the minerals (Fig. 2.38b). The opening of a clay mineral system is generally due to the temperature rise that causes the loss of argon. If the loss is complete, the measured age t_2 will be that of the new closure (cooling period) of the system following its opening. If it is partial, the measured age will be a weighted mean between the accumulation rate of argon during the formation of the mineral and that produced by the new closure of the system. The interpretation of this age becomes impossible in terms of geological history.

2.4.2 The K–Ar Apparent Age of Authigenic-Detrital Mineral Mixtures

2.4.2.1 Evidence of Mixtures of Detrital-Authigenic Illitic Phases

The clay fractions isolated from samples of diagenetic rocks are most often heterogeneous. They are at times composed partially of detrital minerals that are

Fig. 2.39. K–Ar dating of samples of diagenetic shales from two series and composed of a mixture of detrital illite and newly-formed illite whose proportions have been determined by X-ray diffraction (from Pevear 1992). *Points, circles, black squares* and *white squares* correspond to different size fractions. The two straight lines have been obtained by linear regression

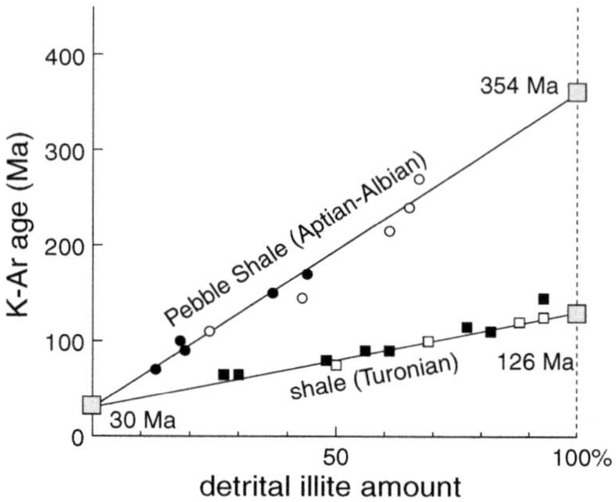

older than the sedimentary bed and of more recent authigenic minerals. The K–Ar age of the clay fraction depends on the different proportions of detrital and authigenic minerals. This has been shown by Pevear (1992) for the shales of the Albian-Aptian and the Turonian of Arkansas (USA). The K–Ar dating has been performed on several size fractions of the shales (2.0–0.2 μm; 0.2–0.02 μm; < 0.02 μm). The detrital illite fraction has been measured by X-ray diffraction using the NEWMOD calculation code (Reynolds 1985). The results (Fig. 2.39) show that both series of analyses can be represented by two straight lines (simple linear regression) that converge for 0% detrital illite at about the same age: 30 Ma. This age corresponds to that of purely neogenetic illites extracted from bentonites interbedded in the Albo-Turonian series. This is a diagenetic age, as opposed to ages 354 and 126 Ma respectively, which are detrital ages. Moreover, Pevear has concluded that the detrital input has changed between the Albian-Aptian and the Turonian because of major tectonic movements at that time.

According to Pevear (1992), the detrital-authigenic K–Ar age mixing function is linear. This mixing law has been discussed in several subsequent papers: Srodon (1999) claimed that the mixing function gives highly curved lines if the end-member compositions contain different % K_2O values. On the contrary, Ylagan et al. (2000) considered that the curvature may be greatly reduced if the smectite content is corrected and the ionic substitutions of K^+ by Na^+ or NH_4^+ are taken into account. A linear relationship was established for $^{40}Ar/^{39}Ar$ dating for detrital mica and neoformed illite in low-grade metamorphism conditions (Jaboyedoff and Cosca 1999).

Whatever the model used, the results of age determinations of radioactive decay products are made uncertain because they are based on several over-simplifying assumptions, the most important of which is that, once an illite or potassium-bearing crystal is formed, it remains present. In a closed chemical system such as a shale, progression in the smectite-to-illite conversion will be accomplished through the dissolution of a high portion of mixed-layered minerals (potassium-bearing) which will be dissolved to form new illite on other crystal surfaces. Illite layers in illite-smectite mixed-layer minerals (I/S MLM) series cannot be simply reduced to an illite component which will provide a given K–Ar age subsequent to its formation. Illite precipitates on I/S minerals and as illite crystallites during long periods of time in pelitic diagenetic environments. Such a system is not static but dynamic. The measure of the quantity of authigenic illite is difficult for two reasons: the possible presence of older detrital "illite" (most often micas) and the presence of discrete illite particles and I/S MLM in the authigenic fraction which are constantly forming from older material.

2.4.2.2 *Identification of the Crystal Growth Effect in Presence of Detrital Inheritance*

Mixing Effect: Δ_{age} Diagenetic shales can be mixtures of detrital and authigenic phases.The former are usually concentrated in the coarse fractions and are older than the age of deposition. The latter phases are concentrated in the finest fractions and are younger, as mentioned previously. Pevear (1992) showed that determining a "diagenetic" age is made possible using the linear relationships between % detrital illite and K–Ar ages measured from different clay size fractions. Taking the depositional age (mean stratigraphic age: 100 Ma and 90 Ma for Albian-Aptian Pebble and Turonian shales, respectively) as the reference, a new parameter may be calculated for Pevear's data: $\Delta_{age} = age_{K\text{–}Ar} - age_{strati}$. The variation of this parameter vs % detrital of total illite, as determined by NEWMOD computation of XRD patterns, shows that K–Ar age for each clay fraction is controlled by the mixture effect and the neogenesis process (Fig. 2.40a). The age difference between the depositional stage and the neogenetic clay fractions (Δ_{cryst}) depends on the burial and thermal history. This means that the absence of any detrital contribution imposes that the mixing line never crosses the reference line (depositional age). Thus, any mixture of authigenic phases must respect the following condition: $\Delta_{age} < 0$.

Neogenetic Age The value Δ_{cryst} is the clay fraction age less the depositional age. The question is how does Δ_{cryst} vary for shale in sedimentary basins of different ages? Using available data (Table 2.8), it seems that the Δ_{cryst} values decrease with increasing depositional age (Fig. 2.40b). The decrease of Δ_{cryst} means that the period of time during which the authigenic illitic minerals are formed is in the same order of magnitude of the experimental error. In addition to the dating error (1σ), other error sources have to be taken into account such as that relative to the measurement of % illite.

Ages by Size Fraction: Δ_{frac} The Δ_{cryst} parameter calculation is based on the definition of the "diagenetic age" which is obtained from extrapolation to 0% of the linear relationship between K–Ar age and % detrital illite. Theoretically, as it is considered to be an end-member in the detrital-authigenic system, it gives the age of the diagenetic illite component. This is classically admitted in the "illite age". There is an analysis procedure recommended by Pevear (1992). However, this "illite age" is questionable since diagenetic illite is not a homogeneous phase but is itself a mixture of several particle populations

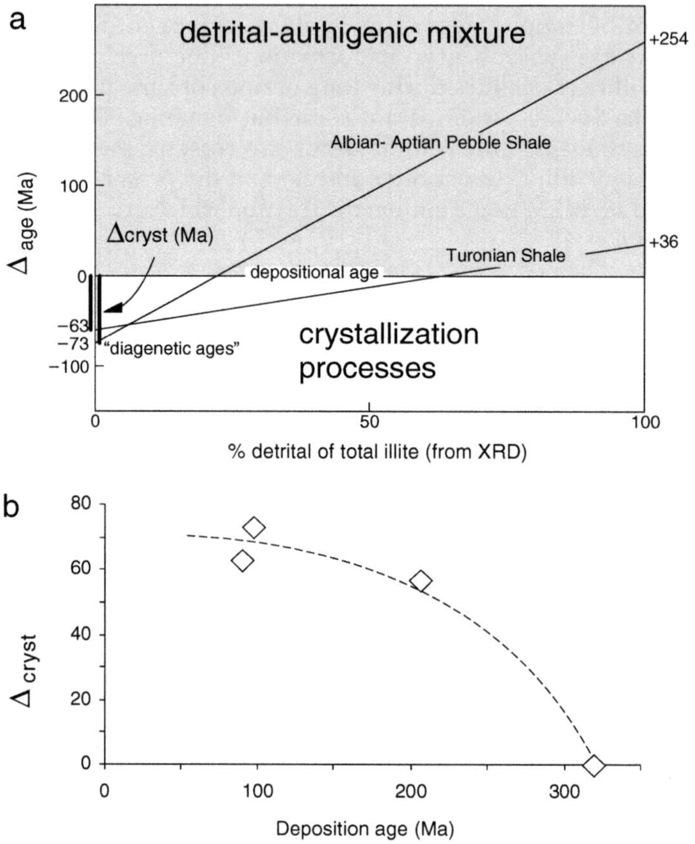

Fig. 2.40a,b. K–Ar age measurement for diagenetic rocks. **a** The linear relationship between apparent K–Ar ages and % illite measured from different clay fractions. $\Delta_{age} = age_{K\text{-}Ar} - age_{strati}$; $\Delta_{cryst} = age_{diag} - age_{depos}$. **b** Variation of Δ_{cryst} (diagenetic age–depositional age) for shales vs depositional ages

Table 2.8. Values of Δ_{cryst} (diagenetic age – depositional age) for shales of different age

Depositional age (Ma)	Calculated diagenetic age (Ma)	Δ_{cryst} (Ma)	σ (Ma)	References
100	27	−73	±1–2%	Pevear (1992)
90	27	−63	±1–2%	Pevear (1992)
205	148	−57	±3–4	Velde and Renac (1996)
320	318–320	0	–	Hofmann et al. (1974)

which have been identified by XRD decomposition as I/S, PCI and WCI or as fine or large, lath-shaped or isometric particles by TEM observations (Lanson and Champion 1991 among others).

Published K–Ar data refer frequently to particle size fractions. Using those which describe only the authigenic phases and not mixtures with detrital micas ($\Delta_{age} < 0$), it is possible to relate the K–Ar ages of the different size fractions to crystallization processes. Indeed, the question is to determine how the radiogenic potassium is accumulated in the authigenic phases through the nucleation and crystal growth processes. This problem is too complicated from a theoretical point of view to be easily solved using a crystal growth model based on a simple layer addition because the mass balance between the growing and dissolving phases is not calculable in the referenced data used here. However, this difficulty could be overcome by using a simple parameter comparing the K–Ar age of any size fraction to that of the finest one: $\Delta_{frac} = \text{K–Ar}_{fraction} - \text{K–Ar}_{finest}$.

2.4.3 Patterns of K–Ar Accumulation During Illite Growth Processes

2.4.3.1 Long-Lived Diagenetic Processes: Ripening Process Dominant (Shales and Clayey Sandstones)

Burley and Flisch (1989) have shown that the I/S MLM of the mudstones and sandstones of the Kimmeridgian have increasing illite contents (hence increasing K_2O contents) with the depth of burial. In these rocks, the diagenetic reactions occur in closed or nearly closed chemical systems. The K–Ar age of these minerals becomes younger as depth increases in the sandstone (Table 2.9, Fig. 2.41). The paradox in dating of clays in diagenetic environments is that frequently the older the bed of the buried sediment, the younger the illites and I/S MLM they contain (Aronson and Hower 1976; Glasman et al. 1989; Morton 1985; Mossman et al. 1992; Renac 1994; Rinkenbach 1988).

The paradox can only be explained by the continuous rejuvenation of authigenic minerals, even if a part of the new material stems from the dissolution of detrital minerals. The question then is how does illitization of I/S MLM proceed in diagenetic environments? Obviously, this question is not answered simply

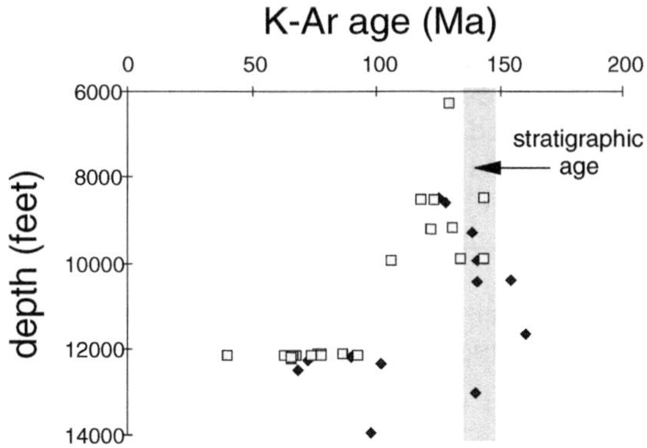

Fig. 2.41. Variation in the K–Ar age of clays extracted from shales (*squares*) and sandstones (*lozenges*) of the Piper and Tartan formation, Outer Moray Firth (Burley and Flisch 1989). The *shaded area* represents the stratigraphic age of these formations (Kimmeridgian)

Table 2.9. Values of the K–Ar age of clays extracted from shales (squares) and sandstones (lozenges) of the Piper and Tartan formation, Outer Moray Firth (Burley and Flisch 1989)

Mudrocks Depth (feet)	K–Ar age (Ma) $\pm 1\sigma$	Sandstones Depth (feet)	K–Ar age (Ma)$\pm 1\sigma$
8482	125.3±1.5	8506	143.8±1.6
8608	128.5±1.5	8515	118.2±1.3
9285	139.1±1.6	8536	123.5±1.4
9948	140.8±1.6	9189	131.1±1.5
10391	154.2±1.7	9210	122.3±1.4
10444	141.2±1.9	6260	129.2±1.5
11671	160.7±1.8	9897	134.4±1.8
12195	89.9±1.4	9899	143.4±1.8
12286	72.6±0.8	9933	105.8±1.6
12355	102.2±1.1	12120	86.8±1.1
12490	68.5±0.8	12139	76.8±0.8
13039	140.6±1.6	12170	73.5±0.8
13980	97.7±1.2	12236	65.9±0.9
		12139	77.7±0.8
		12166	62.9±0.7
		12180	40.6±0.6
		12152	92.7±1.4
		12160	77.8±0.9
		12169	67.7±0.8
		12179	66.4±0.8
		12189	66.0±0.7

by considering that illite layer growth or accumulation on I/S MLM particles is the unique process at work in diagenetic environments. Because the systems are generally of approximately the same mass and chemical composition throughout their diagenetic history (approaching closed chemical systems) the increasing illite content of crystallites in a sample (not total illite content nor total potassium content) is related to a maturation process in which unstable particles dissolve and more stable particles grow. Such a process was thought to be similar to an Ostwald ripening (Inoue et al. 1988; Eberl and Srodon 1988; Lanson and Champion 1991) and has been assumed to exist in shales under conditions of diagenesis because a steady state of crystallite distribution seemed to be reached (the particle size distribution is given by a single log-normal function in reduced coordinates). In fact the Ostwald ripening process cannot be applied to a diagenetic environment because of several missing conditions among the following (Baronnet 1991):

- the system must be isothermal in space and time, i. e., no change in temperature which is not possible in a burial sequence,
- the solid-fluid system is composed of a single stable mineral phase, which is not the case in the smectite-to-illite crystallization sequence,

- the system is closed (isochemical) and mass-conservative, which does not seem to be the case for shales at least in the Gulf Coast USA environment (Berger et al. 1997),
- the oversaturation of solutions with respect to the stable phase must be low but never equal to zero or negative,
- nucleation is not active, i. e., all nuclei are produced at about the same time at the beginning of the evolution of the system.

However, if Ostwald ripening is not possible, strictly speaking, maturation by crystal growth is undoubtedly acting in shale diagenesis. For example, the proportion of lath-shaped particles (I/S MLM) was observed to decrease with depth in samples from a sedimentary basin while the population of poorly crystallized illite (PCI) and then well-crystallized isometric illite (WCI) particles increases (Lanson and Champion 1991; Varajao and Meunier 1995). The clay particles formed under given burial conditions at an instant t become too unstable when burial progresses i. e. when temperature and time ($T \times t$) increase, either because their composition is inappropriate (smectite % too high) or because they are too small (outer surface/volume ratio too big). Therefore the smallest particles of I/S or illite are constantly dissolved; the matter thus released contributes to the growth of particles of greater size which have a greater illite content (Fig. 2.42a). At each dissolution-crystallisation stage, the potassium released in solution is fixed in the crystal lattice of the growing particles but the radiogenic argon associated with potassium escapes from all fixation and can migrate out of the reaction zone. The apparent change in the K–Ar age depends on the ratio between the mass of the growing minerals and dissolving minerals, respectively. Losses of argon are greater than accumulation and thus the apparent age decreases as the particle size increases (Fig. 2.42b).

2.4.3.2
Short-Lived Systems: Nucleation Dominant (Illitization of Kaolinite or Dickite in Sandstones)

The ripening-type process is active during long periods of time in diagenetic series. As they are greater than the dating experimental error (2σ), the K–Ar ratios do not give a geological age but only an integrated measure of the K-accumulation period. Consequently, as the rate of dissolution and crystal growth increases with temperature, the greater the burial depth, the more recent this integrated age. Thus, coming back to definitions given in Fig. 2.37a, the K–Ar ratio gives a geological (diagenetic) age if the K accumulation period of time in illitic phases is shorter than 2σ. At present, we do not know how much time is needed for an illite particle to be formed. Thus, the question is: is it possible to identify the particles which have experienced a sufficiently fast crystallization rate to give a geological age?

Crystal habits depend on the growth rates. It is known that wiskers (hairy illite) are formed in short periods of time while perfect platy hexagonal crystals grow slowly. Wilkinson and Haszeldine (2002) showed that the growth of

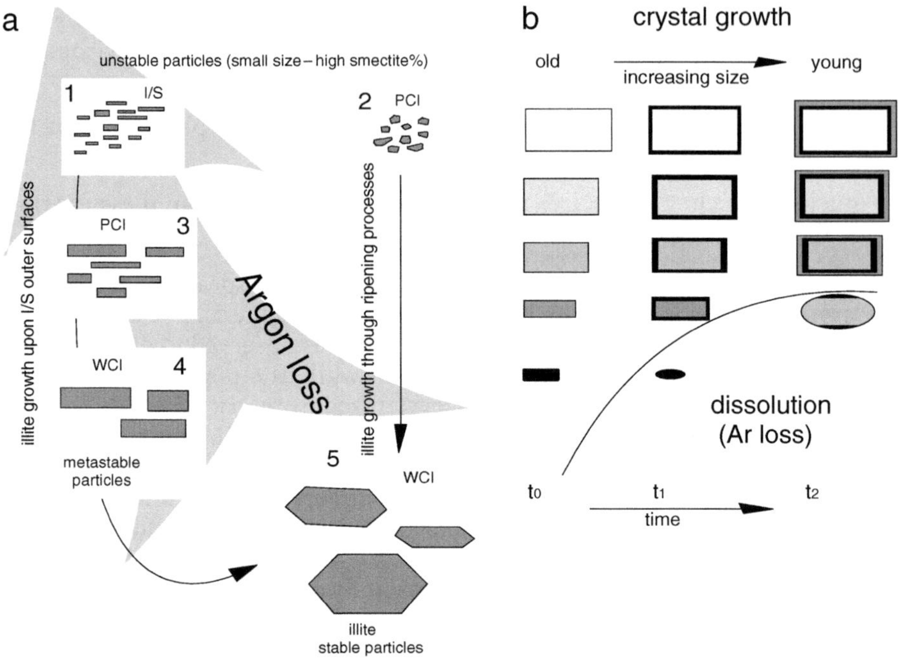

Fig. 2.42a,b. Rejuvenation of the potassium stock by a ripening process. **a** The growth of I/S MLM or illite particles and the increase in the illite content depend on a dissolution process of the unstable particles whose elements are used for the growth of illite layers. PCI: poorly crystallized illite; WCI well-crystallized illite. **b** Rejuvenation effect of a ripening process (no nucleation): the K–Ar age decreases with increasing particle size

fibrous illite is limited by the kinetics of nucleation in abnormal conditions such as high pore fluid supersaturation, high pore fluid velocities, high temperature or presence of a catalyst. The rapid growth is favoured by these abnormal conditions and produces elongated fibrous crystals.

Most of the illitization events in porous sandstones are related to "hydrothermal-like" episodes. This is particularly the case for the lower Permian Rotliegend sandstones in which illite crystals grow either on kaolinite-dickite or quartz overgrowth surfaces. Whatever the shape from fibers to isometric pseudo-hexagonal crystals, illite was shown to be formed during the intense fracturing period of the Kimmerian tectonics at 155 Ma (Lanson et al. 1996). Illitization of kaolinite or dickite requiring high oversaturated solutions with respect to illite, the K-feldspars cannot be the source of K^+ ions. Such fluids were assumed to come from laterally adjacent Zechtein evaporite formations in the vicinity of faults. Because of high oversaturation, the fluids trigger the nucleation and growth of I/S MLM or illite crystals during periods of time shorter than 2σ. They can be considered as instantaneous phenomena compared to the progressive illitization of shales. In the Rotliegend sandstones,

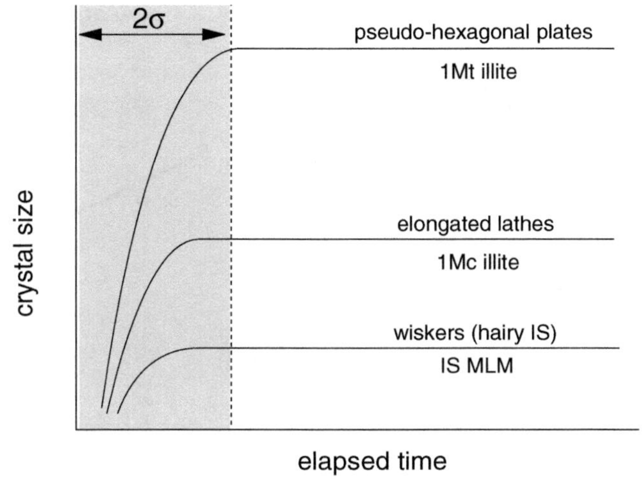

Fig. 2.43. Theoretical representation of the closure of the K–Ar system for three illite types forming in the Rotliengend sandstones (modified from Joesten 1983)

the mean age is about 155 Ma, independent of the composition (I/S from 20 to 5% smectite or illite) or the crystal habits (whiskers, laths, pseudo-hexagonal plates). Composition and crystal habits are dependent on local temperature conditions and fluid oversaturation degree. The I/S MLM or illite crystal growth process which determines the argon accumulation period is similar to that of quartz grains in metamorphic aureoles (Joesten 1983). It is shorter than 2σ (Fig. 2.43).

2.4.3.3
Intermediate Systems: Instantaneous Formation and Continuous Growth of Illite

The sandstones of the upper Brent and upper Skagerrak formations in the northern Viking graben (North Sea area) have been extensively studied. They contain abundant pore-filling illite with almost no I/S MLM in contrast to the adjacent shales. Matthews et al. (1994) showed that the average layer charge and K^+ content increase with depth from 0.69 to 0.80 K^+ per $O_{10}(OH)_2$ no matter the depositional age of the hosting formation. Illitic phases in the lower Jurassic sandstones contain more potassium than the late Triassic ones Here, the K–Ar age increases with K_2O contents (Fig. 2.44a) and with increasing depth (Fig. 2.44b).

This indicates that there is very little or no rejuvenation of the K_2O stock in the system. Consequently, ages increase with depth. The increasing K_2O content with depth implies a continuous growth of illite layers on illite particles during burial (Fig. 2.44c). Matthews et al. (1994) deduced that the formation of illitic clays from various precursors (kaolinite and dickite) in this particular area of the North Sea was rapid but the age increase with depth shows that growth induced by increasing burial conditions was slow.

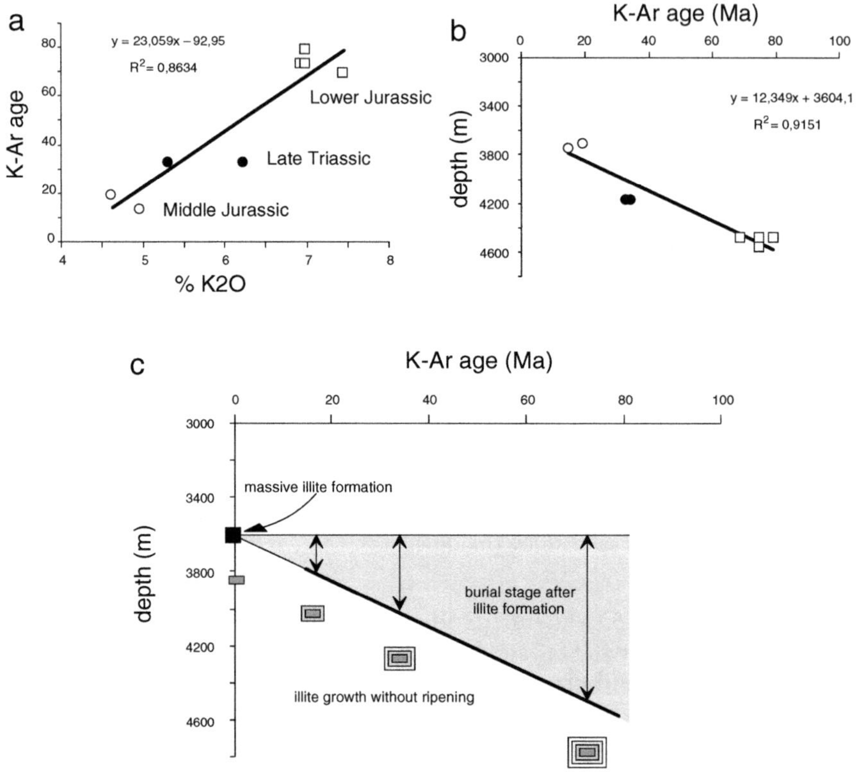

Fig. 2.44a–c. Increasing K–Ar ages with depth in the Rotliengend sandstones (data from Matthews et al. 1994). **a** Plot of K–Ar ages vs K_2O content (per Si_4O_{10}). **b** Variation of K–Ar ages with depth. **c** A continuous growth model explaining the K_2O and K–Ar age variations with depth

The example of the upper Brent and upper Skagerrak formations shows that, in some cases, illite forming in porous sandstones experiences a more complicated thermal history than that recorded by a simple precipitation during hydrothermal-like events. Two steps may be distinguished during its crystallization period of time: 1) a rapid nucleation and growth process ($< 2\sigma$) which is temperature-chemically controlled; 2) a slow growth ($> 2\sigma$) of illite layers on I/S or illite particles whose rate depends on temperature rate increase during burial. Theoretically, the K–Ar dating obtained from clay samples may change according to the relative importance of these two steps:

- if the former predominates, all I/S MLM or illite particles will have the same K–Ar age as shown by Lanson et al. (1996);
- if the second predominates, the age will increase with depth, i. e. temperature. The slope of the line in Fig. 2.44c depends on the rate of K_2O accumulation during the illite growth period.

Table 2.10. K–Ar illitic phase dating from two or three size fractions of clay samples from the Brent sandstones, Heather field, North Sea (data from Glassmann et al. 1989)

Depth (feet)	Size (μm)	%K	K–Ar age (Ma)±1σ
11211	< 0.1	5.303	27.7±0.6
	0.2–0.1	6.273	33.3±0.8
11225	< 0.1	4.405	31.1±0.7
	0.2–0.1	6.281	35.3±0.8
9720	< 0.05	3.126	39.9±1.0
	< 0.1	4.163	44.9±1.0
10973	< 0.1	5.485	50.9±1.2
	0.2–0.1	6.196	50.3±1.1
	0.5–0.2	6.545	56.9±1.3
11711	< 0.1	4.756	26.8±0.7
	0.2–0.1	6.208	39.3±0.8
11778	< 0.1	4.866	34.8±0.8
	0.2–0.1	6.032	45.7±1.1

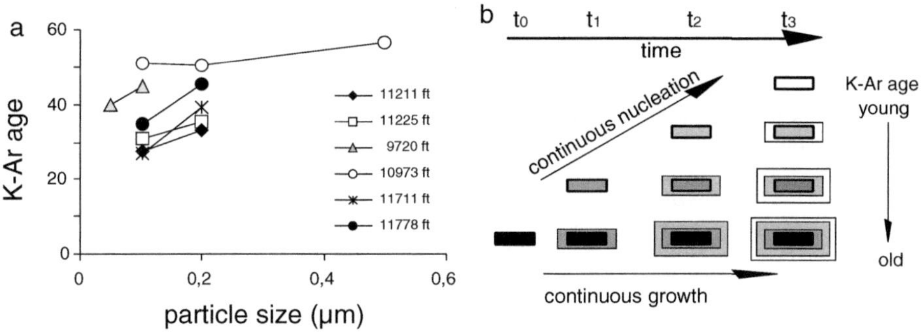

Fig. 2.45a,b. Effects of a continuous nucleation-growth process on the fraction size – K-Ar age relation. **a** K–Ar age variation between fine and coarse fractions from clay minerals sampled at different depths from Glassmann et al. (1989) **b** Schematic representation of a nucleation-growth process continuously active during diagenesis

This seems to be the case of illite forming in the Brent sandstones from Heather field, North Sea (Glassmann et al. 1989). Illite and quartz overgrowth post-date kaolinite cementation. K–Ar ages (Table 2.10) vary in much of the Paleogene (55 to 27 Ma). The coarser fractions are 5 to 10 Ma "older" than the finest ones (Fig. 2.45a). Glassman et al. (1989) suggested that, contamination being negligible, this age difference has to be related to the length of growth history experienced by illitic particles. The fact that the finest fractions are the youngest ones induces a continuously active nucleation during the sandstone burial stage. The coarser fractions grow on old cores and are always older than the finest ones (Fig. 2.45b).

Summary In this section one can see that the relations of illite formation and age determination are strongly interrelated. It is not possible to interpret illite

K–Ar ages without considering how the illite layers on or in clays form. In closed or nearly closed chemical systems, where most of the material in the rock is re-cycled in order to form illite from smectite, continuous change in apparent age is most likely to occur. By contrast, in sandstones or open systems where illite forms from rapid, total recrystallization of non-illitic or smectitic material, the age obtained can, under favorable conditions indicate the age of a given geologic event. It is always possible to obtain a radiometric age, but it will be difficult to interpret its meaning as far as geological events (punctual phenomena) are concerned.

2.4.4 Diagenesis of Bentonites

2.4.4.1 Evolution of the Illite Growth Process with Time

K–Ar data of illitized bentonites from different depositional ages (Moe et al. 1996; Srodon et al. 2002) are given in Table 2.11. It appears that the coarser fractions may have the same age for recent deposits (Oligocene), or older

Table 2.11. Variation of the Δ_{frac} parameter (Δ_{frac} = K–$Ar_{fraction}$ – K–Ar_{finest}) with time in diagenetic bentonites

Size	K–Ar age (Ma)	Δ_{age} (Ma)	2σ
Oligocene bentonites (Srodon et al. 2002)			
< 0.02	17.9	0	1.1
0.02–0.05	18	0.1	1.7
0.05–0.1	17.5	−0. 4	1.6
0.1–0.2	17.4	−0. 5	1.2
< 0.02	18.6	0	2
0.02–0.05	19.5	1.1	2.5
0.05–0.2	20.1	1.5	2.6
Ordovician bentonites (Srodon et al. 2002)			
< 0.02	363	0	10
0.02–0.05	294	−69	8
0.05–2	335	−28	9
< 0.02	360	0	9
0.02–0.05	341	−19	9
0.05–2	346	−14	9
< 0.02	382	0	10
0.02–0.05	353	−29	9
Proterozoic bentonites (Moe et al. 1996)			
< 0.2	606	0	8
0.2–1	675	69	9
1–2	807	201	11

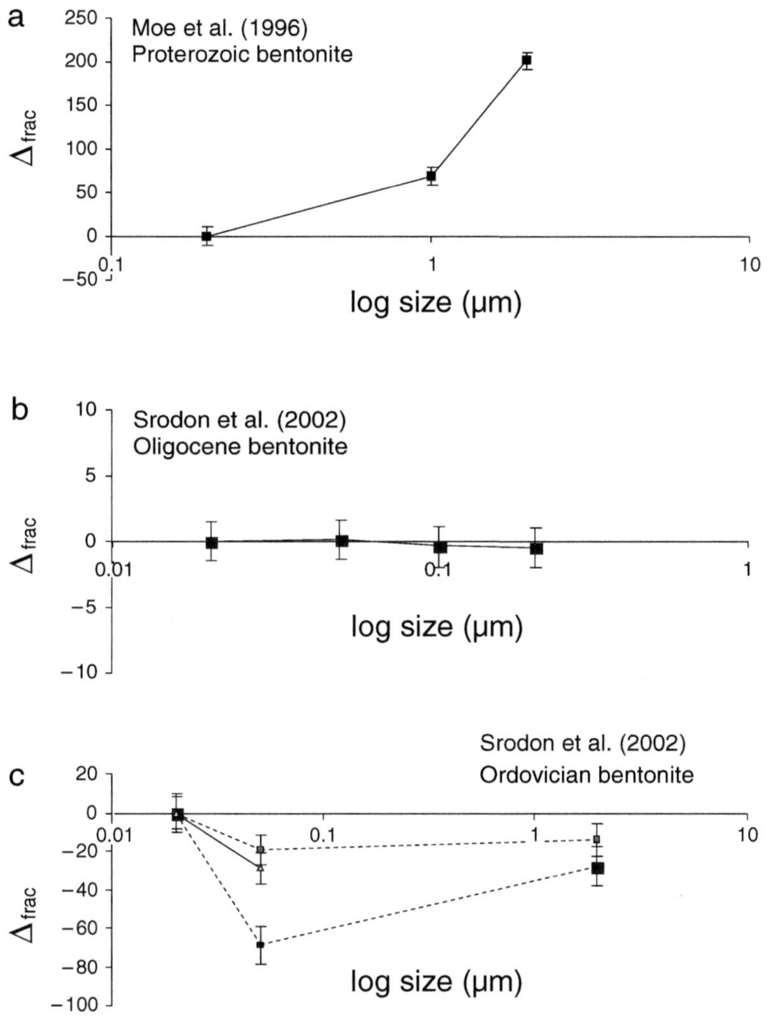

Fig. 2.46a–c. Variation of the Δ_{frac} parameter (K–Ar age difference between a given size fraction and the finest one) versus the fraction size in bentonite deposits of different depositional age. **a** Proterozoic (Moe et al. 1996). **b** Oligocene (Srodon et al. 2002). **c** Ordovician (Srodon et al. 2002)

(Ordovician) – (Proterozoic). The three trends are shown in a: Δ_{frac} vs log size plot (Fig. 2.46a–c). What do these trends mean? Three possible pathways can be envisaged:

- *same ages for coarse and fine particles*: this is possible when the nucleation and crystal growth are sufficiently rapid to produce a fine to coarse-grained particle population in a period less than the dating experimental error ($< 2\sigma$);

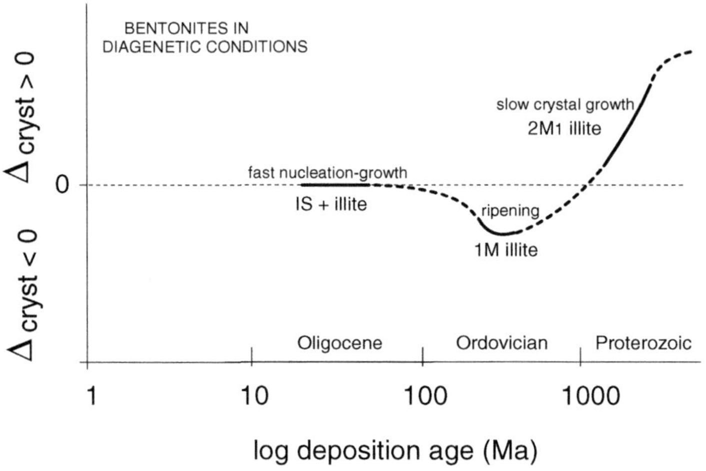

Fig. 2.47. Hypothetical steps of the illitization process of bentonites in diagenetic conditions according to the variation of Δ_{cryst} with time

- *coarser particles younger than the fine ones*: this results typically from a ripening process in which the finest particles are the older ones and are continuously dissolved while the coarser particles continuously grow;
- *coarser particles older than fine ones*: this implies concomitant nucleation-crystal growth. Thus, the internal core of the coarser particles are the oldest fraction.

The illitization of bentonite is a diffusion-controlled process (Velde and Brusewitz 1982; Altaner et al. 1984). This factor could be variable with time. However, even if we ignore the thermal history experienced by the Oligocene, Ordovician or Proterozoic bentonite deposits during their burial stage, their respective K–Ar trends could be considered as steps in a continuous crystallization process. Indeed, the first step which is a fast nucleation and crystal growth process, is the early diagenetic step during which smectite is transformed into illite + I/S MLM (Oligocene). Then, (step 2) for longer periods of time, the K–Ar ages of the different size fractions are controlled by a ripening process leading to the formation of 1M illite (PCI). For Proterozoic deposits (step 3), the dominating process is a slow crystal growth forming large, "perfect" $2M_1$ crystals (Fig. 2.47). All the three processes, i. e. smectite illitization, ripening and crystal growth may coexist at a given time but one is dominating and controls the apparent K–Ar ages.

2.4.4.2
Conclusion

We have seen that K–Ar age is not simply ruled by nucleation and growth of illite layers on I/S MLM. In some cases the biggest particles are younger than the finest ones while the reverse is true in other cases. This means that illite crystal growth should be considered in more detail before interpreting K–Ar analyses in terms of radiometric age. The model proposed by Srodon et al. (2002) may be successfully applied for geological series in which the size distribution of illitic particles results from a continuous nucleation and growth in an open system (bentonites). This is inappropriate for particle distributions ruled by a maturation process. It is also inappropriate for systems in which quick precipitation happens, giving the same radiometric age to all particles independent of their size.

What then is the signification of a "diagenetic age"? Srodon et al. (2002) have clearly shown that the diagenetic age depends closely to the thermal history experienced by the sediments, the thermal history being a function of the burial rate and the local value of the geothermal gradient. We must consider that it depends also on the way that illite grows during illitization.

The illite growth processes depend mostly on two factors: 1) the chemical system characteristics (open or closed); 2) the duration of the crystallization. The example of the bentonite deposit diagenesis shows that at more or less equivalent physico-chemical conditions, the crystal habit and the polytype change with time. Consequently, the argon accumulation mechanism changes also when particles grow and change of crystal habit from hairy to lath and from lath to platy hexagonal.

2.5 Summary

2.5.1 What is Illite?

An initial working definition for the mineral illite was given at the end of Chap. 1 which was a summary of the accumulated knowledge concerning the mineral readily available in the literature. Chap. 2 is designed to rework this data in the context of the geological occurrences of illite and to give new insight into the problem of the definition of illite. Chemically speaking, illite was considered in the past to be a slightly potassium-deficient dioctahedral mica found in low-temperature environments. Charge deficiency originates in the octahedral and tetrahedral sites of the structure. In many ways illite is similar to phengite mica except in the interlayer charge and subsequent potassium content. However, illite is a clay mineral and hence concentrated in the fine fraction ($< 2\ \mu m$) of rocks and other materials in which it is present. It was known that the XRD peak typical of illite is not symmetric, but the physical consequences of this assymetry were not considered, apparently, to be important.

The information presented in Chap. 2 has shown us that illite is clearly a low-potassium mica (i.e. with a 10 Å basal spacing) and it is found in the fine fraction of samples. The potassium content for a unit cell of $O_{10}(OH)_2$ anionic configuration is near 0.9. Mg and Fe content vary depending upon the chemical system in which the illite forms which suggests a solid solution of illite compositions which respond to the chemistry of the system in which it forms. The chemical substitutions in the 2:1 structure of illite are those of celadonite and muscovite producing interlayer charge and pyrophyllite which maintains charge neutrality.

In using curve-decomposition methods of analysis for the peak envelope in the 10 Å region of X-ray diffractograms, one finds that the "illite" component of a sample is most often in fact two components, one with a slightly large band (> 0.6 °2θ Cu Kα width at half height, FWHM) and a position near 10.3 Å and another, more narrow peak (> 0.3 °2θ Cu Kα, FWHM) at very near 10 Å. The first peak is called PCI (poorly crystallized illite), the second WCI or well crystallized illite. This last phase normally has an average of greater than 14 coherently diffracting layers in crystallites present. It is well crystallized. This WCI is composed uniquely of non-expanding layers. Its polytype can be 1M or 2M as far as one can determine at present. The type of WCI polytype might well depend upon physical-chemical conditions of formation. It does appear that in older rocks or hydrothermal systems of high temperature and high potassium activity, the 2M polytype is most likely the ultimately stable form of WCI.

WCI is Illite Most of what is identified as PCI has a small but persistent smectite content (< 5% according to XRD characteristics). It does not develop large, thick, coherently diffracting domains. PCI is not illite.

However, the composite peak attributed to illite in diffraction diagrams, most often composed of both PCI and WCI contributions, is most often asymmetric to the high spacing (low °2θ Cu Kα) PCI side which was assumed in the past to signal the existence of illite. In fact, the asymmetry is due to the presence of two peaks in different, unequal proportions. In more evolved samples, the WCI (true illite) peak is predominant, giving a peak position (position at greatest intensity) of 10 Å. The evolution of the different proportions of PCI and WCI determine most often the values of illite crystallinity. Greater asymmetry indicates the presence of illitic I/S minerals in the illite sample.

From these observations it is evident that illite samples most often contain several crystalline phases, PCI or a high illite content I/S mineral as well as WCI. Illite itself (WCI) has a specific chemical composition, best defined by its potassium content of 0.9 atoms per $O_{10}(OH)_2$. It is a well-crystallized mineral, having more than 14 successive coherently stacked 10 Å layers, which is unusual for a clay mineral. These characteristics can be observed in XRD diagrams when curve decomposition methods are used to separate the different components of the peak complex near 10 Å. Hence we have, to a large extent, a mineral definition based upon crystal structure and chemical composition.

2.5.2
Where Does Illite Form?

The geologic parameters of illite formation are widespread over the realm of conditions in which clay minerals can be found. Illite can be found in soils, forming or being destroyed according to specific chemical conditions. Illite can form in shales and sandstones during burial and diagenesis. Illite can be produced in high temperature hydrothermal environments. The range of physical conditions of illite formation are from 20 °C, surface soil conditions, to below 300 °C, under diagenetic or hydrothermal environments. It is present in virtually all environments where clays are found. It is limited by chemical parameters, based upon potassium availability. Illite is one of the most common clay minerals. Illite (WCI) is rarely the only potassium-rich phyllosilicate present in a geologic sample.

2.5.2.1
How Does Illite Form?

Illite is a potassic mineral and hence its stability depends upon potassium availability. This is in part governed by hydrogen ion activity. If few hydrogen ions are present, a "non-hydrous" potassic phase will be present such as feldspar or zeolites whose water content is not in the form of hydroxyl ionic units. However, in most clay mineral environments water and hydrogen ions are available and hence illite is most often favored when potassium is available, this at the expense of potassium feldspar, for example. From these considerations it is evident that illite can form through chemical potential change. In soils one sees illite forming among other clay phases under conditions of high potassium activity (fertilizer or manure applications). This is most likely due to dissolution-precipitation mechanisms at the expense of less potassic minerals present in the soil. In hydrothermal environments illite is observed to form in cracks and voids, from precipitation of material in solution. The same is true of some sandstone occurrences of illite where illite replaces kaolinite or is precipitated in pore spaces. Thus one mechanism of illite formation is through dissolution – precipitation reactions under conditions of high potassium activity. In these cases there is no specific precursor mineral present in the assemblage in which the illite forms.

A second possibility for the formation of illite is by mineral transition and transformation, the so-called smectite-to-illite transformation. It is shown earlier in this chapter that the transformation of smectite to illite, through mineral growth and mineral dissolution, results in a PCI, mixed-layer mineral with a very low but persistent smectite component. The PCI becomes more important in shale and other closed systems with time and increase in temperature. However, as PCI forms from a small smectite substrate, WCI crystallizes also. Eventually WCI or true illite becomes the major potassic mineral in the clay fraction given enough time or high enough temperatures.

In open, diffusing systems such as bentonite layers in sediments, illite-rich minerals will form in the same manner, but only after very long periods of time, more than 500 Ma. Thus the formation of illite is through precipitation from solution of components forming the potassic clay mineral which has only 10 Å layers present.

2.5.2.2
Why Does Illite Form?

Illite is stable, thermodynamically, at low temperatures. It replaces muscovite or phengite. Even at high potassium content, a chemico-geologic low-temperature system will form illite. The chemical composition of illite will vary as a function of the chemical constraints of the system (bulk composition and the presence of other phases which will define the composition of illite). Illite forms because it is stable.

The process of illite formation can take various paths, precipitation from solution, overgrowth on smectite-illite crystallites, or pseudomorphous replacement of micas as noted above. The fact that illite has a low potassium content constrains the mineral structure to have a neutral site in the succession of octahedra-tetrahedral negatively charged layers. This pyrophyllite component is added to the muscovite and celadonite molecular substitutions in this aluminous mica-like mineral (see Sect. 1.2.4). Three substitutional types give a high degree of liberty in the possible modes of ionic distribution in the mineral structure, ordered, disordered and segregated. As temperature is increased in a geologic system, this highly substituted mica is replaced by a less substituted one, muscovite-phengite with only two substitutional types. Eventually, at very high temperatures, only muscovite substitutions occur. In the thermal sequence, there is a lessening of disorder as temperature is increased. The existence of three substitutional types certainly leads to initial problems of structural accomodation, with the result that the illite crystallites are initially of small coherent diffracting domains, having numerous stacking incompatibilities resulting in stacking faults. This gives us the small coherent diffraction domain PCI, $1M_d$ crystallites. With time, these more disordered types are replaced by crystals with more order and larger coherent diffracting domains, most likely by dissolution and crystal growth precipitation mechanisms. 1M (one 2:1 layer) and eventually 2M (two layers) structures form after very long periods of time or at higher temperatures, near the limits of illite stability. If the system has a very high potassium content (for example in agricultural soils treated with KCl) one can find large illite crystallites. Hence chemical potential can aid in overcoming energy barriers (activation energy necessary to recrystallize a disordered illite) that impede the formation of an ordered illite (WCI).

Illite then forms when a chemical-geological system has sufficient potassium available under conditions of low temperature. The content of alumina must dominate the system. However, illite is rarely found as a pure portion of the potassic phyllosilicates, I/S minerals and PCI are normally present. This is due to the comparatively high energy barriers which impede its formation. However given enough time, illite (WCI) will be the remaining potassic phyllosilicate in a system. As indicated in the phase diagram from Montoya and Hemley (1975),

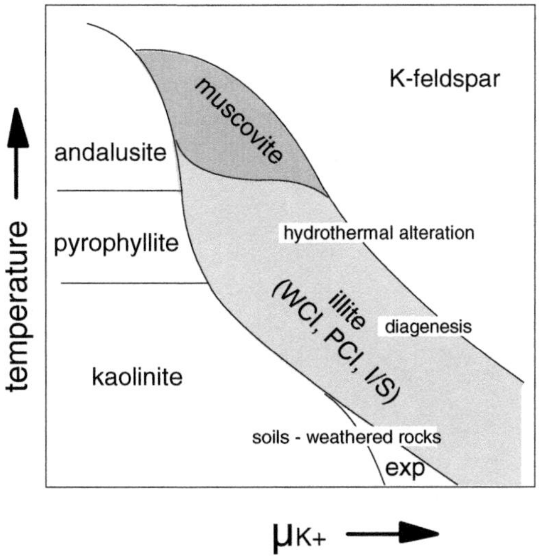

Fig. 2.48. Simplified temperature vs potassium chemical potential phase diagram (inspired from Montoya and Hemley 1975). Illite is the low-temperature micaceous phase. Time is not taken into account here in spite of the fact the crystalline structure of illite depends on the formation rate (kinetics)

a lowering of the temperature and an increase of the potassium chemical potential will trigger the formation of illite (Fig. 2.48). Independent of kinetic constraints, one can consider that illite forms because it is the stable potassic mineral in an aluminous system.

Illite (WCI) is the inevitable low-temperature micaceous mineral present in aluminous rocks, soils and sediments. It is slow to form in closed systems, such as argillaceous sediments, as they are transformed into shales during burial metamorphism. Rapid recrystallization and geothermal pulses produce PCI. However, it can be formed much more rapidly in chemical systems of high potassium activity. The effect of chemical potential on the formation of illite is a key to many geological and ecological problems.

CHAPTER 3

Dynamics of the Smectite-to-Illite Transformation

The occurrence of illite in sediments and sedimentary rocks is closely related to the transformation of smectite to illite under conditions of burial metamorphism. Our objective at this point is to identify the origins of illite (PCI and WCI) in a context of changing temperature. This process is related to the effects of time (t), temperature (T), and composition or chemical activity of certain elements (x) in sediments affected by burial diagenesis. This transformation reaction is the origin of a large portion of the illitic materials found in different geological contexts.

Much recent research has then been directed to a better understanding of the smectite-to-illite transformation which could be used to determine paleotemperatures. These values could then be used to estimate the conditions of burial in sedimentary basins as they affect the evolution of the organic matter they contain and hence the production of oil and gas in sedimentary rocks. More recent interest in the transformation of smectite to illite has been based upon the mineralogical and thermodynamic stability of smectites which have been chosen for their ion exchange and assumed physical plastic characteristics in a use for nuclear waste deposit barriers. The thermal stability of smectite is extremely important for the strategy of smectite use in radioactive waste containment. Hence much time and energy has been spent on determining the reaction characteristics which are responsible for the formation of illite via the transformation of smectite. This use of smectite stability is discussed in Chap. 4.

We divide the analysis of the smectite-to-illite transformation into two parts: 1) studies performed in the laboratory where pressure and temperature are the variables acting on silicate materials and 2) studies of clay evolution in diagenetically altered rocks, essentially in relatively undisturbed sedimentary basins.

3.1 Experimental Studies

The transformation of smectite to illite under various laboratory conditions has been the subject of much study in the last several decades. The smectite content of the interstratified illite/smectite (I/S) minerals interpreted from X-ray diffraction diagrams (XRD) has been taken to indicate the composition of the clay minerals found in diagenetic sedimentary sequences The change in smectite content of the I/S mineral has been related to the conditions of burial in diagenesis, hence the variables of time and temperature (Perry and Hower 1970). Following these concepts, several series of experiments have been performed on synthetic materials (Velde 1969; Eberl and Hower 1976) and on natural clays with the aim in mind to establish the time–temperature relations in the clay transformation. In kinetic studies on natural clays (Eberl and Hower 1976; Howard and Roy 1985; Robertson and Lahann 1981, Huang et al. 1993; Small 1993; Chermack 1989) kinetic relations were determined which differ greatly, by a factor of 10. Such differences indicate that some variables and interpretations have not been made in the same manner by the different authors. There is no reason to think that the experiments were not carried out correctly. We will attempt to show the coherent and incoherent aspects of this considerable amount of experimental information. First we consider the identification and description of the reactions using XRD methods.

3.1.1 The Run Products in Whitney and Northrop's Experiments Using Bentonite

A number of the experiments performed under conditions of relatively high temperature (above 200 °C) and at high pressures (1–2 kb) are similar and contained similar materials, natural smectite minerals, as starting phases. The most complete set of identification procedures is given in Whitney and Northrop (1988) which is used as a starting point here.

3.1.1.1 Experimental Results and Assumptions

Whitney and Northrop (1988) demonstrated that there are two reactions in the transformation of smectite: a more rapid rate for short periods of time and a much slower rate for longer times (Table 3.1). A plot of the results on $\ln(a/(a+x))$ vs. time co-ordinates, where a is the initial 100% smectite concentration and x the amount of non-expanding layers in the potassium-saturated state clearly shows the differences in reaction rates (Table 3.1; Fig. 3.1) This plot follows the usage of Howard and Roy (1985). Whitney and Northrop based their interpretation assuming that no trioctahedral magnesium-rich phyllosilicate phase was formed with the R0 I/S MLM. Chlorite appears only when I/S MLM becomes ordered.

3.1.1.2 Smectite Composition Heterogeneities

The differences in response to glycol treatment for samples in the Na- and K-saturated states indicate that smectites of varying properties are interstratified with illite in mixed-layer phases in all run products. In most of the runs, a variable amount of layers collapse in the potassium-saturated state and re-expand after saturation with Na. This treatment is commonly assumed to

Table 3.1. Run products in the experiments for smectite-to-illite conversion reaction from Whitney and Northrop (1988). High-charge smectite is vermiculite

Sample	Temp. °C	Time (d)	Exp. K	Ord. Type	Exp. Na	High charge	Illite	Low charge	Non-clay min.
SWY-1	–	0	100	–	100	0	0	100	Q, M
A132	250	7	83	R	92	9	8	81	Q
A133	250	14	78	R	91	13	9	78	Q, M
A134	250	30	57	R	91	34	9	57	Q, M
A135	250	60	52	R	90	38	10	52	Q, M
A136	250	120	45	R+O	90	45	10	45	Q
A137	250	220	44	R+O	89	45	11	35	Q
A141	300	1	70	R	92	22	8	70	Q, M
A142	300	7	58	R+O	91	33	9	58	Q, M
A143	300	14	50	O+R	90	40	10	50	Q, M
A144	300	30	45	O+R	90	45	10	45	Q, M
A146	300	120	35	O	82	47	18	35	Q
A147	300	220	35	O	78	43	22	35	Q, C
A151	350	1	47	O+R	90	43	10	47	Q, M
A152	350	7	41	O+R	89	48	11	41	Q, M
A153	350	14	34	O	88	54	12	34	Q
A154	350	30	27	O	87	60	13	27	Q, C
A155	350	60	24	O	80	56	20	24	Q, C
A156	350	120	21	O	76	57	24	19	Q, C
A161	400	1	45	O+R	90	45	10	45	Q, M
A162	400	7	28	O	84	56	16	28	Q, C
A163	400	14	23	O	81	57	19	24	Q, C
A164	400	30	14	O	68	54	32	14	Q, C
A166	400	120	12	O	64	52	36	8	Q, C
A171	450	1	23	O+R	89	66	11	23	Q, C
A172	450	7	12	O	84	72	16	12	Q, C
A173	450	14	8	O	8	0	92	8	Q, C
A174	450	30	5	O	5	0	95	5	Q, C
A175	450	60	4	O	4	0	96	4	Q, C
A176	450	120	4	O	4	0	96	4	Q, C

characterize high-charge smectite layers, i. e. vermiculite layers (Schultz 1969; Howard 1981). The layers in the I/S mineral that remain expandable in both potassium- and sodium-saturated states are low-charge forms. On the contrary, the layers that remain collapsed at 10 Å after Na saturation are considered to be "illite". In summary, we will consider the following three types of layers:

1. Low-charge smectite: expandable after Na and K-saturation
2. High-charge smectite: expandable after Na saturation and then collapsed after K saturation
3. Illite: collapsed after K and then Na saturation

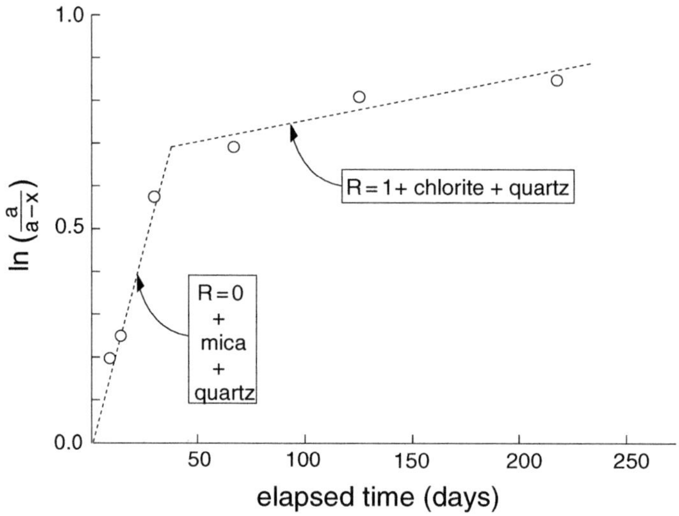

Fig. 3.1. Plot of synthetic hydrothermal mineral reaction progress $\ln(a/(a + x))$ of smectite to nonexpanding layers in the potassium-saturated state according to the data of Whitney and Northrop (1988). The two rates are apparent, that quantifiying the R=0 structural I/S MLM type and that quantifying the R=1 structural type I/S MLM. The run assemblage is different for the two apparent conversion rates. Temperature conditions were 250 °C

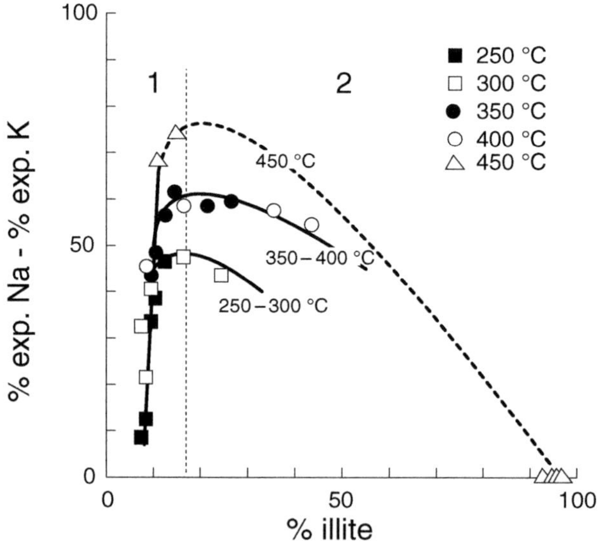

Fig. 3.2. Plot of high-charge layers (non-expanding in the potassium saturation state but re-expanding in the sodium-saturation state) to illite (non-expanding layers in the sodium saturated state) according to the data of Whitney and Northrop (1988). 1 & 2: reactions 3.1 and 3.2 (see text on page 152 for detailed comments)

In the estimations of illite content Whitney and Northrop based their interpretation only upon potassium-saturated samples (non-expanding upon K saturation) which were largely overestimated because much of this material re-expands with Na saturation. This situation is frequently the case for the random I/S structures produced in experiments in which high-charge smectite becomes very important. The amount of these high-charge smectite layers varies with the temperature of the experiment and its duration (Table 3.1, Fig. 3.2). All of the experiments produce high-charge smectite layers while the amount of illite layers does not exceed 11% (random I/S). This reaction proceeds in the presence of quartz and mica which is present in the initial starting materials.

3.1.2 The Different Possible Interpretations of the Experiments

3.1.2.1 A Chemiographic Interpretation of Whitney and Northrop's Results and Hypothesis

In spite of the fact that the chemical composition of the different phyllosilicates produced in each run is not known with precision because the runs are multiphase, it is possible to determine the trends of the reaction using a graphical projection in the M^+–4Si–R^{2+} system where M^+ = Na, K, 2× Ca, R^{2+} = Fe^{2+} + Mg and 4Si = Si content/4 (Meunier and Velde 1989). As the experiments were performed in a closed system, it is possible to estimate how the phases change with temperature and time in order to determine the reaction kinetic parameters. The starting material is a mixture of Wyoming bentonite (i.e. low-charge montmorillonite) containing a small amount of quartz and mica (Whitney and Northrop 1988). Quartz is produced by the reactions as indicated by XRD patterns which show an increase of the quartz peak intensity. This is also the case for other experiments using bentonites (Eberl and Hower 1976; Eberl,1978; Eberl et al. 1978; Inoue 1983; Howard and Roy 1985).

According to the initial assumptions of Whitney and Northrop who did not consider that the tetrahedral charges could increase in the smectite layers, the loss of silica results from the formation of illite layers. Consequently, the R^{2+} component (Mg and Fe) is considered to remain within the smectite structure. Thus, the high-charge smectite layers must always be of a montmorillonitic type. According to this hypothesis, beidellite or beidelletic substitutions are excluded by the existence of the mica–quartz tie-line in the phase diagram (see Fig. 3.4).

The composition of the high-charge montmorillonite layers should be between 0.45 and 0.66 charges per $O_{10}(OH)_2$ (see Proust et al. 1990) because they collapse upon potassium saturation. Assuming that a high-charge smectite layer will have 0.66^+ charge per $O_{10}(OH)_2$, a low-charge layer 0.33^+ and an illite (10 Å non-expandable layers) 0.75^+ or more, it is possible to plot the run products as determined by XRD identification in chemical co-ordinates as in Fig. 3.3a. As an example, the co-ordinates of run products A132 in Whitney and Northrop (1988, Table 3.1) are the following: 9% high-charge montmorillonite

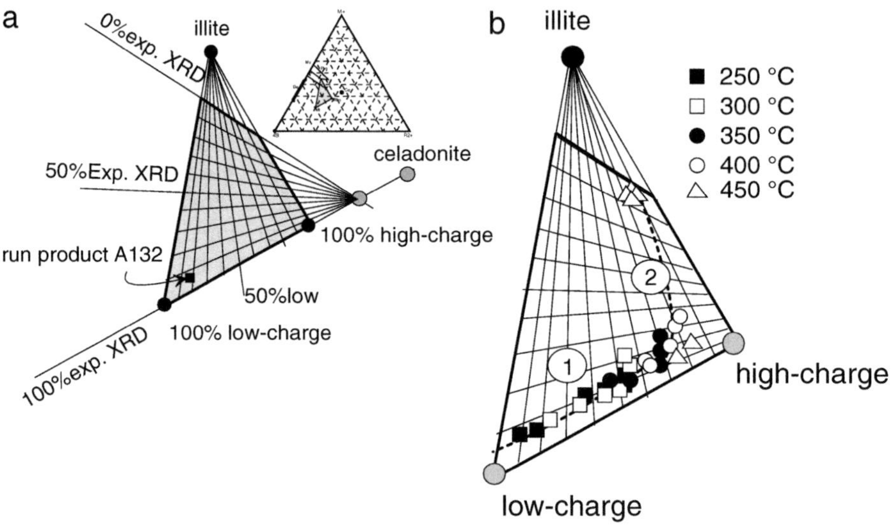

Fig. 3.3a,b. Plot of run results for the data of Whitney and Northrop (1988) according to the phase identifications (XRD determinations of expandable layers) and the starting composition of the reacting clay (SWY). **a)** coordinates are M^+ = (K, Na and 2Ca) content, 4Si and R^{2+} (Mg + Fe^{2+}) as developed in Meunier and Velde (1989); mu: muscovite; be: beidellite; mont: montmorillonite; il: illite I/S: illite-smectite mixed-layered minerals; numbers indicate the layer charge per $O_{10}(OH)_2$; square: plot of the A132 run (see text for detailed explanation); **b)** Plots of the runs corresponding to reactions 3.1 and 3.2

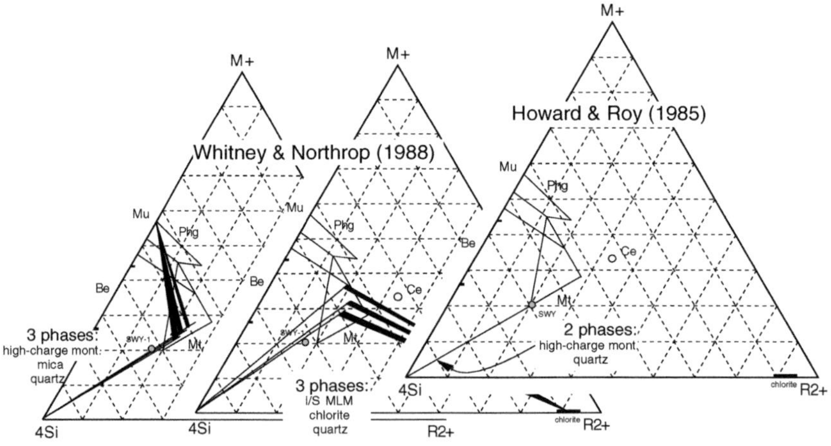

Fig. 3.4a–c. Representation of the phase assemblages in M^+–4Si–R^{2+} co-ordinates for the two portions of the smectite to illite reaction. **a)** Three phase assemblage is imposed by the stoichiometry of the hydrothermal reactions in the experiments of Whitney and Northrop (1988). **a)** Formation of high-charge montmorillonite layers + quartz (+ mica?). **b)** Formation of ordered illite-montmorillonite mixed layered minerals + quartz + chlorite. **c)** The stoichiometry in the experiments of Howard and Roy (1985) allows only a two-phase assemblage: high-charge montmorillonite + quartz

(expanding with. Na only); 81% low-charge montmorillonite (expanding with Na-expanding with K); 8% illite (non-expandable). All of the points for montmorillonite reaction products are given in Fig. 3.3b. It can be observed that the reaction trend from low-charge smectite to illite is deviated towards the high-charge smectite pole.

In the second step of the reactions investigated by Whitney and Northrop (1988) the random illite/montmorillonite mixed-layered mineral is replaced by an ordered I/S. The compositions of the I/S components are assumed to be illite and high-charge montmorillonite (Fig. 3.3b). Some details of the mineral assemblages change: mica disappears when chlorite is produced. The phase association of the reaction shows that the micaceous phase present in the initial starting material disappears when the illite content in the I/S mixed layered phase increases beyond 11%. The mineral which replaces mica in the triphase assemblages is chlorite which contains Mg and Fe ions indicating a relatively high Al/R^{2+} content of the smectitic starting material in the experiments. These I/S minerals formed in longer experiments and at highest temperatures are more illitic as the amounts of high-charge smectite layers decreases (Fig. 3.2, part 2). Therefore, in the second reaction, the charge increase is due to the production of illite layers in the I/S minerals via increased tetrahedral Al.

3.1.2.2
Reinterpretation of the Reaction Sequence Based on the Whitney–Northrop Hypothesis

Whitney and Northrop (1988) stated that there are clearly two mineral reactions observed in their experiments. The first was considered to produce a random I/S in which many of the 10 Å layers were not true illite because interlayer K^+ ions were not fixed and were in fact exchangeable ions. These authors consider that I/S crystals are produced by a "transformation" process (opposed to a dissolution–precipitation process). It is important to note that Whitney and Northrop (1988) showed that there is a difference in the exchange of oxygen between aqueous solution and solids (through oxygen isotope measurements) in the two different series of run products. This suggests that there are two different reaction mechanisms, reaction products and reaction rates.

In spite of the fact that they observed a non-illitic 10 Å component, they concluded that there is a simple first reaction involving random I/S mixed layers (see Figs. 9 and 12, pp 85–86). This is confusing because the progress of the reaction is not directly proportional to the %I in I/S. The first reaction products are not simple random I/S minerals but more complex mixed-layered minerals which are composed of about 10% illite, interstratified with variable amounts of high-charge (vermiculite) and low-charge smectite layers. Reinterpretation of the X-ray patterns, can be made in the following manner:

1. Crystallization of the R=0 I/S MLM

Low-charge montmorillonite (+mica) →	**high-charge montmorillonite (vermiculite) + quartz**	
100% expandable (K sat.)	50% collapsed (K-sat.) thus 50% (*vermiculite* + *illite*)	(3.1)
100% *smectite*	50% expandable (K-sat.) thus 50% *smectite*	

In the second reaction of Whitney and Northrop (1988) the random illite/high-charge montmorillonite mixed-layered mineral is replaced by an ordered I/S. The compositions of the I/S minerals, assumed to be illite and high-charge montmorillonite are plotted in Fig. 3.3b. Conversion of the initial I/S minerals results in a new I/S phase of approximately 50% "illite" and 50% low charge smectite.

2. Crystallization of the R=1 I/S MLM

High-charge montmorillonite → (vermiculite) (I/S)	**ordered I/S**	
50% collapsed (K-sat.)	10% expandable (K-sat.) thus 10% *smectite*	(3.2)
50% expandable (K-sat.)	40% expandable (Na-sat.) thus 40% *vermiculite*	
	50% collapsed (Na-sat.) thus 50% *illite*	

This mineral assemblage changes as mica disappears when chlorite is formed (Fig. 3.4b). The phase association of the reaction shows that the micaceous phase present in the starting material disappears when the illite percent in the I/S mixed layered phase increases beyond 11%. The mineral which replaces mica in the triphase assemblages is chlorite which contains Mg and Fe ions, indicating a relatively high Al/R^{2+} content of the smectitic mineral. These I/S minerals formed in longer runs and at highest temperature *are more illitic* than the percentage of high-charge smectite layers decreases (Fig. 3.2, part 2). Therefore in the second reaction, the charge increase is due to the production of illite layers via increased tetrahedral Al.

3. Illitization of the R=1 like I/S MLM

high-charge montmorillonite → (vermiculite) layers ordered I/S	**illite layers + chlorite + quartz ordered illitic I/S (equivalent to PCI)**	
10% expandable (K-sat.) thus 10% *smectite*	10% expandable (K-sat.) thus 10% *smectite*	(3.3)
40% expandable (Na-sat.) thus 40% *vermiculite*	90% collapsed (Na-sat.) thus 90% *illite* (*PCI*)	
50% collapsed (Na-sat.) thus 50% *illite*		

4. Illite Formation

Ordered illitic I/S	→ illite (equivalent to WCI)	
10% expandable (K-sat.)	0% expandable (Na-sat.)	
3 thus 10% *smectite*	thus 0% *smectite*	(3.4)
90% collapsed (Na-sat.)	100% collapsed (Na-sat.)	
thus 90% *illite*	thus 100% *illite*	
PCI	WCI	

3.1.2.3
A Possible Alternative

Despite very careful mineralogical determinations, Whitney and Northrop have not clearly established that the smectite component in the random and the ordered I/S mixed layers conserve the montmorillonitic composition of the original Wyoming clay. The stoichiometry of mineral reactions in a closed system gives an alternative. Another reaction should be envisaged for the first step of the illitization process: the low-charge montmorillonite could have been transformed into a high-charge beidellite-saponite-quartz mineral assemblage as indicated in Fig. 3.5a. Both beidellite and saponite are smectites and difficult to distinguish by XRD methods. In such a case, the Fe and Mg chemical components are concentrated into a separate phase as was shown in experimental synthesis of clays in the Mg-Al-Si system (Grauby et al. 1993). In

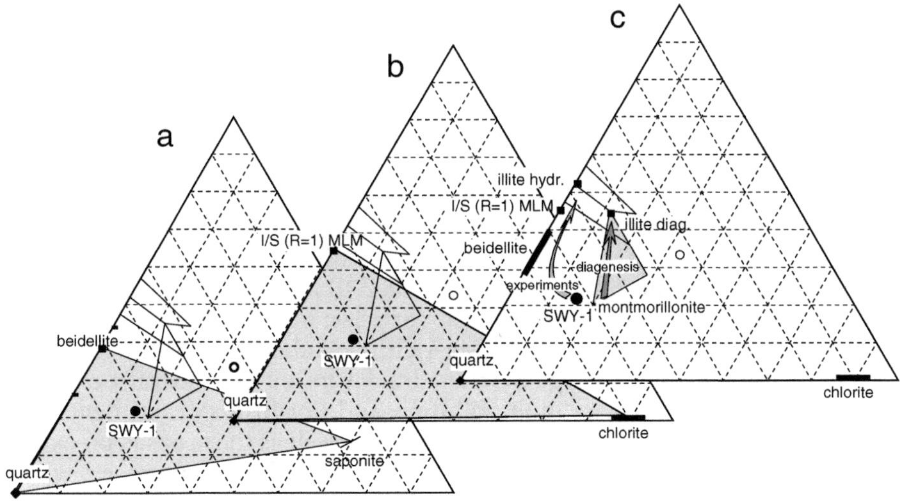

Fig. 3.5a–c. A possible alternative interpretation of the Whitney and Northrop experiments. **a**) The first reaction is: low-charge montmorillonite→high-charge beidellite+saponite+quartz. **b**) the second reaction is high-charge beidellite+saponite→ordered I/S MLM+chlorite+quartz. **c**) Experimental montmorillonite alteration simulates the active hydrothermal boiling zones rather than the diagenetic illitization process

this case, the smectite component in the ordered smectite-illite mixed layers is beidellitic; I/S chemical composition must be located on the beidellite-muscovite join rather than on the montmorillonite-illite join (Fig. 3.5). In such a case, the first reaction would be written as follows:

1. Crystallization of New R0 I/S MLM

low-charge montmorillonite → high-charge beidellite + saponite + quartz (3.5)
dioct. smectite trioct. smectite

Such a reaction was studied by Yamada et al. (1991) and Yamada and Nakasawa (1993) who showed that the saponite-beidellite assemblage crystallized from smectites of the montmorillonite-beidellite series under hydrothermal experimental conditions. Consequently, the second reaction which produced chlorite and ordered I/S mixed layers must be significatively different from that envisaged by Whitney and Northrop. The presence of a trioctahedral smectite is difficult to identify from XRD patterns of oriented preparation because its peaks are superimposed on those of the dioctahedral species. Besides, its relative quantities are very small because the amount of Mg^{2+} ions available after dissolution of the starting montmorillonite is low: 0.3 per O_{10} $(OH)_2$, or a tenth of the Mg amount necessary to form a saponite molecule. The illite component resulting from beidellite destabilization is richer in Al than the illite resulting from montmorillonite as observed in diagenetic series (Meunier and Velde 1989). Besides, the random-to-ordered transition processes in I/S are not identical for montmorillonite and beidellite because of the location of the layer charge.

2. Crystallization of the R1 like I/S MLM

high-charge beidellite → ordered I/S MLM + chlorite + quartz (3.6)

This alternative explanation has been verified by experimental alteration of a Na-saturated montmorillonite from Crook County, $(Ca_{0.02}\ K_{0.02}\ Na_{0.28}\ (Al_{1.51}\ Fe_{0.20}\ Mg_{0.29}Mn_{0.01})\ [Si_{3.95}\ Al_{0.05}]\ O_{10}\ (OH)_2)$ with water in presence of quartz and sanidine at 100 and 200 °C (Beaufort et al. 2001). The clay fraction of the run products was composed of two types of particles: high- and low-charge dioctahedral smectite mixed-layer and Mg-rich trioctahedral smectite (saponite or stevensite). The high-charge dioctahedral layers are tetrahedrally charged. If such a reaction took place in the Whitney and Northrop's experiments as is highly probable, this implies that their results as well as those based on experimental alteration of bentonites do not simulate the illitization process which is active in diagenetic environments in pelitic rocks. In smectite transformation products in pelitic rocks only low-charge smectite and illite have been identified, in both disordered and ordered I/S minerals. Normal identification procedures (Sr, Mg or Ca saturation) of run products in the Whitney–Northrop experiments would show a fully expandable mineral which transforms directly into an ordered 50% I/S mineral.

3.1.2.4
Summary

If the experiments using bentonite and montmorillonitic smectites give reaction products which are not found in pelitic rocks (significant high-charge smectite layers in the disordered I/S minerals). The question then is how can kinetic data obtained from such experimental alteration of bentonites for the last three decades be used to describe smectite-to-illite transformations in shales? First of all, even if they cannot pretend to reproduce the actual mineral reactions involved in the illitization process in natural diagenetic shale environments, they must be taken into account for the following reasons:

- they are, until now, the only way to determine the order of magnitude of basic clay mineral reactions involving smectite and illite;
- they allow one to separately investigate the effects of temperature, time and chemical parameters (K^+ activity for instance);
- they may be compared to kinetic parameters estimated directly from observed smectite-to-illite transition in natural environments (see Velde and Vasseur 1992) or other experiments using different starting materials.

3.2
Kinetics of Experimental Transformations (Natural and Synthetic Starting Materials)

Kinetics values are of great importance as predictive tools for the transformations of clays in natural settings subjected to temperatures above ambient levels. For this reason we discuss here how these kinetic values are derived from experiments of different sorts.

3.2.1
Kinetics of Illite Formation Using Synthetic, Chemical Compositions

The experimental synthesis has been used for more than a century to determine the stability fields of the constituent minerals of natural rocks. The example selected here is based on the study by Eberl and Hower (1976) dedicated to the progressive formation of illite from the transformation of a glass of determined chemical composition under controlled conditions of temperature and time. The glass is made from a stoichiometric combination of components forming potassic beidellite: $K_{0.34}$ (Al_2) $[Si_{3.66}$ $Al_{0.34}]$ O_{10} $(OH)_2$. The starting materials contain no magnesium nor iron, and hence are a simplification of a natural system. Experimental data are given in Table 3.2.

The experiments carried out with the K-glass show the following sequence of mineral reactions:

Table 3.2. Kinetic data of the experimental transformation from smectite into illite-smectite mixed layers (after Eberl and Hower 1976). Kaol.: kaolinite; Qz: quartz; Felds.: alkaline feldspar; Pyr: pyrophyllite

Temperature (°C)	Duration (days)	Smectite %	Other products
152	101	No reaction	
260	99	85	Kaol.
260	266	65	Kaol.
300	31	85	Kaol.+Qz+Felds.
300	88	70	Kaol.+Felds.
343	5	90	–
343	23	80	Kaol.
343	88	35	Qz+Felds.
343	99	25	Qz?
393	3	70	Qz
393	14	70	Qz?+Pyr.?
393	23	35	Qz+Pyr.?+Felds.
393	169	15	Qz+Pyr.

1. glass → 100% expandable smectite
2. smectite → illite–smectite mixed layers+kaolinite (or pyrophyllite) + quartz + feldspar

The formation of illite layers in the second reaction can be described by a first-order kinetic equation with the following integration:

$$\ln \frac{a}{a-x} = kt \qquad (3.7)$$

a: initial concentration of smectite (100%)
x: percentage of smectite layers transformed into illite.

Therefore, $a - x$ represents the percentage of smectite layers measured by the degree of expansion on the X-ray diffraction patterns X (column smectite % in Table 3.3), so

$$\ln \frac{100}{\%\ \text{smectite}} = kt \qquad (3.8)$$

In this case, the rate constant k is expressed in days^{-1}. If the order of the reaction is one, the Eq. 3.27 shall be represented by a straight line with slope k in a diagram ln (100% smectite) as a function of time for each of the temperatures selected for the experiment (Fig. 3.6a). The activation energy of the reaction is calculated from the Arrhenius equation:

$$E_a = \frac{\Delta \log_{10} k}{\Delta \frac{1}{T}} \times 2.303\, R \qquad (3.9)$$

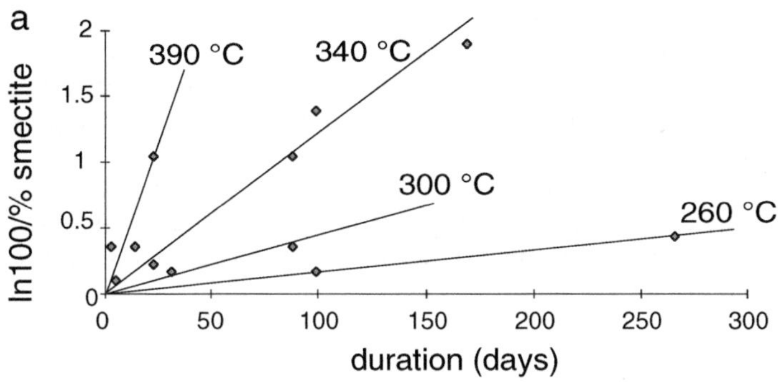

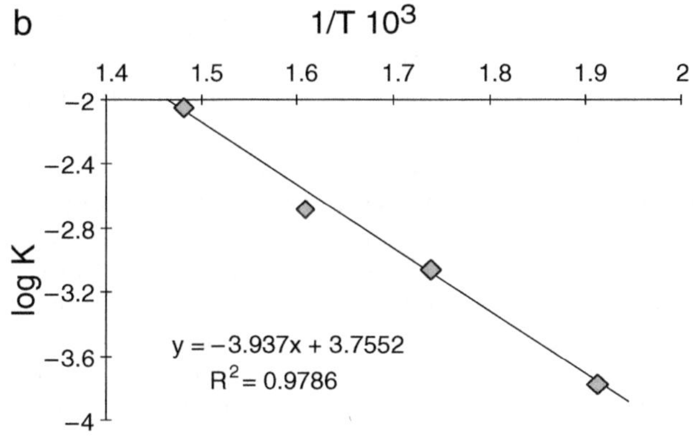

Fig. 3.6a,b. Kinetics for illite formation (from Eberl and Hower 1976). **a**) Graphical representation of (3.8): $\ln(100/\%\ \text{smectite}) = kt$. The rate constant for each temperature corresponds to the slope of the various straight lines. **b**) Determination of the activation energy (E_a) using an Arrhenius plot

The equation above is represented by a straight line in a diagram $\log_{10}k$ as a function of the inverse of temperature (Fig. 3.17); its slope is equal to: $E_a/(2.303\,R)$. One can deduce the value of E_a for the illitization reaction of smectite under experimental conditions. In the case presented: $E_a = 81.9 \pm 14.6\,\text{kJ}\,\text{mol}^{-1}$ (Fig. 3.6b).

What does the value of E_a then represent? In reality, the mechanism enabling smectite to disappear and produce illite is not well known. Several explanations may be proposed. For instance, one can imagine that each smectite layer is transformed individually into illite. In this case, E_a represents the energy necessary for breaking the Si–O bond in tetrahedra so that Al may substitute for Si. One can also imagine that the illite layers grow on certain smectite layers by heterogeneous nucleation and "feed" on the chemical elements in solution resulting from the dissolution of the other smectite layers. In this case, two mechanisms take place in parallel: dissolution and growth. The slower one controls the rate of the complete reaction. If it is dissolution, E_a then represents the energy necessary for breaking all the chemical bonds within the smectite crystals. In contrast, if it is growth, E_a represents the energy necessary for each ion or group of ions to find its place during the growth steps, i. e., crystal incorporation.

Again, one should be reminded that this study is a simplification of natural mineral assemblages in which important major elements such as Na, Mg, Fe are not present. However, these data can be compared to others in synthetic or natural systems.

3.2.2
Kinetics Using Natural Smectite Minerals

The study of Whitney and Northrop (1988) discussed above is the most complete concerning the range of experimental conditions and the analysis of the run products (250–400 °C, 1–220 days duration) using natural minerals as starting materials. It gives us the greatest insight into the kinetics of the smectite illitization reaction produced under experimental conditions.

As mentioned above, the smectite-to-illite transition has been observed to occur in two parts, where two transformation reactions change the apparent illite content of the I/S minerals to more illitic forms. The most striking observation which can be made on these data is that, from the tables of run results and XRD diagrams presented in the studies, the reaction products are not the same for these two reactions. In the earlier, rapid reaction, the product is a *three-component random mixed-layered mineral* composed of illite, low-charge and high-charge smectite. This complex phase is associated with quartz and small amounts of mica. Mica and quartz were considered as components of the starting material by the authors. Nevertheless, the XRD patterns of run products from the 300 °C and 400 °C series (Whitney and Northrop 1988, Fig. 2, p 79) show an increase of the 3.34 Å peak intensity, suggesting that quartz is produced during the hydrothermal reaction.

Experiments by Eberl and Hower (1976), Eberl (1978), Eberl et al. (1978), Inoue (1983), and Howard and Roy (1985) using natural smectite minerals as starting materials show exactly the same sequences of mineral phases under hydrothermal treatment. It is difficult to determine if the quantity of the mica phase present in the run short products remains constant or even increases, nevertheless, it is obvious that the mica phase disappeared from the longer run products in which chlorite was formed. Mica and quartz can thus be considered to be part of the first reaction product assemblage.

The second reaction sequence, found after approximately 40 days at 250 °C and higher temperatures, produces an R=1 (ordered) MLM of increasing illite content composed of 10 Å non-expandable layers in the potassium-saturated and in the sodium-saturated states and smectite layers (expandable in the Na-saturated state) associated with quartz and chlorite. In the first reaction sequence, the non-expanding layer mineral is a high-charge smectite or vermiculite. In the second reaction the non-expanding mineral is an illite. The two reactions do not give the same non-expandable mineral. Only in the second

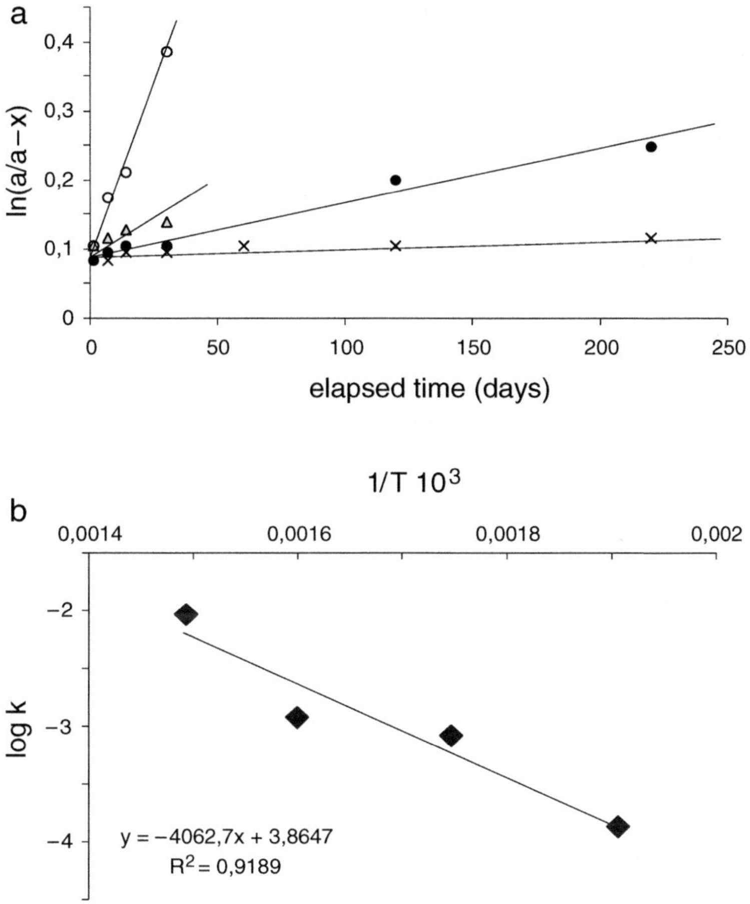

Fig. 3.7a,b. Smectite transformation experiments from Whitney and Northrop (1988). **a**) Variation of $\ln(a/a-x)$ with time (see text for details). **b**) Arrhenius plot

reaction involving an ordered, mixed-layered mineral does one observe a true, non-expanding illite component. Thus the smectite-to-illite reaction observed in the laboratory experiments produces illite only in the second segment of the reaction sequence.

The kinetics of vermiculite or high-charge smectite formation from reaction (3.1) can be described in the experiments of Whitney and Northrop (1988). Indeed, it is possible to calculate the activation energy of the reaction which produces the 10 Å non-expandable layers (i. e. illite layers) from the run corresponding to part 1 in Fig. 3.2. The value of ln $(a/a-x)$ where a is 100% smectite and x the illite layer percent (nonexpandable layers in the Na-saturated state, are linearly correlated with time (Fig. 3.7a). The lines do not intercept at the origin but at ln $(a/a-x)$ = 0.9 which correspond to a value of x = 8%. This means that the starting material which is apparently homogeneous (100% expanding layers at 25 °C) is heterogeneous at elevated temperature in the experimental conditions: 8% layers collapse to 10 Å before reaction 1 proceeds. The Arrhenius plot (Fig. 3.7b) gives an activation energy of $18.7 \pm 2.2\,\mathrm{kcal\,mol^{-1}}$. This value is close to that calculated by Eberl and Hower (1976) in experiments using natural smectite.

3.3 The Bulk Composition Effect (K_2O)

3.3.1 Natural Minerals

Most studies of smectite stability designed to reproduce the smectite-to-illite transformation use smectites of bentonitic origin, and in fact a type called Wyoming bentonite is used almost exclusively. It is a montmorillonitic smectite, the charge on the layer originates through substitutions in the octahedral layer site. The charge is relatively low, near $0.3/O_{10}\ (OH)_2$ anionic charge. The following analysis of experimental results is based on three experimental studies; Robertson and Lahann (1981), Howard and Roy (1985), Whitney and Northrop (1988). Other studies have been performed using natural clays (Eberl and Hower 1976; Inoue 1983), but the data do not cover the same range of experimental conditions, or use precisely the same starting materials as those in the three quoted above. All studies used montmorillonitic smectites (Wyoming and Chambers type) treated in autoclaves at 250 °C for time periods of 40 days. These are the conditions which are common to all the studies. XRD identification methods were similar, thus there is a homogeneity of information and method.

If we consider experiments with the same duration at the same temperature (up to 40 days, 250 °C) in the three following studies (Robertson and Lahann 1981; Howard and Roy 1985; Whitney and Northrop 1988), it is evident that there is a great difference in the quantity of low-charge smectite transformed to high-charge montmorillonite for the same periods of time (Fig. 3.8). This difference has led to different estimations of the activation energy necessary to initiate the reaction. Investigating the experimental procedure in each study more closely, it is apparent that although the starting materials are all natural montmorillonite smectites, the amount of potassium available compared to the solids is very different in each study. Table 3.3 shows the different clay starting materials used by the various authors. The samples of Howard and Roy (1985) and Whitney and Northrop (1988) are near montmorillonites while those of Robertson and Lahann (1981) are montmorillonites-beidellites.

Howard and Roy (1981) use entirely nonpotassic materials, Whitney and Northrop (1988) use potassium-saturated smectites and Robertson and Lahann (1981) use Na or Ca-saturated clay with a solution which contains 800 ppm potassium ions. If one calculates these differences to percent K_2O of total non-aqueous elements, it is obvious that the experimental charges are not identical. As it turns out, the higher the potassium content, the faster the reaction proceeds, if all other variables are held constant (Fig. 3.9b).

Table 3.3. Compositions of smectites used by various authors in the smectite-to-illite conversion experiments. Potassium content is not that of experiments (see text)

Smectite	Unit formula per $O_{10}(OH)_2$	References
Chambers	$Ca_{0.29}$ $(Al_{1.33}$ $Fe_{0.22}$ $Mg_{0.45})$ $[Si_{3.85}$ $Al_{0.15}]$ O_{10} $(OH)_2$	Robertson and Lahann (1981)
Polkville	$Ca_{0.28}$ $(Al_{1.87}$ $Fe_{0.32}$ $Mg_{0.29})$ $[Si_{3.78}$ $Al_{0.22}]$ O_{10} $(OH)_2$	Robertson and Lahann (1981)
Wyoming	$Ca_{0.08}$ $K_{0.02}$ $Na_{0.13}$ $(Al_{1.59}$ $Fe_{0.16}$ $Mg_{0.21})$ $[Si_{3.98}$ $Al_{0.02}]$ O_{10} $(OH)_2$	Howard and Roy (1985)
Wyoming	$K_{0.30}$ $Na_{0.02}$ $(Al_{1.52}$ $Fe_{0.23}$ $Mg_{0.27})$ $[Si_{3.90}$ $Al_{0.10}]$ O_{10} $(OH)_2$	Whitney and Northrop (1988)

Using the relation $k = ln(a/a - x)\ t^{-1}$ where a is the 100% smectite content starting material, x the quantity of illite found in the I/S phase and t the time in days, one can compare the reaction constant values (k) estimated from these experimental results. This is found in Table 3.3, as well. Plotting these values as a function of potassium content (Fig. 3.8) demonstrates the importance of the potassium content in the run product. These results are chemically reasonable in that the reaction which the researchers wished to study was approximately: smectite + K→mica-like layer (in the I/S MLM). Run product tables of the published studies show that the reaction is slightly more complex since other products are present, but the reactants are the same in the studies and follow the reaction equation as written. The amount of potassium present appears to be the factor which controls the observed energy of activation of the

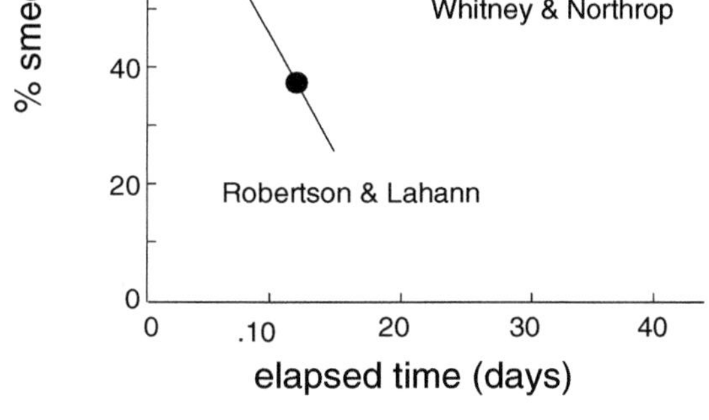

Fig. 3.8a,b. Reaction progress as smectite percent at 250 °C as a function of run time from experimental data (Howard and Roy 1981; Robertson and Lahann 1981; Whitney and Northrop 1988)

reaction produced in the hydrothermal experiments reported by the authors. It is evident that in the experiments performed, the law of mass action governed the rate of change in the smectite content of the I/S minerals as much or more than the temperature conditions.

The important aspect of these observations based upon this experimental data is that there is a different rate of reaction depending upon the reaction conditions. The first reaction, producing a high-charge montmorillonitic smectite layer in the I/S mineral, is more rapid and should have a lower activation energy, while the second is slower and could well have a much higher activation energy. Since the reactions are not the same, one would expect that the reaction rate to be different. Although the conclusions concerning kinetic parameters of the authors of the three studies are apparently contradictory, it is evident that one can assimilate all of the results into the same system which depends upon the bulk composition (K^+ content) of the experiments. Increasing the potassium content of the run charge (mass effect on the reaction as written smectite + K^+ $\rightarrow$ nonexpandable layer) certainly favors the production of an apparently nonexpanding mineral. It appears that the rate constant is roughly linearly dependant upon potassium concentration (Fig. 3.8b). This early phase which collapses at low temperature is in fact a high-charge montmorillonite. The montmorillonite re-expands when sodium-saturated.

The extracted kinetic values of activation energy for the natural smectite $\rightarrow$ illite reaction under hydrothermal conditions are as follows:

4 kcal:	Howard and Roy (1985)
18 kcal:	Whitney and Northrop (1986)
30 kcal:	Robertson and Lahann (1981)

One can then plot the estimated activation energy as a function of potassium content (Fig. 3.9) which demonstrates a reasonably direct relationship.

Observed reaction constant (dx/dt in % change per day) for experiments at near 250 °C and 30 day run duration for three studies: Howard and Roy (1985),

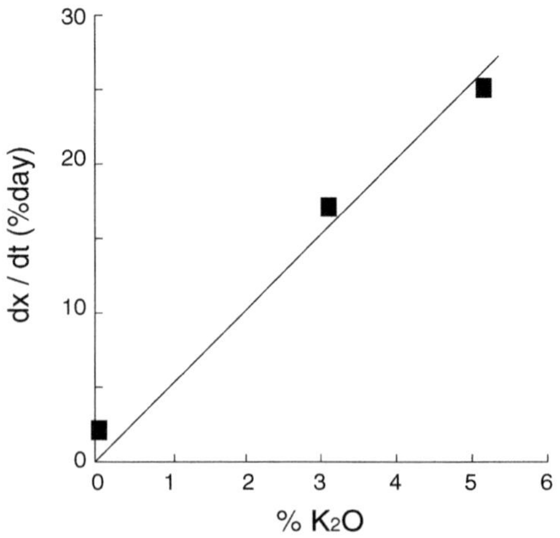

Fig. 3.9. Plot of potassium content of experimental charges against the calculated activation energy for the transformation of smectite to vermiculite (high-charge smectite)

Whitney and Northrop (1988) and Robertson and Lahann (1981) plotted as a function of potassium content of the solids in the experimental charge.

The increase in activation energy then follows an increase in potassium in the system. Huang et al. (1993) have done experiments on various materials, synthetic and natural smectites with varying amounts of potassium present. Their results using natural Wyoming bentonite show that the potassium content of a system strongly influences the reaction rate, as one would expect. The existence of two types of interstratified mineral products, one with both high- and low-charge smectite components (montmorillonite), the other a true I/S mineral with changing illite content, is important to observe. These authors develop a model for smectite-to-illite conversion which takes into account the factors of time, temperature and potassium activity in solution.

3.3.2 Multiparameter Kinetics

The smectite illitization reaction is dependent on the concentration of the dissolved potassium in the aqueous solution of the system. The observations discussed above show that when a natural smectite is used as a starting material the smectite-to-illite transformation is a heterogeneous reaction that produces several different mineral phases other than I/S minerals (quartz and probably a ferromagnesian phyllosilicate: saponite or chlorite). If one simplifies this reaction to the dissolution of smectite to produce illite-rich minerals in a one-step reaction, the general kinetic equation may be written as follows (Pytte 1982; Pytte and Reynolds 1989; Huang et al. 1993):

$$-\frac{dS}{dt} = k\,[K^+]^a\,S^b \tag{3.10}$$

S: molar fraction (smectite %) of smectite in the illite-smectite mixed layer

$[K^+]$: concentration of the dissolved potassium

k: rate constant

a and b: constants defining the order of the reaction.

If the concentration of potassium $[K^+]$ is constant, $k[K^+]=k'$ is constant; the equation above becomes:

$$-\frac{dS}{dt} = k'S^b \tag{3.11}$$

This equation shows that the smectite ratio in I/S MLM decreases with time according to $-dS/dt = k'S^2$ (smectite %). If b=2, the order of the reaction kinetics is 2 and the equation becomes: $-dS/S^2 = k'dt$. The integral gives: $1/S = k'y + I$. I is an integration constant. Considering the limit conditions $S = 1$ and $t = 0$, one obtains: $1/S = k't + 1$. This shows that the quantity $1/S$ is a linear function of t. This has been verified experimentally by Huang et al. (1993), as presented in Fig. 3.10a.

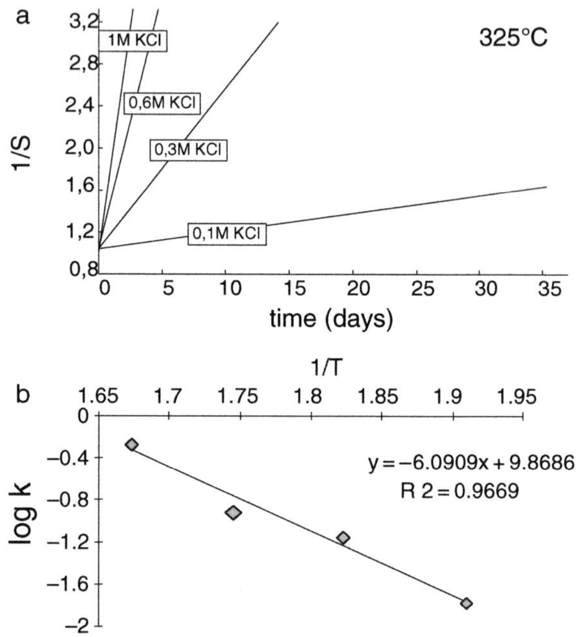

Fig. 3.10a,b. Multi-parameter kinetics of the smectite → illite (from Huang et al. 1993). **a**) Progress of the reaction as a function of time (S being the smectite content in I/S). **b**) Arrhenius diagram showing the experimental data for a second-order kinetics

We know that $k' = k[K^+]$, so $\log k' = \log k + a \log[K^+]$. The value of a can then be obtained by determining the slope of the curves in a diagram $\log k'$ as a function of log $[K^+]$. Huang et al. (1993) have shown that $a \approx 1$. This suggests that the reaction is of first order for the concentration of potassium ($k' = k[K^+]$) and of second order for the molar fraction of smectite (smectite %) in illite-smectite mixed layers. Equation $-\mathrm{d}S/\mathrm{d}t = k\,[K^+]^a\, S^b$ is then written as follows:

$$-\frac{\mathrm{d}S}{\mathrm{d}t} = k\,[K^+]\,S^2 \tag{3.12}$$

k' is determined in diagram log (duration in days) as a function of log $[K^+]$ for the value of log $[K^+] = 0$. When log $[K^+] = 0$, $k' = k$. The results obtained by Huang et al. (1993) are given in Table 3.4. The activation energy E_a can then be obtained from these data which is displayed in an Arrhenius diagram according to the following equation:

$$k = A \exp^{-E_a/(RT)} \tag{3.13}$$

hence

$$\ln k = -\left(\frac{E_a}{R}\right)\left(\frac{1}{T}\right) + \ln A \tag{3.14}$$

The results obtained (Fig. 3.10b) give $E_a = 117\,\mathrm{kJ\,mol^{-1}}$ and $A = 8.08 \times 10^{-4}$ $\mathrm{s^{-1}\,mol^{-1}\,l}$.

Table 3.4. Values of k' for various temperatures according to the experimental data from Huang et al. (1993)

Temperature (°C)	log k (days^{-1})
250	−1.8
275	−1.13
300	−0.91
325	−0.25

This formulation takes into account the chemical parameters of potassium in solution (assumed to be constant in the solution and hence controlled from outside of the rock system), the temperature at which the reaction occurs and the time of reaction under these conditions. Thus this is the general basis for making a model of mineral change in basin sedimentation and burial diagenesis. In diagenesis, such a model must take into account the changing temperature for each layer of sediment and its evolution as a function of time.

3.3.3 Formation of Muscovite at High KOH Concentrations: Shape and Polymorph

Bauer A. (1997) and Bauer et al. (2000) have shown the importance of growth kinetics on the shape and polymorph in synthetic illite. The experiments used natural kaolinite which was dissolved to form illite by precipitation in KOH solutions of high pH (13 and above) at temperatures ranging from 5 to 110 °C. The effect of temperature was to accelerate the reaction of transformation which resulted in different crystal shapes. The run products showed the typical XRD two-band structure, but the peak widths were very narrow. The mineral formed is pure, high-charge muscovite. The peak at 10 Å has a FWHM of less than 0.2 °2θ, and the peak at 10.2 Å has a width of less than 0. 5 °2θ. These peak positions and widths probably only represent mica crystals with different grain sizes. Hence although one would normally designate a WCI and PCI peak, the PCI peak material does not contain any smectite layers, contrary to the PCI crystals in sedimentary rocks.

Fast growth (higher temperatures) promotes the formation of hexagonal books of crystals of the 1 M polytype. Slower growth promotes books of 2 M mica and very slow growth promotes initially lath-shaped mica crystals which are replaced by 2 M hexagonal mica crystals (Fig. 3.11).

These experiments demonstrate that crystal growth rate can influence the shape and polytype of micas.

The applicability of these various experimental results to natural systems is difficult to assess. It is known that I/S minerals found in different environments (bentonite, shale diagenesis, metasomatic and hydrothermal series) have different overall compositional trends for the individual layer components (Velde and Brusewitz 1986; Meunier and Velde 1988). These compositional differences will undoubtedly result in differences in reaction kinetics. In the experiments of Whitney and Northrop (1989), a different reaction product (type of I/S mineral) can change the rate of reaction to a great degree. It is therefore absolutely

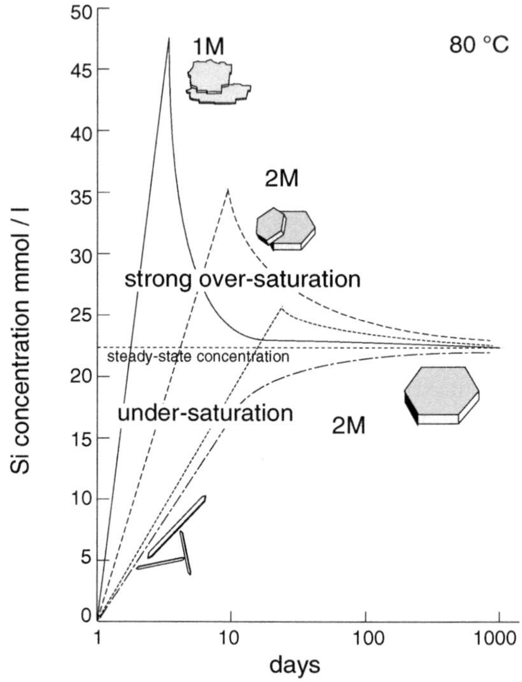

Fig. 3.11a,b. Concentration-time dependence of mica crystal shapes and polytypes from Bauer et al. (2000)

necessary to compare experimental results with minerals which have the same type of interstratified mineral present. Only then can we derive correct time and temperature values of the transformation of smectite to illite so common in diagenesis and other geological processes.

In man-made systems, where the starting material is bentonite smectite, such as in barriers around nuclear waste, one can use the experimental data on Wyoming bentonites to predict the behaviour and life span of the clays when subjected to increased temperature. Since the desired lifetime of such systems is on the order of a million years, the rapid reactions are pertinent and those of diagenesis are not.

3.4 Kinetics of the Smectite-to-Illite Conversion Process in Natural Environments

The sections above deal with experiments which are performed in simplified systems using only one mineral phase as a starting material. This is not the case in natural systems. It is therefore interesting to try to extract kinetic data from natural systems in order to compare this information with data from more pure, simple systems where parasite reactions are kept to a minimum.

The smectite-to-illite transformation has been observed in several different natural environments: those of deep burial at low to medium geothermal gradients (20–50 °C/km) where petroleum is sought, in high-temperature environments (geothermal gradients of greater than 80 °C/km) where high-temperature energy is sought and in various types of thermal pulse environments such as the intrusion of a magma into sedimentary rocks. We will investigate these different types of thermal regimes and their effect on clay mineral transformation as a function of the thermal regime.

3.4.1 Burial Diagenesis

Most authors who have attempted to fit experimental data or observations of clay mineral change in sedimentary basins have used the single-step model described above. It is assumed that the initial stages of I/S mineral transformation in the disordered I/S R0 interlayering and those of the second, more illitic I/S R=1 structure behave in the same manner. However, Velde and Vasseur (1992) have taken into account the different mineral structures assuming that the R=0 structure I/S minerals is a separate mineral phase which transforms into the ordered I/S mineral by dissolution. This assumption is due to the observation that in diagenetic mineral series one sees an R=0 sequence of I/S MLM of different proportions of illite first and an R=1 series of changing illite content. There appears to be a rapid change in relative proportions of the two minerals in XRD patterns near 50% illite composition. Figure 3.12 shows the evidence for this interpretation as observed by XRD in a deep well in Texas (Tertiary in age). The figure indicates a strong disordered I/S peak in one sample, a mixture of disordered I/S and ordered (R1) I/S in another, and a sample dominated by ordered I/S. Both I/S minerals have a composition near 50% illite/smectite.

In the samples, one sees large irregular crystals and thin laths together. The laths are identified with ordered I/S (see Lanson and Champion 1991). Based upon these observations, a two-step model of smectite-to-illite transformation was proposed (Fig. 3.13).

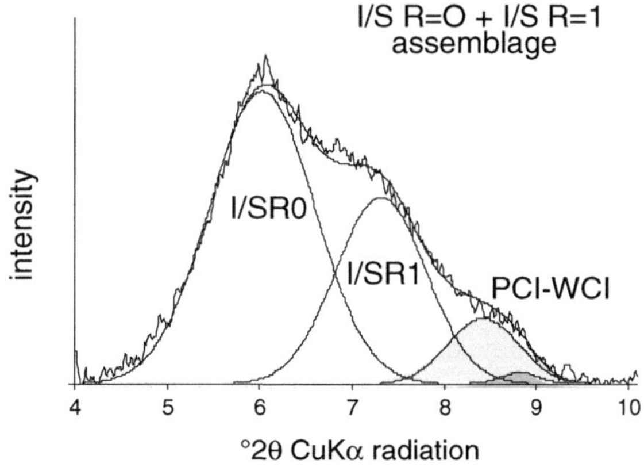

Fig. 3.12a,b. Background-subtracted, decomposed XRD patterns of Sr-saturated, air-dried < 2 μm fractions of a shale samples from a deep well in Texas (Tertiary) **a** Random I/S **b** mixture of disordered I/S and ordered (R=1) I/S **c**)sample-dominated by ordered I/S Both I/S minerals have a composition near 50% illite/smectite

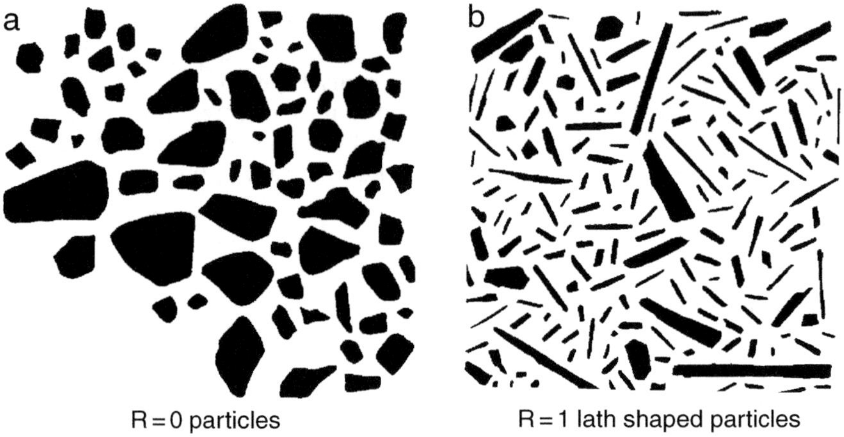

Fig. 3.13a,b. Demonstration of the presence of different particle sizes and shapes in a shale sample containing both R=1 and R=0 I/S MLM. Crystal shapes are derived by tracings from electron transmission micrographs of samples from a single sample in a deep well in the Texas Gulf Coast, Fig. 3.12. The Fig. 3.13a shows the irregular shaped R0 crystals in the sample and Fig. 3.13b shows the lath shaped R1 crystals in the sample

3.4.2 The Dual Reaction Kinetic Model (Velde and Vasseur 1992)

The kinetic model of Velde and Vasseur (1992) has been previously developed as a function of time and temperature variables using data from seven deep wells. It is based on the fact that the smectite-to-illite conversion proceeds through the formation of two mineral structures which evolve in their chemical environment in such a way that they are related to one another by a sequential reaction. The first reaction describes the change of smectite with a disordered stacking (R=0) from fully expandable (100% smectite) to a reaction maximum with 50% smectite. The second reaction starts with the formation of an illite-smectite of 50% illite content which is assumed to evolve to 100% illite. This reaction is dependent upon the first one, for each smectite layer lost in the R=0 mineral structures (seen as increasing illite content) a new crystal is created, the ordered illite-smectite (R=1) phase of 50% illite composition. This crystal is assumed to evolve towards a fully illitic phase which is the term of the second reaction. New crystals of R=1 I/S are created as long as some R=0 phases change composition towards the end-point of the first reaction. The proportion of the R=0 crystals decreases as time proceeds in the reaction sequence.

Velde and Vasseur (1992) propose an empirical kinetic model (Fig. 3.14) based on I/S sequences observed in 7 basins of differing ages (1–200 Ma). Present-day thermal gradients are relatively low, from 25 to 35 °C/km. It is assumed that the present-day gradient pertained throughout the history of observed sedimentation, an assumption difficult to verify. This model is based upon the two-stage smectite-to-illite transformation, corresponding to

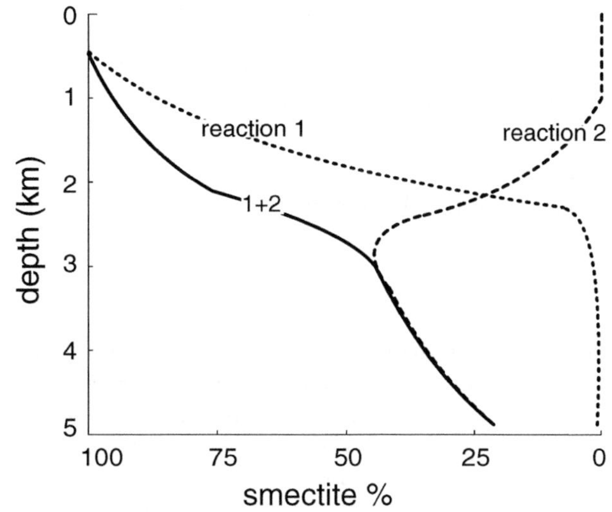

Fig. 3.14. Empirical kinetic model of the smectite-to-illite transformation through two reactions (Velde and Vasseur 1992). Reaction 1: smectite → randomly ordered I/S mineral (R0) with 100 to 50% smectite (dotted curve); reaction 2: R0 → regularly ordered I/S mineral (R1) with 50 to 0% smectite (dashed curve). The continuous curve is the sum of both reactions

the transformation of two mineral structures with changing compositions (% illite). The first reaction describes the transformation of smectite into a randomly ordered I/S mineral (R0) whose illite percentage ranges from 0% to 50%.

The second reaction involves ordered R1 I/S minerals whose illite percentage ranges from 50% to 100%. The second reaction is initially dependent on the first reaction because, for each lost smectite layer in a R0 I/S mineral (relative increase in the illite content), a R1 50% illite crystal is formed. Subsequently, this crystal progresses towards the 100% illite composition, which is the end-point of the reaction (Fig. 3.14). New R1 I/S crystals are produced as long as there are R0 I/S minerals with a composition tending towards the end of reaction 1 (50% illite).

The kinetic equations developed from observations of illite components of natural I/S minerals in diagenetic burial sequences take into account the illitization of the I/S minerals but not the kinetics of the R0 to R1 reaction. They are as follows:

Reaction 1 (randomly ordered I/S (R0) with a percentage in smectite component ranging from 100 to 50%):

$$\frac{\mathrm{d}S}{\mathrm{d}t} = -k_1 S \quad \text{with} \quad \log(k_1) = \log(A_1) - \frac{E_1}{RT} \tag{3.15}$$

$$\log(A_1) = 24.4\,\mathrm{Ma}^{-1} \quad \text{and} \quad E_1 = 76.8\,\mathrm{kJ\,mol}^{-1}$$

Reaction 2 (regularly ordered I/S mineral (R1) with 50 to 0% smectite)

$$\frac{\mathrm{d}M}{\mathrm{d}t} = k_1 S - k_2 M \quad \text{with} \quad \log(k_2) = \log(A_2) - \frac{E_2}{RT} \tag{3.16}$$

$$\log(A_2) = 7.2\,\mathrm{Ma}^{-1} \quad \text{and} \quad E_2 = 37.4\,\mathrm{kJ\,mol}^{-1}$$

This model gives a rather good description of the diagenetic transformations of the I/S minerals in shales. Its major drawback lies in an early prediction (too shallow depth) of the end point at 100% illite in certain sequences of sediments (Varajao and Meunier 1995).

3.4.3
Changes in Reaction Kinetics

3.4.3.1
Old Basins

Elliot and Matisoff (1996) and Hillier et al. (1995) have attempted to use the two-step model and also one-step models of smectite-to-illite transformation. The results of comparisons of different models indicate that the younger the basin, the greater correspondence with a one-step reaction fit. This is verified to be more the case when the geothermal gradient is high (55 °C/km). In fact, when the experimental data is fitted to a kinetic model, the kinetic parameters change, indicating a slowing of the reaction (observations of the present authors) in older basins (> 250 Ma). If one remembers the observation in Chap. 2 that in older basins, the illite (WCI) component is greater than that in younger basins, one can suspect that there is in fact a third reaction occurring in these closed diagenetic systems. This is the crystallization of WCI illite which forms at the same time as the ordered I/S mineral is adding illite layers. In this case the two phases I/S and WCI are competing for dissolved illite components. The slower the reaction (lower temperatures), the more competition there is for illite components and the slower the I/S reaction rate going toward the PCI end-member will be.

It could be expected that a three-step reaction model would be more appropriate for the description of clay mineral change in pelitic rocks of sedimentary basins under conditions of low-temperature transformation. This observation reinforces those presented in Chap. 2: illite is WCI, and not a product of smectite-to-illite transformation or evolution.

3.4.3.2
Kinetics of the Smectite-to-Illite Conversion Process at High Temperatures

Elliot and Matisoff (1996) and Lanson and Velde (1993) have attempted to model the observed change in I/S composition found for the Salton Sea (Yau et al. 1987). Here it is clear that a one-step transformation is more appropriate for the reaction kinetic formulations (Fig. 3.15).

Summary

The kinetic reaction formulations for the smectite-to-illite transformation in an essentially closed system (pelitic sediments and shales) appear to indicate two major types of reaction progress. In high-temperature sequences (geothermal and other regimes) a one-step process seems to be most appropriate. In

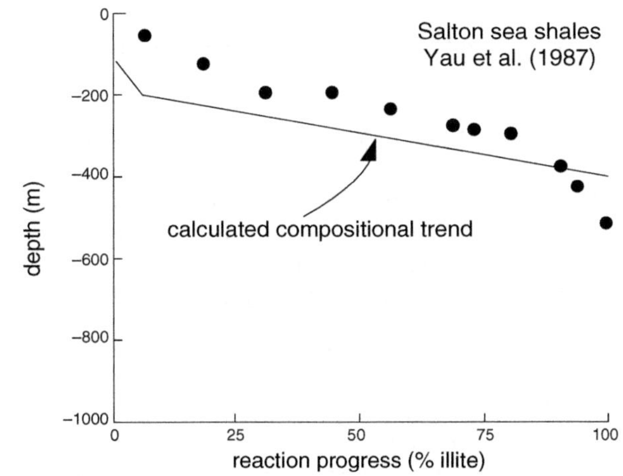

Fig. 3.15. Velde Model compared to experimental data for Salton Sea (CA) well determinations (data from Yau et al. 1987, calculated curve from Lanson and Velde 1993)

any event the data for reaction progress can be best fit using a one-step model. In older, lower-temperature series (< 80°/km geothermal gradient) a two-step reaction formulation seems to be more appropriate. In the instance of rapid reactions, the illitic end-member attained is PCI in type, with a small amount of smectite present and a rather small coherent diffracting domain for the crystallites. This has been presented in Fig. 1.6 where young sediments reach an advanced stage of maturity.

Two-Step Process

The older basin materials show the formation of PCI and WCI (Figs. 1.6c and 3.12) and a two-step kinetic formulation is best employed in the older sequences (Mesozoic). In these samples it is apparent that the presence of WCI should be included in the formulation and one would be better off using a three-step reaction. Thus one can propose a series of models to account for the different growth dynamics of clay minerals in the smectite-to-illite transformation:

1. One-step: high temp. Young basin and high temperatures
2. Two-step: model in sequences of moderate temperature and ages from early Tertiary to mid-Mesozoic.
3. Three-step: reaction in old sequences formed at low temperatures in older rocks, Mesozoic and older.

Kinetic data can be deduced for the one- and two-step models but as yet none is available for the three-step model where WCI becomes a major component of the potassic clay phases.

3.5
Success and Failure of the Multiparameter Models

3.5.1
The Kinetic Model of Pytte and Reynolds (1989) (Thermal Metamorphism)

Pytte and Reynolds (1989) proposed a more general kinetic model for smectite-to-illite conversion in shales which have experienced either long-time–low-temperature conditions (burial diagenesis) or short-time–high-temperature ones (contact metamorphism) They used the following equation:

$$-dS/dt = S^a (K/Na)^b A \exp(-E/RT) \tag{3.17}$$

a and b represent integers which, when summed up, give the reaction order.

The best fit of % smectite versus temperature for conditions ranging from contact metamorphism to burial diagenesis is obtained with a sixth-order kinetic expression, fifth-order with respect to smectite content and first-order with respect to the ratio K/Na (Fig. 3.16). The main advantage of a high-order expression is to drastically decrease the reaction rate as the I/S composition approaches 100% illite.

Pytte and Reynolds (1989) have discussed the weaknesses of their model. Indeed, the illitization process is not convincingly fitted in many diagenetic series (see Fig. 8.4, p 138). The authors consider that their empirical formulation (fifth-order with respect to smectite content) is difficult to reconcile with the physico-chemical process which governs the dissolution of smectite and the precipitation of illite.

3.5.2
Drawbacks of Multi-Parameter Kinetic Models

Several models based upon kinetic laws have been proposed, either from observation of natural I/S sequences (Bethke and Altaner 1986; Pytte and Reynolds 1989; Velde and Vasseur 1992) or from experimentation (Eberl and Hower 1976; Huang et al. 1993). Values of the activation energies and pre-exponential factor vary according to whether illitization is considered as a simple reaction (smectite → illite) or as a stage reaction (smectite→R0 I/S minerals → R1 I/S minerals...).

In the smectite → illite simple reaction, the law of mass action is considered to affect the reaction rate. Huang et al. (1993) propose a rate formulation in which the relative K^+ content acts as a linear factor:

$$\frac{-dS}{dt} = A \exp\left(\frac{-E_a}{RT}\right) \times [K^+] \times S^2 \tag{3.18}$$

S: % smectite in the I/S minerals
t: time (s)
E_a: activation energy (28 kcal mol^{-1})
T: absolute temperature (°K)
S^2: indicates a second-order reaction
A: frequency factor (8.08×10^{-4} s^{-1})
R: perfect gas constant (1.987 cal °K^{-1} mol^{-1})
$[K^+]$: concentration (molarity) of K^+ in solution

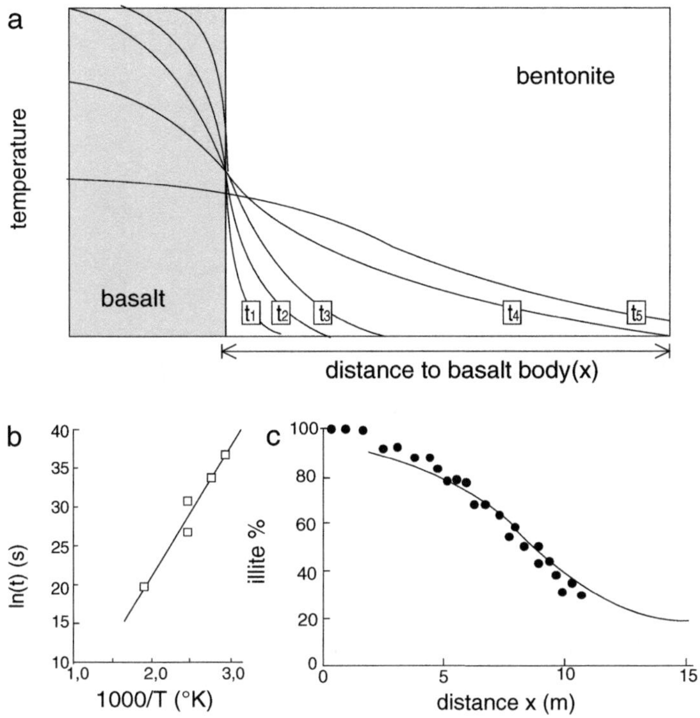

Fig. 3.16a–c. Thermometamorphism of shales (Pytte and Reynolds 1984). **a**) One-dimensional schematic representation of the effects of thermal diffusion around a basaltic vein crosscutting smectite-rich shales : temperature variation versus distance for different increasing time values from t_0 (date of the magmatic intrusion) to t_5. **b**) Time-temperature relationship of the formation of more than 90% of illite/smectite mixed layers with more than 80% illite in the clay fraction. **c**) Variation in the illite content in shales with distance to the basaltic vein (*full dots*). The *solid line* represents the curve calculated from the kinetic equation using the fifth-order with respect to smectite content

By comparing the results of experiments performed under the same temperature conditions (180 °C) for different K^+ concentrations (Eberl and Hower 1976; Howard and Roy 1985; Robertson and Lahann 1981; Whitney and Northrop 1988), Meunier et al. (1998) have shown that a linear relationship exists between the potassium content and the reaction progress (dS/dt).

The major difficulty in this type of approach (simple reaction) is the selection of the reaction order, which is determined by its mechanism. What does a second-order (Huang et al. 1993) or a sixth-order reaction (Pytte and Reynolds 1989) mean? To avoid this problem, other kinetic approaches consider that illitization is a stage reaction. Bethke and Altaner (1986) have attempted to describe the I/S sequences observed in shales by successive reactions forming SSI then ISI structures from smectite S. Each of the forms S, SSI and ISI is characterised by a specific reactivity coefficient: 0.5, 1 and 1.1. The three reaction constants k_1, k_2 and k_3 are calculated as follows:

$$k_1 = A \exp\left(\frac{-E_1}{RT}\right) \tag{3.19}$$

$$\frac{k_0}{k_1} = \exp\left(\frac{-\left[E_0 - E_1\right]}{RT}\right) \tag{3.20}$$

$$A = 10^{-3}\,\mathrm{s}^{-1}$$

$$E_0 - E_1 = 0.5\,\mathrm{kcal\,mol}^{-1}$$

$$E_2 - E_1 = 1.7\,\mathrm{kcal\,mol}^{-1}$$

This complicated velocity rule has only been applied to those shales which have undergone a deep burial. Indeed, this rule does not properly describe less advanced diagenetic transformations.

3.6 Stability Controls (T, t, μ_x)

3.6.1 Comparison of Experimental Models and Natural Systems

Despite the scatter in estimations of activation energy, it is possible to compare the experimental data with those observed for similar reactions in natural settings. Essentially two sets of data can be used for natural settings, those of high thermal regimes (over 100 °C/km), geothermal areas such as the Salton Sea, and those of burial diagenesis. The data summarized and commented upon by Velde and Lanson (1993) for the high-temperature regime in the Salton Sea (CA, USA) area can be used to estimate the reaction progress to be expected when heating occurs over a time span of about 50,000 years (Fig. 18). These authors conclude that a one-step reaction describes the transformation of smectite to illite described by Yau et al. (1987).

In contrast, reactions at lower temperatures (less than 30 °C/km) are most often best fitted using a two-step reaction (Velde and Vasseur 1993). Reactions of intermediate thermal regime can be fitted either using a two-step reaction (Hillier et al. 1995) or one-step reaction (Elliot and Matisoff, 1996).

3.6.2 Kinetic Parameter Values

One can compare the results discussed above concerning the activation energy derived by different methods, which determines the reaction rate as a function of temperature (Table 3.5).

Here we see that the determinations using different methods based upon natural mineral series are reasonably close, in the range of 20 to 30 kcal/mol. In fact most values are near 26 kcal/mol. This suggests that the conversion of smectite

Table 3.5. Activation energy values for the formation of illite (mica) from smectite or kaolinite from experimental studies and modelling of natural I/S series

Experiments		Estimations through the modelling of natural IS series	
Natural smectite → illite		**Illitization of bentonite by thermal effect**	
4 kcal	Howard and Roy (1985)	25 kcal	Pusch and Madsen (1995)
18 kcal	Whitney and Northrop (1986)	27 & 7 kcal	Esposito and Whitney (1995)
28 kcal	Huang et al. (1993)	25 kcal	Pytte and Reynolds (1989)
30 kcal	Robertson and Lahann (1981)	**Illitization of shales in a geothermal field**	
Synthetic beidellite → illite		28 kcal	Velde and Lanson (1992)
20 kcal	Eberl and Hower (1976)	**Illitization of shales by diagenesis**	
kaolinite → muscovite		27 & 9 kcal	Velde and Vasseur (1992)
30 kcal	Chermarck (1989)	20 kcal	Bethke and Altaner (1986)
28 kcal	Small (1993)	30 kcal	Elliott et al. (1991)

to illite in pelitic rocks is rather similar concerning its response to temperature. As we have discussed above, the use of natural bentonites in experiments produces phases which are not those observed in natural pelitic rock assemblages which can account for some of the scatter in experimental results. Also one should be reminded that not all of the experiments have the same starting composition as far as potassium content is concerned. The experiments using variable potassium content (Huang et al. 1993) give an activation energy of 28 kcal/mol which is near those values derived from natural mineral suites.

3.6.3 Importance of Mineral Reactions

There is a second, important factor in the normal kinetic formulations, which is the frequency factor, a pre-exponential constant. This value describes the type of material acted upon by increase in temperature and could well be modified depending upon the specific sediments, or rocks being altered by chemical influences.

In general, when comparing the laboratory data with those for natural bentonites, one could assume that the bentonites are much more stable than they would be predicted to be in extrapolating laboratory data. Since the natural bentonite is initially a nearly pure smectite mineral, and because the element crucial for the layer composition, potassium, must be brought into the immediate system, it can be deduced that at least a portion of the difference in reaction rate can be attributed to a slightly different configuration of materials and minerals between laboratory and natural setting. The bentonite layer in a sedimentary basin, a nearly pure smectite, is a reasonable analogue for the propositions of clay barriers in high-energy waste storage facilities. Hence one could expect that bentonite smectites would be stable in one meter or less thicknesses in sedimentary rocks for several millions of years at 150 °C, as is the case of the samples of Sucha et al. (1998) in the East Slovack sedimentary basin.

3.7 Summary

We can observe from the material presented in this chapter that the kinetics of change of smectite material which will eventually produce illite depends upon an increase in temperature which is necessary to destabilize and activate the smectite from its initial stable state. Chemical influences are very pronounced. In laboratory experiments, pure mica can be produced from kaolinite in several month's time but only at very high pH conditions, in the presence of high amounts of potassium. However, similar reactions have been observed in sandstones where kaolinite is transformed into illite (see Sect. 2.2.5, Chap. 2). In general, it takes very long periods of time to produce a significant amount of illite under normal pelitic rock compositions in closed systems at low temperatures (less than 100 °C). This time is shortened by increased temperatures. But, for the most part, the short time span, high-temperature reactions produce PCI instead of WCI. Thus the true, pure mica-like, low-charge phase is not favoured in closed chemical systems.

However, in systems where potassium is readily available, one can produce WCI in significant amounts and pure mica at very high potassium and pH conditions. Thus the chemical potential is more important for the production of illite (WCI) than temperature or time.

3.8 Application of Kinetics to K–Ar Dating

3.8.1 The Problem for K–Ar Dating of Illite from Shales

Extracting K–Ar ages from diagenetic illites is still debated because the models proposed until now over-simplify the evolution of I/S particles during maturation in diagenetic series. These problems are discussed in Chap. 2.4. Indeed, even the most recent model proposed by Srodon et al. (2002) is still based on a continuous nucleation and growth of 2-nm-thick fundamental particles which does not take the dissolution of the pre-existing unstable particles into account. The preceding discussion on crystal growth and the dynamics of mineral change in closed systems where the smectite-to-illite transformation occurs can be used as a guide to a more precise interpretation of age dating of illitic clays. At this point it might be useful to forget previous models and to come back to practical observations and measurements. Some results of age-dated sedimentary sequences could be helpful.

1. In relatively young stratigraphic sequences K–Ar age decreases with particle size no matter the stratigraphic age (Hofmann et al. 1974). For the clay fraction, the difference between K–Ar and stratigraphic ages increases with depth (Aronson and Hower 1976), as the apparent age determined by radioactive decay products decreases with depth and stratigraphic age. This indicates a loss of radiogenic argon from the system which might indicate

that the amount of "recycled" potassium, that gained from the dissolution of pre-existing minerals to form more illitic minerals increases with depth, releasing radio-decay argon from the potassic mineral system (Aronson and Hower 1976; Burley and Flisch 1989; Glasman et al. 1989; Morton 1985; Mossman et al. 1992; Renac 1994; Rinkenbach 1988).

2. Velde and Renac (1996) showed in a study of upper Carboniferous–Trias-Hettangian age rocks, that the fine-sized clay fraction (< 0.2 µm) is richer in K_2O and exhibits lower K–Ar ages than the coarser fraction (0.2–2 µm). This indicates that stable pure illite crystals are forming in this fraction and that they do not recrystallize.

The two points discussed above lead to the following apparent paradoxes:

- the first implies that the older the bed of the buried sediment, the younger the clays (illites and ordered illite/smectite mixed-layer minerals) they contain. If we refer to XRD observations in which different types of I/S and PCI, WCI minerals are present in varying proportions as a function of the burial history of the sample; in the I/S minerals which contain more illite, the rejuvenation of the radiogenic potassium in the clay "reservoir" is probably related to the dissolution of unstable I/S particles;
- the second observation shows that the most illite-rich particles are the finest. This implies that small and thin crystals (less than 6 layers) result from a nucleation step. In these old rocks, these crystals contribute to the PCI band intensity and eventually to the WCI peak.

In a constant or approximately constant fixed mass system, an increase in the mass of one particle can only be possible by the loss of another. Why do some I/S particles dissolve while others grow? Coming back to the Ostwald ripening theory (Baronnet 1976), we learn that the particles of a size r_0 at a given time t_0 begin to grow. Because oversaturation of the solution decreases during their growth, they reach a critical size (equilibrium). Then they dissolve to the benefit of particles of a larger size. The size distribution is stable in reduced coordinates (Fig. 3.17a). However, as growth proceeds, the mode is displaced to larger values (Fig. 3.17b). The shape of the distribution curve of crystallites is constant.

Considering the reaction steps in the smectite-to-illite transformation in shales (see Sect. 2.2.2), the maturation process which is active under diagenetic conditions makes the smectite-rich I/S particles unstable. The components released during their dissolution contribute to the growth of larger, more isometric I/S particles with a greater illite content. Illite precipitates on the surfaces of pre-existing crystallites. At each dissolution-crystallisation stage, the potassium released in solution is fixed in the crystal lattice of the growing particles, but the radiogenic argon escapes and migrates out of the reaction zone. Thus, the apparent change in the K–Ar age of a shale sample depends on the mass ratio between the growing and dissolving particles.

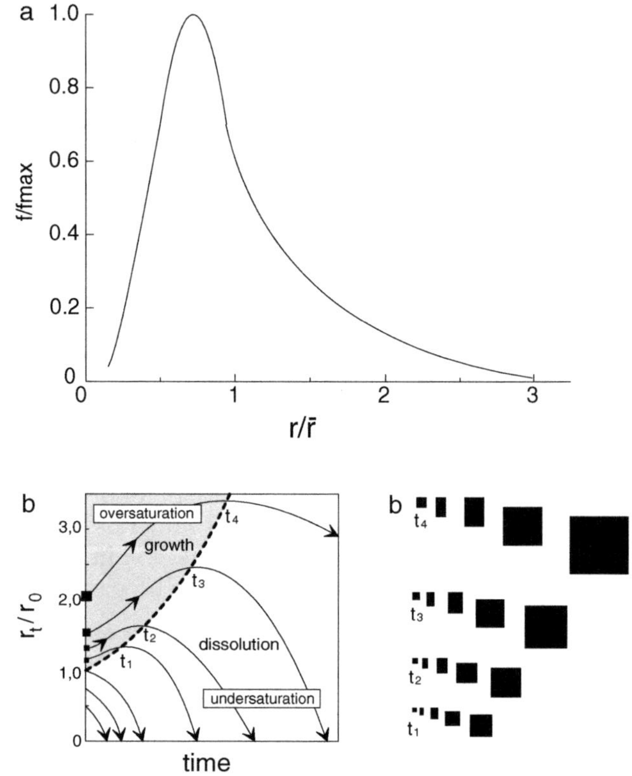

Fig. 3.17a,b. The ripening process. **a)** The size-distribution curve in reduced coordinates. **b)** Schematic representation of the maturation effect on the particle size distribution (from Baronnet 1976)

3.8.2 K–Ar Age and Mass Transfer During Smectite-to-Illite Conversion

The K–Ar dating method for illitic clays is dependant upon an interpretation of the recycling rate of potassium which controls the argon lost through the dissolution-nucleation-growth processes (Fig. 3.18). Argon loss depends on:

- the rate of dissolution of detrital minerals, should they be present,
- the rate of dissolution of unstable I/S particles as they dissolve to form more illite layers on other, larger grains.

Argon gain depends on:

- the rate of nucleation and growth of the ultimately stable illite particles (WCI),
- the rate of illite layer growth on I/S particles, which eventually form PCI

Thus, the K–Ar ratio of a given sample is the summation of the remaining part of the Ar in the old stock (undissolved I/S) with the amount of K–Ar fixed in overgrown illite layers. If one wishes to interpret this information on the

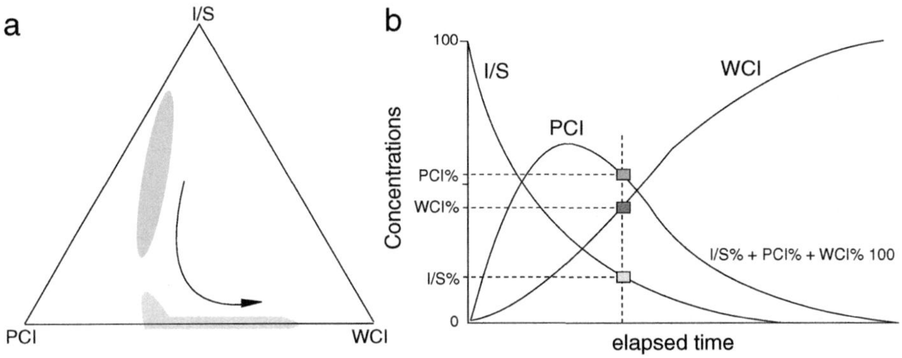

Fig. 3.18a,b. Variation of relative proportions of I/S, PCI and WCI sub-populations. **a**) Diagenesis to epizone transition (from Gharrabi et al. 1998). **b**) A theoretical representation of the variations of I/S, PCI and WCI quantities vs elapsed time in a simple chain reaction model for a fixed temperature. For any time I/S% + PCI% + WCI%=100

growth of illite layers as a function of time, the question now is to choose the most adapted variable to represent the rejuvenation of I/S:

- *the bulk K_2O amount of a given clay fraction.* No matter the size range of the clay fraction, the chemical analyses integrates the old and young stock of potassium. This total potassium content of changing old/young proportions is compared to the radiogenic argon which is by definition related to the old potassium in the minerals. The result is at best a continuous trend with increasing diagenetic conditions which may be used to extrapolate a "diagenetic age";
- *the % illite calculated from XRD patterns.* This parameter is classically used to determine the illite age. In most studies; the calculation of the % illite is based on the peak position, considering first, that the I/S MLM particle population is homogeneous and second, that peak position varies continuously in a 100%–0% expandable layer range. Usually the amounts of K_2O show a continuous trend with increasing diagenetic conditions;
- *the WCI+PCI proportions given by XRD decomposition.* The decomposition of XRD patterns showed systematically that when ordered I/S MLM series occur they are composed of three particle populations: I/S, PCI (<5% smectite layers+small free illite particles) and WCI (0% smectite layers), the relative proportions change with depth (Lanson and Champion 1991; Renac and Meunier 1995;Varajao and Meunier 1995).
- With increasing depth, the relative proportion of WCI increases and the PCI peak position shifts toward 10.0 Å and the peaks become sharper (increasing mean CSDS). Consequently, at a given depth, the young K_2O stock is concentrated in the overgrowth illite layers on I/S-PCI particles and in the pure illite crystals which have nucleated freely. These layers are, in their turn, concentrated in the WCI+PCI particle populations whose proportions can be measured using diffraction peak areas.

What are the main changes induced by increasing diagenetic conditions?

1. WCI is the ultimate product which is produced during the smectite-to-illite conversion. PCI is the end product of the smectite-to-illite transformation (Lanson et al. 1996; Gharrabi et al. 1998);
2. the mean CSDS of each particle population increases. It was shown that the CSDS and junction probabilities of I/S depend on the geothermal gradient, but not on the age of the sediments: the higher the gradient, the higher the % illite in I/S
3. smectite content of I/S particles decreases to about 5% illite layers (PCI);
4. I/S amount (peak area) decreases and disappears at the anchizone-epizone boundary.

According to the observations discussed above, the smectite-to-illite conversion process is undoubtedly controlled by the dissolution of I/S particles to the benefit of PCI and WCI. Thus, in spite of the fact that I/S, PCI and WCI are not mineral phases *sensu stricto*, the conversion process could be considered as a simple chain reaction (Lasaga 1981): I/S→PCI+WCI using the following simplifying assumptions:

- composition change in I/S and PCI phases can be ignored since it is related to the growth of illite layers which accumulate in WCI;
- the small free illite particles (mean CSDS<6 layers) which contribute to the broad PCI peak intensity are not distinguished from larger particles having less than 5% smectite layers;
- the two reactions (1) I/S → PCI and (2) PCI → WCI are assumed to be first-order reactions; thus, their respective apparent rate constant k_1 and k_2 are expressed in time^{-1} units.

At a given temperature, (given depth in the sediment pile), the variation of "phase" concentration with time may be written as follows:

$$\frac{\mathrm{d}C_{\mathrm{I/S}}}{\mathrm{d}t} = -k_1 C_{\mathrm{I/S}} \tag{3.21}$$

$$\frac{\mathrm{d}C_{\mathrm{PCI}}}{\mathrm{d}t} = k_1 C_{\mathrm{I/S}} - k_2 C_{\mathrm{PCI}} \tag{3.22}$$

$$\frac{\mathrm{d}C_{\mathrm{WCI}}}{\mathrm{d}t} = k_2 C_{\mathrm{PCI}} \tag{3.23}$$

If $C^0_{\mathrm{I/S}}$ represents the initial proportion of I/S (100%), by integration, Eq. 3.21 gives:

$$C_{\mathrm{I/S}}(t) = C^0_{\mathrm{I/S}}\,\mathrm{e}^{-k_1 t} \tag{3.24}$$

Thus, inserting (3.24) in to (3.22):

$$\frac{dC_{\text{PCI}}}{dt} = k_1 C^0_{\text{I/S}}\, e^{-k_1 t} - k_2 C_{\text{PCI}} \tag{3.25}$$

$$\frac{dC_{\text{PCI}}}{dt} + k_2 C_{\text{PCI}} = k_1 C^0_{\text{I/S}}\, e^{-k_1 t} \tag{3.26}$$

Following Lasaga's recommendation for solving these equations using substitutions:

$$C'_{\text{PCI}} \equiv C_{\text{PCI}}\, e^{k_2 t} \tag{3.27}$$

Then,

$$\frac{dC'_{\text{PCI}}}{dt} = e^{k_2 t}\left(\frac{dC_{\text{PCI}}}{dt} + k_2 C_{\text{PCI}}\right)$$

Introducing Eq. 3.25:

$$\frac{dC'_{\text{PCI}}}{dt} = e^{k_2 t} k_1 C^0_{\text{I/S}}\, e^{-k_1 t}$$

By integration $\int_0^{C'_{\text{PCI}}} \text{d'}_{\text{PCI}} = \int_0^t k_1 C^0_{\text{I/S}}\, e^{(k_2-k_1)t} dt$, one obtains the variation of the PCI concentration with time at a given temperature:

$$C_{\text{PCI}}(t) = C^0_{\text{PCI}} e^{-k_2 t} + \frac{k_1 C^0_{\text{I/S}}}{k_2 - k_1}\left(e^{-k_1 t} - e^{k_2 t}\right)$$

As C^0_{WCI} and C^0_{PCI} are equal to zero and $C^0_{I/S}$ to 100%, $C_{\text{PCI}}(t) = \frac{100 k_1}{k_2 - k_1}\left(e^{-k_1 t} - e^{k_2 t}\right)$.

Lasaga demonstrates how $C_{WCI}(t)$ can be obtained:

$$\begin{aligned} C_{\text{WCI}}(t) &= C^0_{\text{WCI}} + C^0_{\text{PCI}}[1 - e^{-k_2 t}] \\ &\quad + C^0_{\text{I/S}}\left[1 - e^{-k_1 t} - \frac{k_1}{k_1 - k_2}\left(e^{-k_1 t} - e^{-k_2 t}\right)\right] \end{aligned} \tag{3.28}$$

Thus

$$C_{\text{WCI}}(t) = C^0_{\text{I/S}}\left[1 - e^{-k_1 t} - \frac{k_1}{k_1 - k_2}\left(e^{-k_1 t} - e^{-k_2 t}\right)\right] \tag{3.29}$$

The graphical presentation of the three equations $C_{\text{I/S}}(t)$, $C_{\text{PCI}}(t)$ and $C_{\text{WCI}}(t)$, representing the simple chain reaction for a fixed temperature is given in Fig. 3.18b. At a given elapsed time, the apparent K–Ar age will depend on the relative proportions of K_2O component in the old (I/S%) and the young (PCI% + WCI%) reservoirs. The transfer from the first to the second increases with time.

3.8.3
An Example: The Balazuc Series (Renac 1994)

If one accepts the analysis discussed above, it should be possible to test the hypotheses on a series of measurements of natural samples. The following case study is presented.

K–Ar dating has been performed on two-size clay fractions extracted from samples in the Carnian to Hettangian sedimentary pile of a the Ardèche passive margin from southeastern France (Renac 1994). Major data are presented in Table 3.6 and Fig. 3.19. The calculation of K_2O component transfer in a diagenetic series is made difficult for two reasons:

- the temperature increases with depth which changes the apparent reaction rate constants: $k = A\,e^{-E_a/RT}$ where A is the pre-exponential factor and E_a the activation energy for the reaction;
- the composition of the unstable "mineral phases" (I/S and PCI) changes during the reaction because of illite layer addition and nucleation of free illite particles (small mean CSDS). Thus, their respective K_2O content increases with time.

This is typically the case here where, paradoxically, the < 0.2 μm fraction has more K_2O than the 0.2–2 μm one in spite of the fact that its WCI amount is smaller. This means that small free illite particles (mean CSDS < 14 layers) are within the PCI peak due to their small size and increase the PCI% (Fig. 3.19).

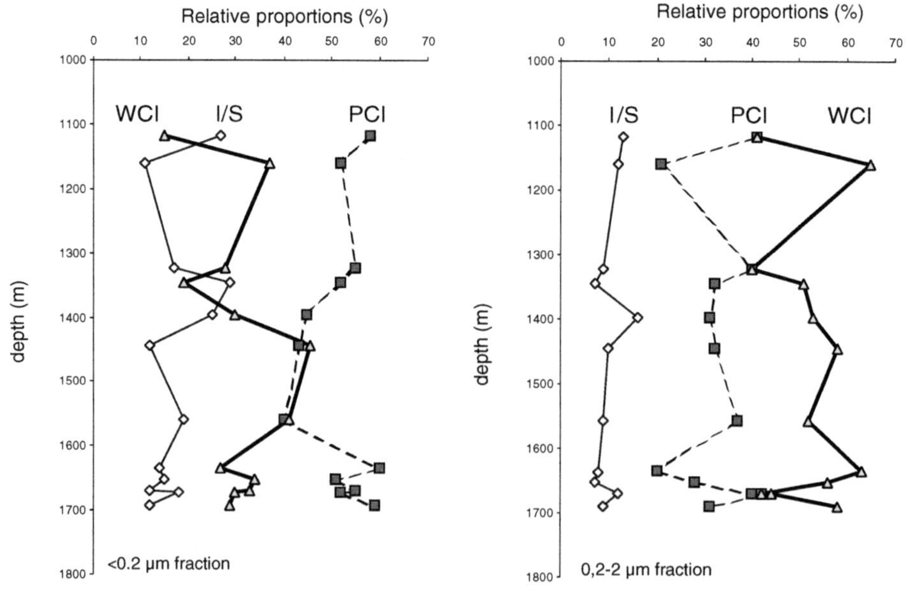

Fig. 3.19. Variation of I/S, PCI and WCI % of two size fractions (< 0.2 and 0.2–2μm) with depth in the Ardèche passive margin (southeastern France)

Table 3.6. Diagenetic series of the Ardèche passive margin. Composition of the < 0.2 and 0.2–2 μm size fraction and apparent K–Ar ages related to depth, i. e., stratigraphical age (data from Renac 1994)

Depth	Stratigr.	0.2–2 μm					< 0.2 μm				
(m)	age (Ma)	K_2O %	K/Ar age (Ma)	I/S%	PCI%	WCI%	K_2O %	K/Ar age (Ma)	I/S%	PCI%	WCI%
1117	203	4.25	175.8±3.9	18	41	41	6.58	141.8±3.2	27	58	15
1160	203	2.86	150.3±4.0	14	21	65					
1220	204	4.61	188.1±4.2	19	41	40	6.82	153.6±3.4	20	55	25
1251	204	5.65	181.2±4.0	20	46	34	6.88	150.4±3.4	25	55	20
1281	205						6.82	147.0±3.8	16	59	25
1323	209						7.59	137.1±3.2	17	55	28
1420	215	5.90	182.5±4.0	15	52	33	7.49	141.8±3.2	24	46	30
1444	216	7.13	190.4±4.0	11	37	52	7.59	155.3±3.5	12	43	45
1540	221						8.82	136.2±3.0	9	36	55
1558	228						7.85	156.6±3.5	19	40	41
1653	239						8.20	138.2±3.1	15	51	34
1666	244						8.74	132.3±3.0	10	57	33
1671	297						7.68	175.0±3.8	18	52	30

In a first approximation, the above-mentioned difficulties may be overcome for a given sample (fixed temperature) if the relative proportions of I/S, PCI and WCI and K–Ar age are determined. Then the elapsed time may be estimated using the difference between the stratigraphic and K–Ar ages, which we call Δage. Indeed, the simple reaction chain equations are adjusted until they fit with the I/S, PCI and WCI proportions for a given Δage giving an indirect determination of k_1 and k_2. This empirical approach was attempted for a diagenetic series of the Ardèche passive margin which gave the data set below (Table 3.6).

K–Ar ages were measured on the < 0.2 and 0.2–2 µm fractions (Table 2.4, Renac 1994). Results obtained for the two fractions are considered to be isochrons rather than mixing lines. Indeed, the detrital micas from magmatic and metamorphic rock bodies which neighbour the passive margin are much older than the 145.5 ± 4.6 (MSWD: 4.06) and 199.3 ± 5.2 Ma (MSWD: 0.66) deduced ages from < 0.2 and 0.2–2 µm fraction isochrons, respectively. The K–Ar age of both fractions are always younger than the stratigraphic ages (Δ age). This means that, if present, detrital mica quantities should be identical in all the samples analyzed in the sedimentary pile, i. e., a constant input during the Carnian to Hettangian period (about 25 Ma). This is highly improbable. The authigenic characteristics of I/S, PCI and WCI particle populations are reinforced by the Δ age increase with depth (Fig. 2.27a). Then, a question rises: what do the isochrons mean for a mixture of authigenic illitic phases?

The K–Ar age deduced from isochrons gives a mean date of the Ar accumulation. Thus, as the 0.2–2 µm fraction is the oldest one, this means that Ar was accumulated earlier than in the finest fraction. In other words, the coarser fractions stop growing or grow much more slowly than the finest one. This is typical of non-isothermal growth processes and has been demonstrated for quartz crystals in a contact metamorphic aureole (Joesten 1983). This implies that the K–Ar age difference between the fine and coarse fractions should be constant through the sedimentary pile. Indeed, as the growth rate depends on temperature conditions, the accumulation rate of K_2O in the two fractions must vary concomitantly with depth. Thus, the age difference between the two fractions must be constant. This is the case, the age difference between the fine and the coarse fractions is about 35 Ma (Fig. 3.20).

The following simplifying assumptions have been used to calculate the k_1 and k_2 rate constants:

1. the illitization of ordered I/S is considered to be a simple chain reaction I/S(1) $\rightarrow$ PCI(2) $\rightarrow$ WCI,
2. I/S, PCI and WCI quantities are considered to be proportional to the intensities of their corresponding decomposed bands,
3. time duration has been calculated for each sample using the difference between the total burial period obtained by modeling of the burial history (Renac and Meunier 1995) and the K–Ar age measured on clay fractions (duration = sediment age – K–Ar age).

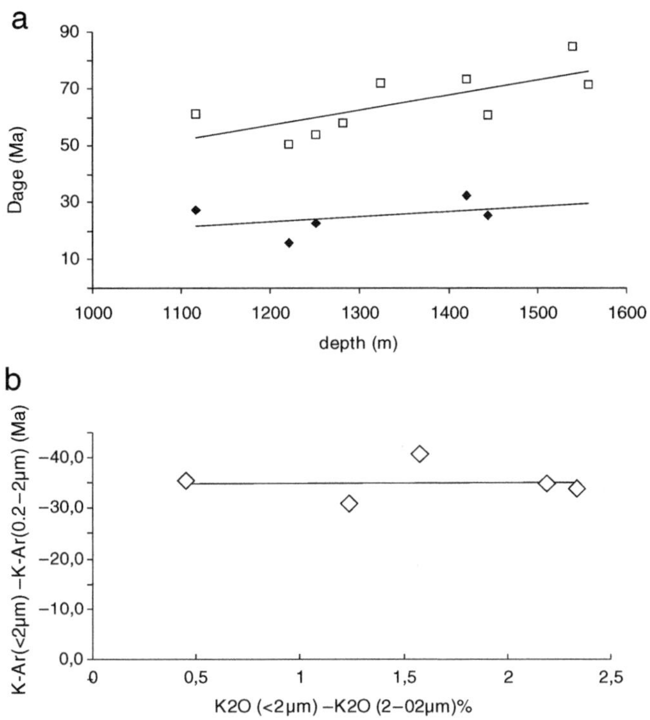

Fig. 3.20a,b. K–Ar dating of the <0.2 and 0.2–2 μm fractions from clay samples of the diagenetic series in the Ardèche passive margin (data from Renac 1994). **a**) Variation with depth of the difference between stratigraphic and K–Ar ages for the two size fractions. **b**) Relationship between potassium content ($K_2O_{<0.2\,\mu m}$–$K_2O_{2-0.2\,\mu m}$ and age differences: ($K–Ar_{<0.2\,\mu m}$-$K–Ar_{0.2-2\,\mu m}$)

Thus, the $k_1$and k_2 rate constants have been determined for each sample by fitting the curves on the measured compositions (Table 3.7; Fig. 3.21a). The results show that k_2 is lower than k_1 and that both increase with depth as expected (increasing temperature). It is to be observed that k_2 increases more rapidly than k_1 indicating that the rate of illite nucleation and growth, i. e., rejuvenation rate of the K_2O stock, increases with depth. This is coherent with the observation of decreasing K–Ar ages with depth found by Aronson and Hower (1976).

Can k_1 and k_2 be used as "true" rate constants? If so, as temperature can be measured through organic matter maturation or calculated using the method proposed by Velde and Vasseur 1992), the Arrhenius equation: $k = A\,e^{-E_a/RT}$ may be used to calculate the values of the pre-exponential factor (A) and the activation energy (E_a) for the reactions. This is theoretically possible but the meaning of the results should remain ambiguous since I/S(1) → PCI(2) → WCI is not a real simple chain reaction. The necessity to use simplifying

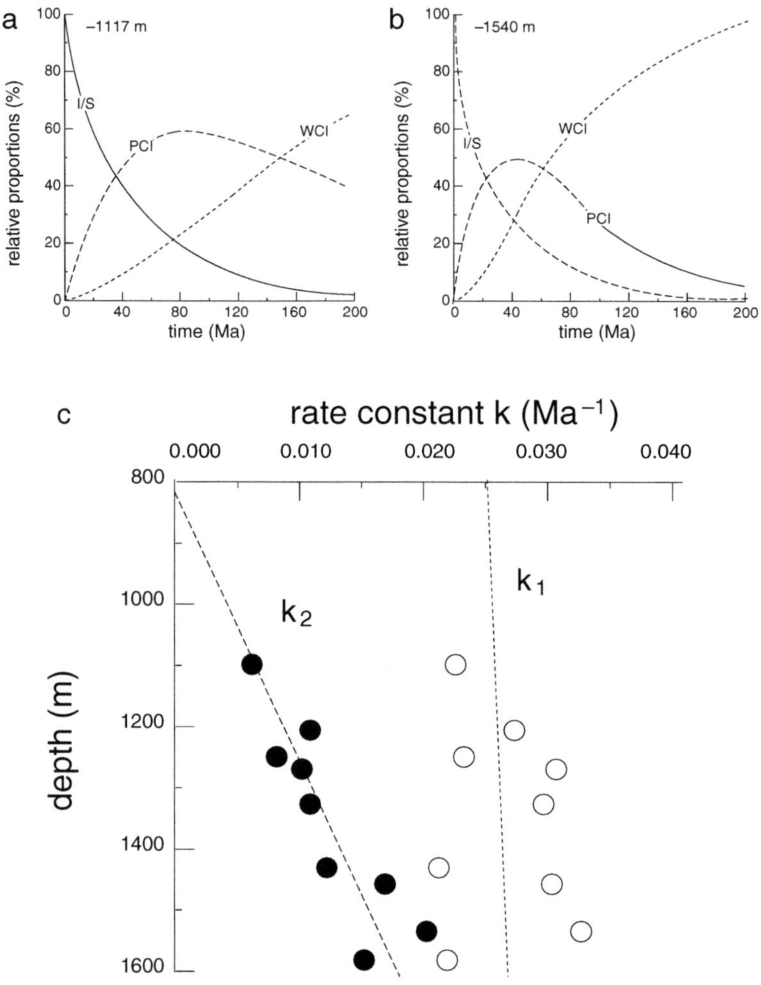

Fig. 3.21a,b. Calculated illitization of ordered I/S MLM assuming that I/S(k_1) →PCI(k_2) → WCI is a simple chain reaction (from Renac 1994). **a**) Determination of the apparent rate k_1 and k_2 constants for two samples by fitting the measured I/S%, PCI% and WCI%. **b**) Variation of k_1 and k_2 rate constants (Ma^{-1}) with depth, i. e., temperature

assumptions (see above) introduces errors. One of the biggest one comes from PCI% which is the summation of the diffracted intensities of illite-rich I/S and small free illite particles. Indeed, I/S should dissolve while free illite particles should grow.

As a temporary conclusion, we can say that K–Ar dating methods should not ignore the rejuvenation effect induced by the dissolution of unstable particles. They have to integrate the old, diminishing Ar-rich K_2O stock with the increasing young Ar-poor stock. The latter is itself dependent on the illite growth rate.

Table 3.7. Values of the two rate constants k_1 and k_2 obtained from the fit of I/S, PCI and WCI proportions (from Renac 1994)

Depth (m)	k_1 (Ma^{-1})	k_2 (Ma^{-1})
1117	0.0228	0.007
1220	0.0276	0.012
1251	0.0237	0.009
1281	0.0312	0.011
1323	0.0299	0.011
1420	0.0207	0.012
1444	0.0302	0.016
1540	0.0319	0.019
1558	0.0217	0.016

The simple chain reaction model proposed here could be a satisfying approach in the future if the rate constants are refined in order to represent a real mineral reaction.

CHAPTER 4

Applications

"Car, comme le sel assaisonne les viandes, ainsi l'argille et le sablon estans distribués ès terroirs par juste proportion, ou par nature ou par artifice, les rendent faciles à labourer, à retenir et à rejeter convenablement l'humidité; et par ce moyen, domptés, aprivoisés, engraissés, rapportent gaiement toutes sortes de fruicts."

Olivier de Serres, 1600

Overview

There are two fields of current human endeavor where illites play an important role. Other clays have been and some are still used in industrial processes, smectite and kaolinite being the chief minerals. Smectites have been used as thixotropic agents to fluidify paints, drilling muds in oil exploration, or as filling agents in hot dogs and other foods. Illite is not adapted to most of such industrial uses. In the past it was a major component of common clays used for the production of cooking pots, plates, and, continuing to this day, especially in Europe, tiles and bricks, but this is a diminishing field of usage. Hence we will not try to find a use for illite in factories. However the importance of illite will perhaps be even greater in the future.

Illite, and potassium clays, are essential for agriculture. In the recent past (since about 1950) use of potassium fertilizer became a very common practice in agriculture. Its origin dates back to the late 19th century, but almost the complete use of this material to increase or maintain plant fertility is more recent. Use of potassium fertilizer is necessary because intensive farming practices deplete much of the available potassium from agricultural soils. If one is to decrease the dependence on artificial means of maintaining fertility one needs to understand how farming techniques in the past kept fertility at a reasonable level without the addition of potassium. For example, the major portion of France has been farmed for approximately the last 4,000 years. The spectacular decrease in fertility is only a recent phenomenon. It has occurred as productivity increased dramatically with the use of hybrid plants and artificial fertilizing agents. If the two are interrelated by necessity, there is no going back to traditional methods of farming or at least in ameliorating these methods. However, if there is a mean of increasing the efficiency of traditional methods of farming and maintaining fertility, there is a great hope for the future of the world's soils. Most of this hope lies in better understanding the release and re-capture of potassium by illites and the role of ammonium, a replacement ion for potassium in illite and the second major fertilizing agent (out of three). The stability of potassium illite layers in soil clay minerals and their relation to the ammonium form is the key to future research in soil clay mineralogy. Increasing fertility and suppressing dependence on nitrate-forming ammonium fertilizer will solve many problems of water resource procuration.

A second, and potentially as important, aspect of illite in man's activities is that of the storage of high energy radioactive waste. Illite has an important role to play here. If one can devise a safe and long-lasting barrier to the migration of nuclear waste elements in the surface environment, the use of atomic energy can proceed and the use of petroleum, the major producer of gas causing the greenhouse effect and major climate change, can be slowed to a reasonable consumption. This is a major challenge at present, and one that will decide the future development of human activity.

Illite has played and still plays a major role in the recovery of primary resources such as petroleum and ores. This exploration potential resulted in a correct identification of illite occurrence which is of primary importance to the efficient discovery of porphyry-type ore deposits. It is also important for evaluating the production potentials of sandstone reservoirs for petroleum deposits and the thermal history of sediments. The aspects of illite occurrence are less pressing, perhaps, but none the less important to the future of resource exploration and exploitation.

The importance of illite to prospecting in geothermal energy fields is also very great. A mastered use of this knowledge can increase the efficiency of finding and using this little exploited energy source which can be used to develop electricity in using the natural emanations of heat and vapour escaping from the earth and not adding to the pollution of the atmosphere.

These are some of the more evident aspects of the use of illite mineralogy in the activity of modern societies. We develop several of these aspects below.

4.1 Exploration and Exploitation of Natural Resources

4.1.1 Geothermal Resources

Prospecting geothermal resource or oil has a similar goal and faces to identical difficulties: to find the reservoirs of the energy-bearing fluids which may migrate into porous or fractured rocks. The identification of clay minerals is commonly used because they are highly reactive (see Chap. 3). Indeed, the crystal structure of illite-smectite or chlorite-smectite MLMs was shown to change gradually with depth, i.e. temperature increase in geothermal fields.

Any "anomaly" in the general trend indicates the predominance of local conditions and are used to locate the hot-fluid reservoirs.

4.1.1.1
Mineral Sequences and Prospecting

The example of the Chipilapa geothermal field (El Salvador) is used here as a case study to illustrate several very important principle aspects of rock alteration under conditions of high geothermal gradients. Rock transformations and clay deposits are the key to understanding the existence of vapour reservoirs and prospecting for them.

The Geology of the Chipilapa Geothermal Field The geothermal field at Chipilapa has developed in calco-alkaline volcanic formations in the recent past (Pliocene to present day) where andesitic lava flows and dacitic pyroclastic deposits alternate (Beaufort et al. 1995). The hydrothermal activity has been active for about 16,000 years but recent crises are attested by phreatic explosion craters. Infiltrating meteoric fluids are heated up to 250 °C in reservoirs and drained by faults that delineate a "thermal plume" (Fig. 4.1a). Two reservoirs are exploited at 600 to 800 m and 1,150 to 1,400 m depth. Drill hole CH 8 is far from the production area; it is used as a reference for the heat-diffusion-controlled illitization process prevalent around the present high-temperature region. For rocks having a low permeability in this area, thermal conduction is the prevailing heat dissipation process. Drill hole CH 7bis is located at the centre of the "thermal plume". It is used as a reference for hydrothermal alterations in high-permeability areas where convection is the prevailing heat-dissipation process.

The clay content of the rocks increases with depth from 0 to 500 m. They form an impermeable clay cap that strongly decreases heat losses. Therefore, the thermal gradient is very high between 300 and 550 m: 60–80 °C over just 250 m or 24–320 °C km^{-1}. Surface reservoir A immediately beneath the clay cap contains high-temperature fluids (> 185 °C) with a significant vapour phase (steam cap). From 650 to 1,100 m, below the clay cap, the permeability of rocks is low, as is the thermal gradient. The amount of vapour of the deeper aquifer (reservoir B) between 1,100 and 1,400 m is small (prevailing liquid phase). Today, the boiling point of fluids are found between 700 and 800 m in depth.

The clay minerals formed under these high-temperature conditions are illite/smectite and chlorite/smectite mixed-layer minerals the composition varies with depth. The I/S MLM smectite content decreases with increasing depth (Fig. 4.1b). The variation is not progressive, but occurs in stages:

1. Far from the plume anomaly (well CH 8), the I/S MLM are randomly ordered (R0) and rich in smectite (>90%). In this well, the major mineralogical change occurs between 400 and 600 m, where the temperature suddenly passes from 60 to 80 °C a local gradient of 100 °C/km. The smectite-rich R0 I/S MLM transforms into regularly ordered (R1) I/S MLM (50% smectite

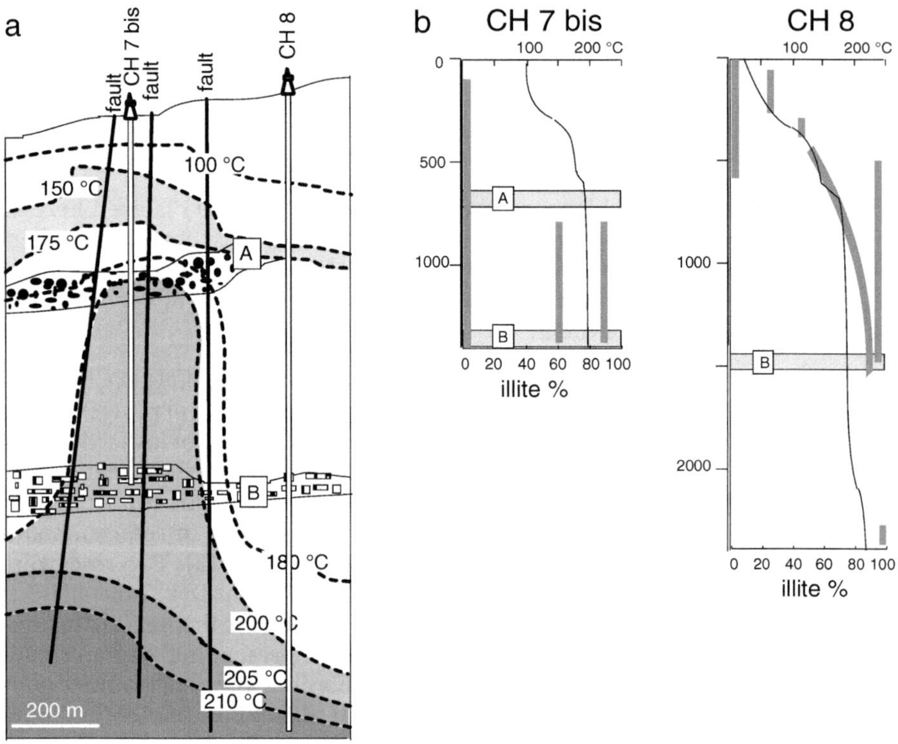

Fig. 4.1a,b. The Chipilapa geothermal field, El Salvador (Beaufort et al. 1995). **a** The "thermal plume" is framed by three faults. Two reservoirs (A and B) are exploited by a series of drill holes. **b** The distribution of smectite, I/S and illite in two drill-holes in the plume (well CH-7bis) and in the conduction zone (well CH-8)

- 50% illite) then quickly into R1 I/S minerals with 60% illite. The illite content then increases steadily up to over 95% (PCI: 10.13–10.18 Å – FWHM = 0.8–1.0 °2θ Cu Kα) at −1,750 m depth.

2. In the thermal anomaly (well CH7 bis), transformations of the I/S MLM take place differently. Here pure smectites and R1 I/S MLM with 60 and 90% illite coexist. There is no steady enrichment in illite in I/S MLM with depth. Clay minerals in reservoirs, notably dioctahedral smectites and I/S MLM, are not the same as those observed in diagenetic series. In the diagenetic smectite-to-illite transformation mineralogy the illite is interstratified with montmorillonite (Mg, Fe in the octahedral site creating a large portion of the layer charge). Beidellites (aluminous smectites) predominate in geothermal fields and illites are much more aluminous (close to potassium-poor muscovite). Accordingly, the montmorillonite → illite transformation is inappropriate for these systems. Once the hydrothermal system is sealed by

the formation of clays, the conductive regime is re-established and smectites transform directly into illite or chlorite. Formed at about 4,000 years ago (t_{4000}) between 180 and 240 °C in the field's active fractures, beidellite and saponite were subsequently transformed into mixed-layer minerals of the saponite-chlorite and the beidellite-illite series (Fig. 4.2). Their transformation towards the end members-chlorite and illite progresses with depth. The disappearance of the expandable components which were transformed into illite or chlorite at increasing depth indicates that during the time interval (4,000 years), the fluid-rock ratio decreases and consequently the conductive thermal effect (mean geothermal gradient) prevailed over convective conditions (local thermal anomaly due to hydrothermal fluid circulation).

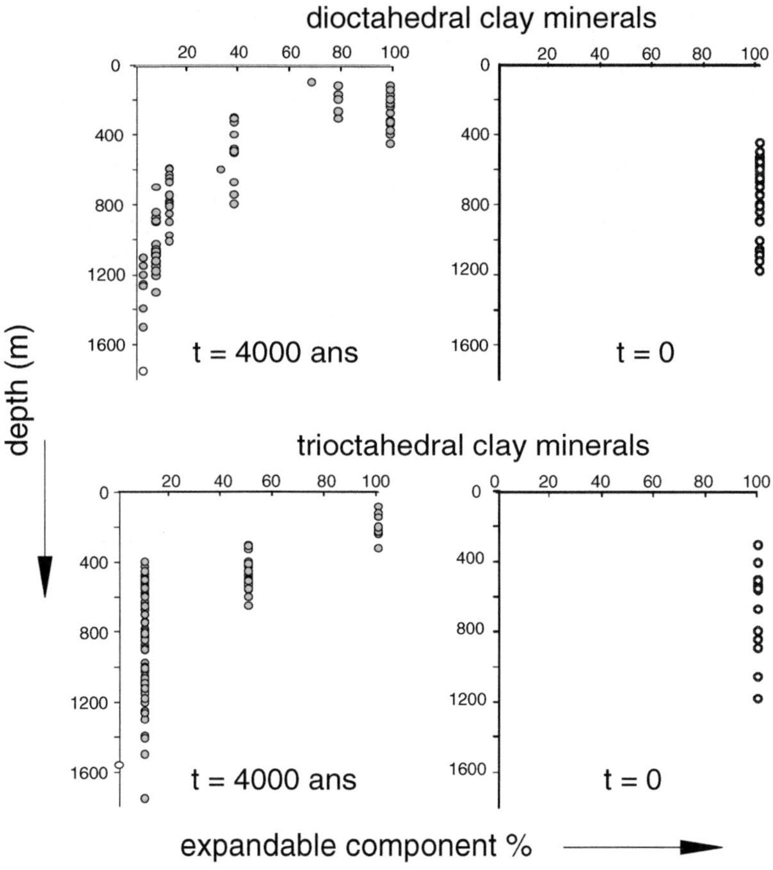

Fig. 4.2. Variation in the mineralogical composition of clays from the di- and trioctahedral series as a function of depth. Time t_0 corresponds to the precipitation of clays in fractures from boiling solutions. Time 4,000 years represents the duration of the local thermal anomaly (4,000 years at Chipilapa) that caused high-gradient diagenetic transformations in the rock

I/S MLM Series and Local Anomalies The pioneer work of Steiner (1968) showed that the illite content of I/S MLM progressively increases with depth in geothermal fields, indicating that there is a relationship between hydrothermal and high-gradient diagenesis. Thus, the kinetic models established for the smectite (montmorillonite)-to-illite conversion (Huang et al. 1993, for instance) describe the complete I/S sequence which is controlled by heat diffusion through the rocks reasonably well. However, these models, as sophisticated as they may be (see Sect. 3.5.2), fail to explain the clay formation in the active zones of geothermal fields in which hot fluids circulate. Since smectitic minerals are formed locally, while the surrounding rocks contain more or less illite-rich I/S MLM, anomalies in the I/S sequence can used as guides for prospecting the hot fluid reservoirs.

The distribution of the fluid-conducting fault network usually changes during the period of hydrothermal activity in a given zone due to fracture-opening by earthquakes or phreatic eruptions. Thus, a given rock may experience successively different clay formation processes, for example "high-temperature gradient" illitization and sudden smectite precipitation which results from a brusk change in the temperature of altering fluids. The thermal plume which is predominantly located in the conducting fault network moves within the geothermal area during its period of activity, changing local thermal gradients and fluid sources. When the plume (high-temperature fluids) is displaced, locally, previously formed smectite can be illitized in its turn due to higher temperatures. The heat conduction becomes the governing factor in mineral transformation (Fig. 4.3).

Thus, two parameters can be used to evaluate the industrial capacities of a geothermal field using clay minerals:

- the heat capacity of the rock sequence can be evaluated by the % illite vs depth curve. The sharper the change in illite content with depth, the higher the heat transfer;
- any local smectite anomaly in the I/S MLM series may be used as a prospecting guide for the hot fluid reservoirs via veins or porous rocks.

4.1.1.2
Mineralogical Problems

Illitization Process from Beidellite or Montmorillonite There are some differences between the I/S MLM series observed in geothermal fields compared to diagenetic smectite-to-illite transformations and it is interesting to compare them:

- the random interstratification (R=0) stage in the beidellite-to-illite conversion series is totally absent in geothermal alteration. The disordered, highly smectitic I/S MLM encountered are in the clay cap of the Chipilapa geothermal field is of the montmorillonite type. Thus the initial stages of rock transformation are similar in character to those in other geological environments,

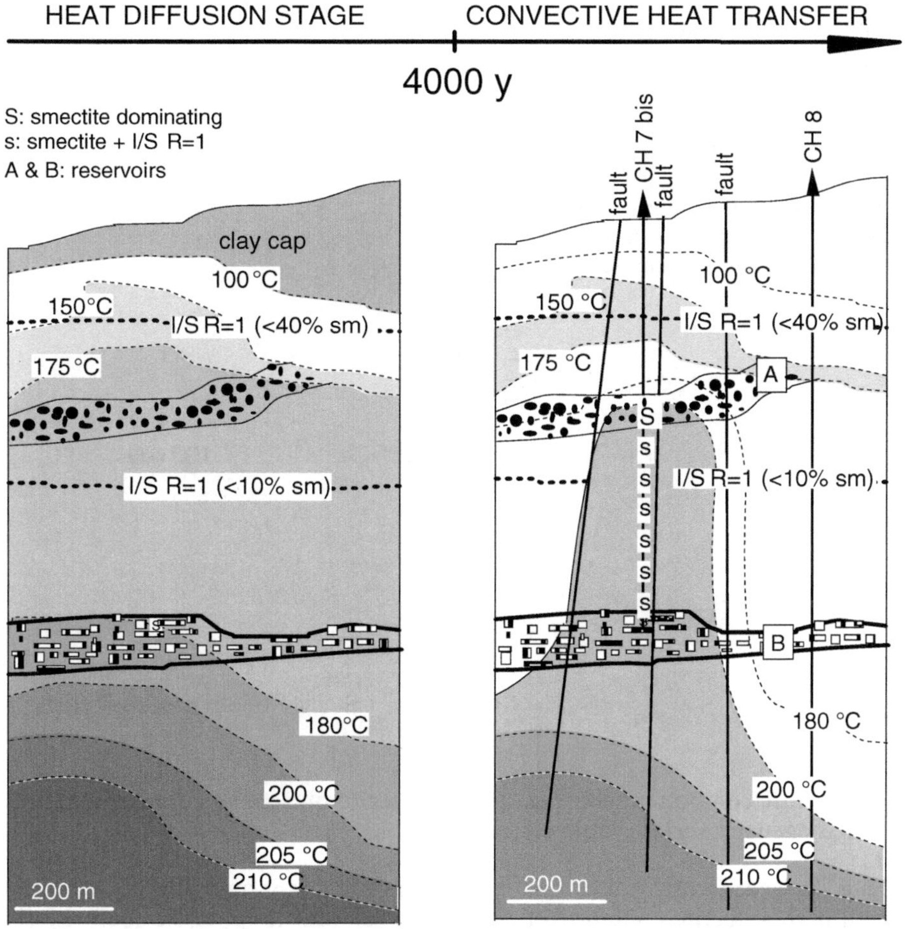

Fig. 4.3. Schematic representation of the two alteration stages in the Chipilapa geothermal field. Smectite presently formed in the hot fluid reservoirs are superimposed to I/S MLM minerals previously formed during the heat diffusion stage

- the I/S MLM sequence, caused by the thermal plume, begins with a potassic rectorite phase (regularly ordered mixed-layer mineral) of 50% illite composition.
- beidellite and not montmorillonite is the initial smectitic mineral, its transformation into illite is a function of the *time* $\times$ *temperature* parameter;
- in deeper zones, the potassic rectorite (50% illite) is present along with an ordered illite-rich I/S MLM, its illite content and coherent domain size increase steadily with depth (Fig. 4.4). The crystal morphology passes from small-sized lath-shaped particles (60–80% illite) to more isomorphic and thicker large-sized particles (> 85% illite).

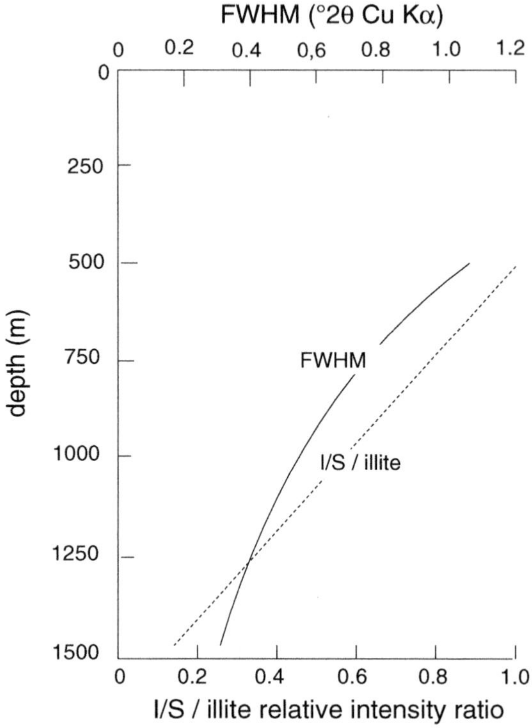

Fig. 4.4. Variation in the full width at half-maximum intensity (FWHM) of the illite (PCI) peak and in the intensity ratio of I/S and illite diffraction bands as a function of depth in the Chipilapa geothermal field. These parameters have been determined after decomposition of the diffraction patterns into Gaussian curves

- the illitic component composition is closer to muscovite compared to illite in diagenetic series (Table 4.1).

Thus, in geothermal zones one encounters high montmorillonite I/S MLM with high smectite content in the clay cap. The initial potassic rectorite (illite/beidelite ordered mineral) in the hydrothermal plume zones and the evolution of an I/S MLM towards illite (PCI?) occurs still in the presence of potassic rectorite.

Unfortunately, the discrimination between coexisting beidellite particles and regular I/S MLMs is difficult. There seems to be no significant difference of size and morphology between them. Therefore, the standard reaction process

Table 4.1. Geothermal field of Chipilapa (El Salvador). Chemical compositions of beidellite and I/S MLM in which the smectite percent decreases. (Patrier et al. 1998)

Beidellite	$Ca_{0.16}\ K_{0.03}\ (Al_{1.895}\ Fe^{3+}_{0.04}\ Mg_{0.085}\ Mn_{0.005})\ [Si_{3.615}\ Al_{0.385}]\ O_{10}\ (OH)_2$
I/S R=1 (≈0%sm)	$Ca_{0.10}\ K_{0.44}\ (Al_{1.655}\ Fe^{3+}_{0.20}\ Mg_{0.16})\ [Si_{3.47}\ Al_{0.53}]\ O_{10}\ (OH)_2$
I/S R=1	$Ca_{0.015}\ K_{0.69}\ (Al_{1.825}\ Fe^{3+}_{0.10}\ Mg_{0.08})\ [Si_{3.325}\ Al_{0.675}]\ O_{10}\ (OH)_2$
I/S R=1	$Ca_{0.015}\ K_{0.775}\ (Al_{1.735}\ Fe^{3+}_{0.13}\ Mg_{0.1585})\ [Si_{3.265}\ Al_{0.735}]\ O_{10}\ (OH)_2$
I/S MLM (<5% sm)	$Ca_{0.01}\ K_{0.84}\ (Al_{1.765}\ Fe^{3+}_{0.165}\ Mg_{0.075})\ [Si_{3.205}\ Al_{0.795}]\ O_{10}\ (OH)_2$

is probably a solid state transformation. The adsorption of K^+ ions brings about the irreversible collapse of high-charge layers. Subsequently, the increase in the illite content of I/S MLMs with depth accompanied by size and morphology changes of the crystallites can only be due to a crystal growth phenomenon through the addition of illite layers over I/S particles that is similar to the one described in diagenesis. Inoue and Kitagawa (1994) have shown that, in geothermal systems, the growth of I/S minerals and illites produces spiral steps originating from the emergence of a dislocation on the (001) faces.

The Randomly Ordered I/S MLM Problem in Geothermal Series The presence of 100% expandable smectite at elevated temperature conditions in hot fluid reservoirs has been shown by Inoue et al. (1991) in different Japanese geothermal fields. Most often, smectite coexists with ordered I/S MLM and occasionally with illite. These authors considered that the complete I/S sequence observed in the low-temperature areas represented a fossil smectite-to-illite conversion stage. However, they described the presence of randomly ordered I/S MLM in zones far from the active reservoirs. At these locations, the geothermal gradient is rather constant in the first kilometer, while it is strongly affected in active zones at the hot fluid reservoir levels.

It appears that the randomly ordered I/S MLMs are formed under two different conditions in geothermal fields:

- in zones where the geothermal gradient is controlled by heat diffusion. This is the case in most of the geothermal fields which are located in andesitic-dacitic volcanic areas. In Japan, the chemical compositions of the complete I/S MLM series including R=0 I/S show that the smectite component is probably a montmorillonite (Meunier and Velde 1989);
- in the clay cap above active zones. This is the case at Chipilapa where montmorillonite; R=0 I/S MLM and beidellite coexist in the first 200 m.

The presence of randomly ordered I/S MLMs appears to be strongly related to the montmorillonite structure of the smectite component either in geothermal fields or in diagenetic series.

4.1.1.3
Summary

Although the presence of smectite (highly smectitic disordered I/S MLM) poses some problems of mineral stability, its presence can be used to identify zones of potential high-energy fluids. In this type of prospecting, one looks at the general sequence of I/S MLM minerals as a function of depth in a well, and when a strong anomaly in smectite content occurs, one can assume that there is a high probability of finding a conducting vein. The more rapid the change in illite content of the I/S MLM minerals in a drill hole sequence, the higher the thermal gradient and the higher the potential of finding a "hot" zone of geothermal fluids.

4.1.2 Clays and Petroleum

Work on clay minerals in sediments and sedimentary rocks has been sustained since the 1960s by hopes of using information on clay change to indicate the thermal histories of sediments in sedimentary basins. It was assumed that one could link the dynamics of clay change to the dynamics of change in organic matter. Since clays are more common than is petroleum, or at least significant deposits of organic matter useful for thermal determinations, such studies were financed by petroleum companies for some time. Relations of clay change are usually determined by the state of transformation of smectite to illite in pelitic rocks (shales). Chap. 3 discusses the dynamics of this reaction. In the following section, we would like to look at some uses of such clay mineral change in relation to determining geologic conditions in the past and comparison with organic matter maturation.

In this section we will look at the change in illite content of the I/S minerals formed by the so-called smectite-to-illite reaction. It must be kept in mind that this reaction is in fact a smectite-to-PCI reaction transformation and hence does not produce WCI or true illite. It describes the addition of illite layers to I/S minerals over a period of time at various temperatures.

4.1.2.1 Smectite-to-Illite Transformation and Geothermal Gradients

Clay mineral change under conditions of increasing temperature during the formation of sedimentary basins is the basis of the usage of clays in the search for petroleum. These studies are founded upon the dynamics of smectite-to-illite transformations. In Chap. 3 the different types of kinetic formulations and different approaches to the interpretation of clay mineral determinations under the dynamics of basin sedimentation are described. If one uses the formulations of change as a function of time and temperature, one should be able to simulate the observed data for a sedimentation series, and with sufficient stratigraphic information (time determinations), there should be a possibility to predict the temperatures at which the clays evolved. This is the theory. However, the dynamics of clay change are slow, occurring over thousands of years at the upper limit of clay stability and over hundreds of millions of years under conditions of shallow basin burial. If the sedimentary regime occurs over temperatures of normal geothermal gradients for sedimentary basins (20–30 °C/km), time is a very important factor, and tens to millions of years are required for smectite-to-illite transformation. However, it is probably unlikely that a sedimentary basin will maintain a constant thermal gradient over such long periods of time. Hence extrapolations over long periods are hazardous. For this reason, clay mineral change is best assessed on a comparative scale. Initial questions of thermal regime can be answered in a comparative manner, from point to point in a basin where the sediments have experienced the same, or at least very similar, burial histories. Several examples of this approach are given below. These studies focus on the formation of illite layers in I/S minerals, considering the end-point of the reaction is PCI.

4.1.2.2
Comparison of Several Wells from the Texas Gulf Coast

Initially, one can observe the type of data usually gathered in studies of basin history. Data from wells drilled in the Texas Gulf of Mexico region are presented in Fig. 4.5. The first two wells contain sediments of slightly different ages (40–157 and 40–113 Ma) while the present-day geothermal gradients are similar (Fig. 4.5a). The overall burial history is the same, where recent erosion has truncated the upper part of the series at 40 Ma. The well with the oldest sedimentary history is further advanced in the reaction progress of the smectite-to-illite transition. The curves of clay composition (illite content of the I/S phases) and depth are similar in shape. There is a difference in the reaction progress due to a longer residence time in the basin. The depth range of the wells is similar and hence the temperatures experienced by the materials are comparable. The factor of time in the reaction kinetics is the principal influence on the differences in reaction progress in the smectite-to-illite conversion. In the second example (Fig. 4.5b), the sediments of the wells are of nearly the same ages but the geothermal gradient (actual) is rather different. The well with the higher present-day gradient shows a greater progress in the smectite-to-illite reaction, as would be expected. In this example, temperature is the major variable in reaction progress with burial history being almost the same in the two wells.

A more complex situation is shown in Fig. 4.5c where two wells are compared, one inland from the Gulf of Mexico and the other 150 km to the south on the

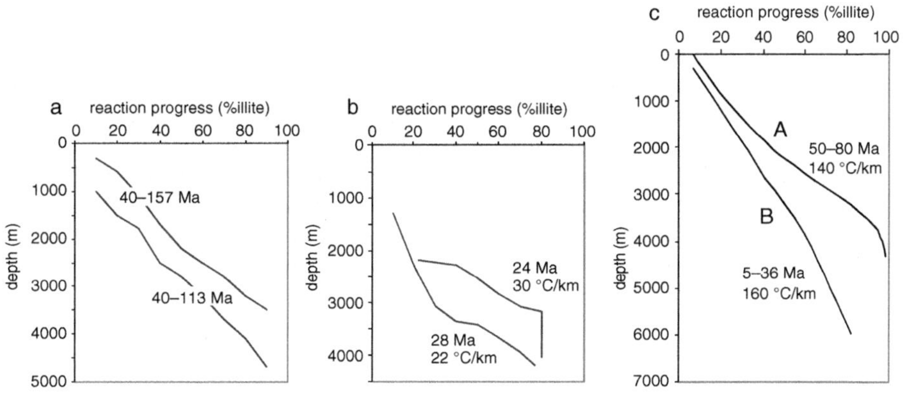

Fig. 4.5a–c. Depth–reaction progress as illite content in I/S minerals. **a** Data for wells from the gulf of Mexico (data from Velde and Vasseur 1992). **b** Reaction progress data for depth in wells reported by Schmidt (1973) and Perry and Hower (1979). **c** Depth–reaction progress data for two wells from the Gulf Coast area (data from Velde and Vasseur 1992). A: offshore well; B: onshore well

edge of the Gulf. The growth and development of the Gulf Coast sedimentary basin is one of subsidence and uplift with zones further offshore subsiding to receive sediments. A series of block-faulted units play one against the other as the active sedimentation moves further into the present-day Gulf of Mexico from the continent. In the example presented in Fig. 4.5c the B well in older sediments (onshore) indicates an older tectonic block while the A well with younger sediments is located in a more recent subsided block.

The A well with younger sediments shows a less mature smectite-to-illite sequence than the B well with older sediments. Thus the maturity relations are as expected in that more time has elapsed during the burial history of the inland B well than in the offshore A well. Thus one would expect a greater transformation of the clays in the well with the older sediments. The estimation of the differences in reaction progress must take into account the entire geologic history of the materials shown by the two wells. In the well with older sediments one finds that it outcrops with rocks of 40 Ma age and hence has experienced deeper overall burial than shown today with subsequent erosion occurring at the top of the series, about 3 km of sediments. Therefore, the lower sediments in the well would have experienced higher temperatures in the past than those prevailing today due to deeper burial. Subsequent exhumation would have lowered the temperature of the deeper sediments as there was movement towards the surface. Further, in sediments in the onshore well there was a long period of deeper burial for most of the sequence. These factors also explain the difference in reaction progress. It is also possible that variations in the thermal gradient occurred during these tectonic events of burial and uplift for the onshore situation. In the example considered in Fig. 4.5c it is very difficult to compare the two wells without making some large assumptions of factors such as rate of erosion, total burial depth, etc. In this case, the use of clay mineral change as a method of indicating paleohistory is limited.

In general, then, it is most safe to compare sedimentary sequences which have experienced the same burial history, or nearly the same history. This means that the burial history (depth and time) events must be the same and comparable. It is also safer to use several wells compared one to the other in order to estimate relative differences. In this way one can attempt to estimate thermal history or the importance of erosion episodes. Then it is possible to use the kinetics of clay transformation (see Chap. 3) to construct models of burial history as they affect the clays. By comparing the calculated effects on clay transformation with the observed ones, one can form an iterative approach to reconstructing the burial and thermal history of a given series of sediments or perhaps a sedimentary basin.

4.1.2.3
Paris Basin Wells

The Paris Basin is often taken as a model of a simple sedimentation sequence in a stable tectonic setting. This is probably true when compared to other basins, but there are some characteristics of the Paris Basin which modify this simple concept. It is true that the basin is rather regular, somewhat circular in present shape, and the Tertiary history is clearly visible. However the pre-Tertiary history is somewhat, though not terribly, more complex. One can use clay mineral transformations in order to estimate the complexity of the geologic history of segments of the basin.

The information reported here is for the most part unpublished (Velde, Lanson) excepting clay mineral determinations for several wells given in Velde and Vasseur (1992). Clay mineralogy in several deep wells was investigated as well that of outcrop samples found on the edges of the basin. This data is the basis for the study described below. The overall objective is to see whether one can compare data acquired from outcrop samples with those from a deep well. In doing so, one compares different parts of the sedimentary basin.

Figure 4.6a shows the geographic extent of the Paris Basin considered here and the sampling sites. A synthetic cross section is given for reference. In the basin two major events are clearly visible, one is an uplift and tilting of the basin in the early Tertiary followed by erosion and deposition of several hundreds of meters of sediment in the early to mid-Tertiary. A second, older tectonic event occurred in the mid-Cretaceous, when tilting occurred and erosion of the early Cretaceous layers occurred differentially from east (little apparent erosion) to the west where more erosion occurred. Using stratigraphic data (Fig. 4.6b) one can deduce that in the southwestern areas studied in the Aquitaine area erosion of early Tertiary is about 156 m, for the Seuil du Poitou 226 m. At the south central site, Bar sur Aube, only 94 m were removed. In the center of the basin (deep wells S,V) no perceptible erosion occurred.

One can use the central deep wells as a reference and attempt to see whether the edges of the present basin have experienced the same burial conditions, specifically thermal gradients. It is clear that the erosion is a factor but the differences in thickness of the layers or their depths after erosion is not very great, 200 m or less. Hence one would expect that the effect of time would be more important than burial depth. If the burial history is similar for different sequences then one can consider the importance of overall thermal gradients pertaining to the burial history of the different parts of the basin. The temperature will be the major factor in determining the state of reaction progress in the smectite-to-illite transformation.

The reference for this comparative study is the reaction progress of clays in two seep wells roughly in the center (greatest depth) of the Paris Basin found at Melun, near Paris itself. These are wells V-S, figured as a continuous line on the graphs of Fig. 4.7. Clay compositions from three other wells are used to compare evolution with depth, one from a site between Orléans and Tours (well Sg) one from near Bar le Duc (well Q) and one from near Troyes (D). These wells cover the Tertiary, Cretaceous, Jurassic and end in the Triassic. Outcrop samples were taken from rocks outcropping in the Cretaceous-Jurassic sequences between Verdun and Sedan (SM) in the east of the Paris Basin, in the region of Bar sur Aube (B/S) in the southeast and the region just north of Poitiers (SP). Also, samples were taken just to the southwest of the edge of the Paris Basin on the northern part of the Aquitain Basin (Aq) in the Saintes-Angouleme region. These sampling sites form a band going from east to southwest in the Paris Basin.

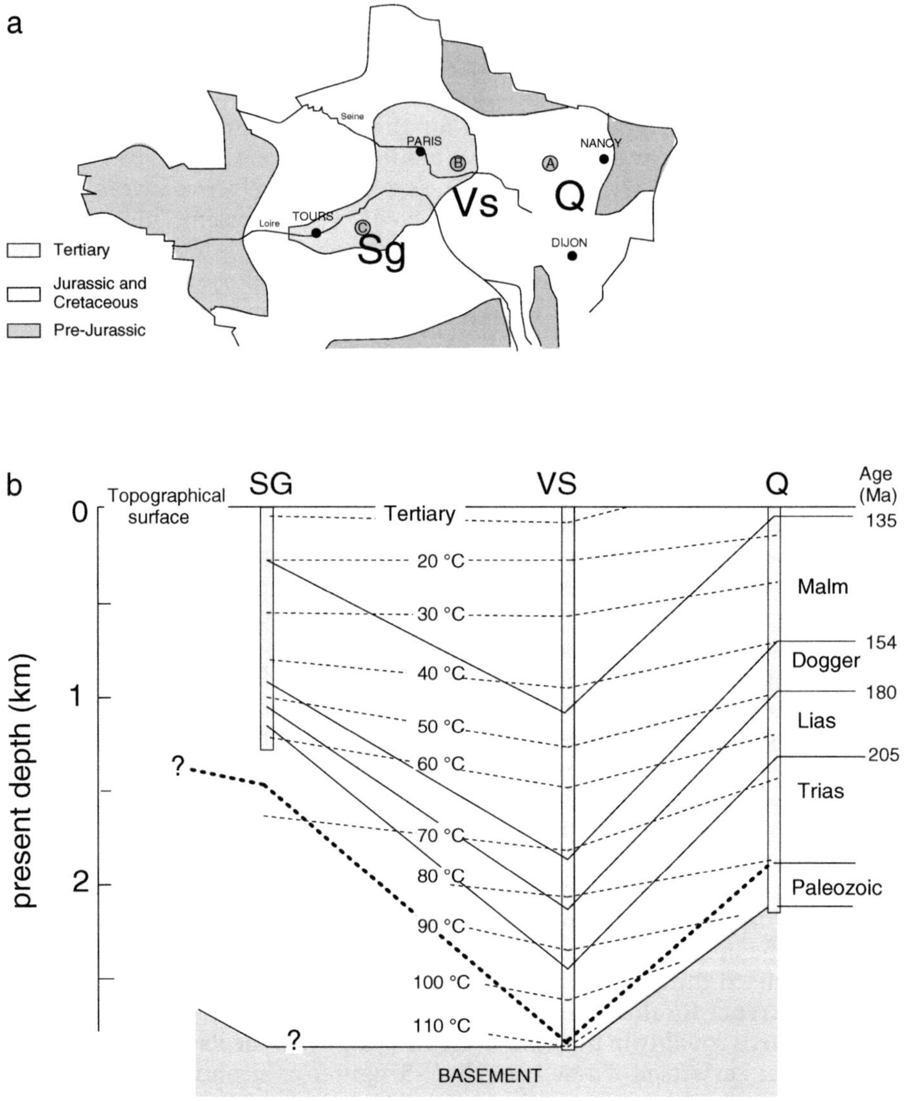

Fig. 4.6a,b. The Paris Basin. **a** General outlines of the Paris Basin outcrop extent in France with positions of deep wells indicated (Q, VS, Sg) and **b** Schematic stratigraphic section from east to west in the Paris Basin

Figure 4.8 shows the clay mineral determinations indicated by the range of values which were observed for the several samples investigated which ranged from 3 to 12 per site in the case of outcrop samples. Of the seven sites compared to the V-S well clay reaction progress determinations, two show very similar advancement states, that at the extreme western edge of the Paris Basin itself

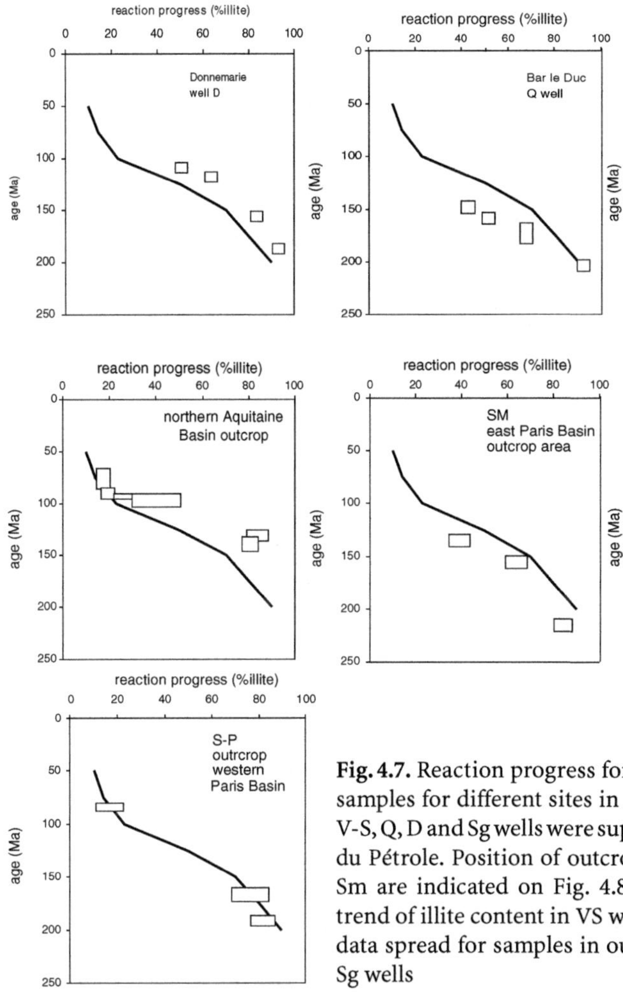

Fig. 4.7. Reaction progress for different wells and outcrop samples for different sites in the Paris Basin. Samples for V-S, Q, D and Sg wells were supplied by the Institut français du Pétrole. Position of outcrop samples, Aq, Sp, B/A and Sm are indicated on Fig. 4.8. The *heavy line* shows the trend of illite content in VS wells, and the *rectangles* show data spread for samples in outcrop samples and the Q,D, Sg wells

(SP) and the well near Tours to the south and west of the site of the reference wells. If the general assumptions are correct (little effect due to the erosion in the Cretaceous and similar histories during the Tertiary uplift and erosion event), one can assume that these three sites experienced similar geothermal gradients throughout most of their burial history. The Tertiary erosion event seems to have little affect on the reaction progress or the Cretaceous erosion event in that this latter episode more strongly affected the Cretaceous sediments in the western part of the Paris Basin. In going to the south of the western edge of the basin, into the Aquitinian Province, one sees that the clays show more reaction progress than those of the V-S reference well. Hence, one can suspect a higher overall geothermal gradient, or at least a higher accumulation of thermal energy over the time of burial.

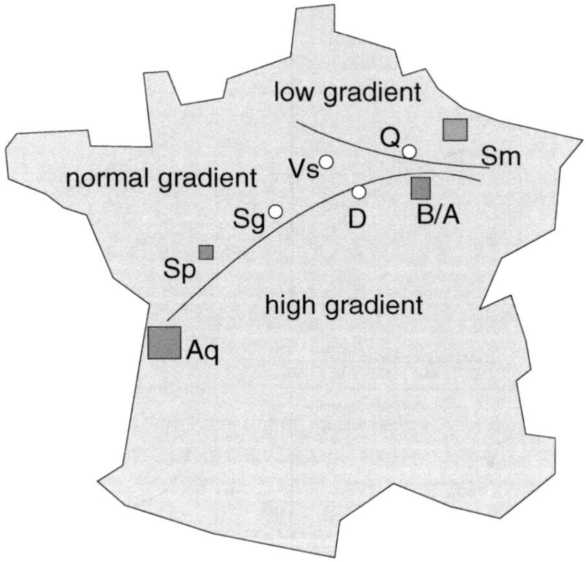

Fig. 4.8. comparison of overall thermal regimes for samples at different points in the Paris Basin. Comparison is made with the V-S well found in the deepest portion of the basin, near Melun. Normal gradient indicates clay compositions similar to the VS deep wells, high and low gradient indicate wells showing greater change in illite content of I/S minerals than the VS wells or less change in illite content as a function of depth. This estimation is only valid for the sedimentary materials in the Paris Basin outlined in Fig. 4.6

If we go to the east of the reference well site, well Q and to the edge of the basin (site SM), the graphs indicate a lower rate of reaction progress than in the reference wells, the % illite is lower for a given age of deposition. However, at the two sites investigated to the south of these (well D and B/A outcrops) the reaction progress is greater for a given age of sedimentation than that of the reference wells S-V. One can establish a sort of crude map of overall paleogradients in these portions of the Paris Basin as shown in Fig. 4.8, considering either higher gradient, same gradient or lower gradient. In general, these data suggest a higher overall paleogradient from Triassic to Tertiary in the southern part of the basin and a low gradient in the northeastern part of the Paris Basin.

It is possible to establish the differences in thermal gradient numerically, utilizing some general assumptions. For this estimation, we use a program developed by G Vasseur (CNRS, Paris) based upon the derived kinetic values from the study in Velde and Vasseur (1992). Simulations assume a 600 m erosion during the Tertiary uplift, and that the apparent hiatus of lower Cretaceous material is that eroded during that uplift event. If we assume constant thermal gradients for the entire length of burial history, the lowest gradient (least advanced smectite-to-illite reaction) indicates an overall gradient of 22 °C km^{-1} (SM section) and the highest near 50 °C/km (B/A). These relatively small differences in reaction progress in the different wells are possible because the entire sections are rather shallow, the deepest well being less than 2 km at maximum depth (Triassic age). Thus even though the differences in geothermal gradient are significant, the shallow burial gives rather small differences in absolute temperature for a given stratigraphic horizon.

4.1.2.4
Local Changes in Thermal Regime

Another use of clay reaction progress in the smectite-to-illite (PCI) reaction transformation is one of following local changes in thermal gradient, e.g. of the site in near the Sg well between Orléans and Tours in central France, in the Paris Basin. Mathieu and Velde (1989) have determined the clay composition (reaction progress in the smectite-to-illite transformation) in two wells found near to the Sg well cited in the preceding section. The Sg well has experienced a "normal" overall geothermal gradient of near 32 °C km^{-1} similar to that of the basin center S-V well according to the stratigraphic information in each well. However, two wells only several kilometers from this reference well for the area show strong alteration in clay composition compared to the reference for the area. Clay composition alteration changes rapidly in the upper portions of the wells at less than 500 m depth (see Fig. 4.9). This is strange in that the stratigraphy indicates little difference from the reference well. Such change indicates strong thermal variations, difficult to imagine on a regional scale over such distances. A closer investigation of the geology showed that the two wells traversed major fault systems in the lower two thirds of their 1.5 km depth length.

If we take the Sg well reaction progress (% illite) line as representing a normal thermal history, one can plot same data for the two adjacent wells, C1 and C2, in the same depth-reaction progress diagram. If we look at the distance of the analysis points from major faults in the last two wells, it is apparent that there is a relationship of reaction progress and distance from faults. The obvious conclusion is that there is a relationship between the faults and the reaction progress which cannot be explained by stratigraphic hiatii but by thermal anomalies. Vertical, distensive faults are obvious candidates for transfer of hot

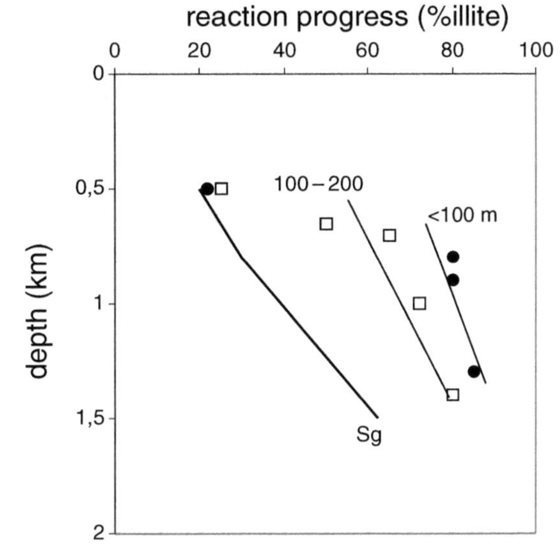

Fig. 4.9. Reaction progress vs depth in three wells situated several hundred metres one from the other in the Paris Basin (data from Mathieu and Velde 1989). Numbers indicate the distance of the samples (in metres) from a major fault as a function of depth

deeper fluids towards upper portions of a geological structure. It appears that the clays in the sediments of the wells near faults in this part of the Paris Basin have experienced significant periods of higher, local temperatures than the normal basin geothermal gradients would have provided.

Here again, thermal effects can be detected in clay assemblages using a comparative method to establish normal and aberrant relations. One must use sediments which have experienced the same burial histories and major tectonic events in order to get a true picture of the effects of the geological variables of temperature and time as they affect the clay mineralogy of sediments experiencing burial diagenesis.

Summary It is clear from the studies discussed above that clay mineral reactions can only be used to determine geologic thermal history in a comparative manner. Since the thermal history of a given segment of a geologic structure can be variable in its history from time to time, one can only observe an overall, integrated, thermal history in studying clay mineral compositions. The clays register the total thermal input in a given rock, and the reaction progress in the smectite-to-illite reaction is then a summation of this input. However, if the comparison is made for geologic units having the same stratigraphic successions, one can determine the differences in relative thermal input from one point to the other and determine a general pattern of thermal history which can lead to usage in petroleum prospection.

4.1.2.5
Clay Mineral Transformation and Transformation of Organic Matter

General Statements The initial premise of this section is that the alteration of clays and that of organic matter in sediments could be correlated. Since both are kinetically controlled, by variables of time and temperature, one should be able to find a method to compare one to the other. As we can deduce from the description of clay mineral genesis and the formation of illite in sedimentary rocks, lithology can strongly influence the type of I/S mineral present. Sandstones can produce illite as a precipitation product under conditions of high potassium activity while shales will be controlled by closed system mineral transformation via dissolution-precipitation mechanisms. Velde and Nicot (1985), Xinhua et al. (1996) and Clauer et al. (1999) have observed such differences in core material in deep wells. Thus, the usage of clay mineral determinations should be accompanied by a strict control of lithologies.

It appears that the same types of problems exist for the determination of the evolution of organic matter. It is well known that three major types of organic matter are found in sediments, types I, II and III where type III is related to terrestrial material with a strong proportion of lignitic components. These materials are the most reliable in determining organic matter evolution (Tissot and Welt 1978; Espitalie 1986). Unfortunately type III materials are less likely to produce abundant petroleum than the others which have an origin in marine environments. Type I and type II organic matter gives little or poor reflectance data (vitrinite determinations) and scant information as to thermal maturity using pyrolysis (Tmax) determinations. Thus the best indicator of thermal history is the less likely to indicate good economic deposits. Further, Carr (2000) indicates different conditions of burial and chemical environment which can affect the evolution of the same organic matter. However, when sufficient reliable data is available one can determine paleothermal gradients of different ages with great success (see Corcoran and Clayton 1999, for example).

It is clear that both methods of determining the burial and thermal history of sedimentary rocks, clay minerals and organic matter, have their drawbacks. It is necessary, as usual, to use multiple observations to determine such events with any precision.

Diagenesis Much work has been performed on problems of mineral change and organic matter maturation during burial diagenesis. This is the realm of petroleum geology. Waples (1980) developed a general method (time-temperature index or TTI) to estimate the conditions under which organic matter in sediments could generate petroleum. Bruce (1984) used this method, combined with estimation of illite content of I/S minerals, to try to determine the correlations between these two sets of information, the TTI and clay mineral transformation estimated by the change from low illite content minerals (near 20%) to high illite content minerals (near 80% illite). Bruce noted a difference in the temperature ranges where clays change depending upon the geologic setting. His observations indicate that the major change in clay I/S composition can occur before the TTI of 15 is reached, conditions necessary for petroleum generation or it can occur after the TTI is reached. Hence there is a clear difference between clay kinetics and those of organic mater transformation which releases oil or gas.

Figure 4.10 presents some data to this aspect, showing the change in maturity of organic matter (measured by the pyrolysis method designated as Tmax, Espitalier 1986) and reaction progress in the smectite-to-illite transformation. Data for wells of Tertiary, Mesozoic and Paleozoic age (unpublished data of the author and Xinhau et al. 1998) are plotted, which indicate that the older the well, the further the illite reaction progresses while organic maturity increases. This shows the effect of time on clay change is much less important for changes of organic matter. Relations of illite content of I/S minerals compared to organic maturity measured by Tmax are shown in Fig. 4.11. Here the range of maturity giving rise to oil and gas is described compared to two stages of clay maturity, 90 and 60% illite content, for wells of different ages having experienced different overall thermal gradients.

In fact, the pyrolysis value of Tmax does not correspond linearly to the evolution of kerogen or oil-producing organic matter in sedimentary rocks. Figure 4.12 indicates the reaction progress of organic transformation and the end of clay mineral evolution (90% illite). It is quite clear that the two materials have very different reaction kinetic functions and they show as a result a curved relation when plotted one against the other. It seems evident from these last diagrams that the direct correlation of clay transformation to organic matter maturation is not possible. The kinetic factors are far too different for such an operation. One can combine the kinetic formulation of Velde and Vasseur

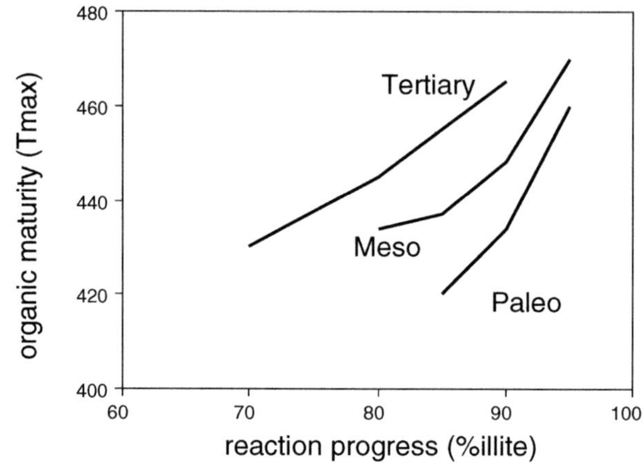

Fig. 4.10. Data for three wells for organic maturity (Tmax pyrolysis index) and illite content in I/S minerals. Tertiary example based on data from Xinhau et al. (1996), Mesozoic and Paleozoic examples based on data from Institut français du Pétrole and unpublished clay mineral determinations (author BV) and Illinois State Geological Survey data and unpublished clay data of the author (BV)

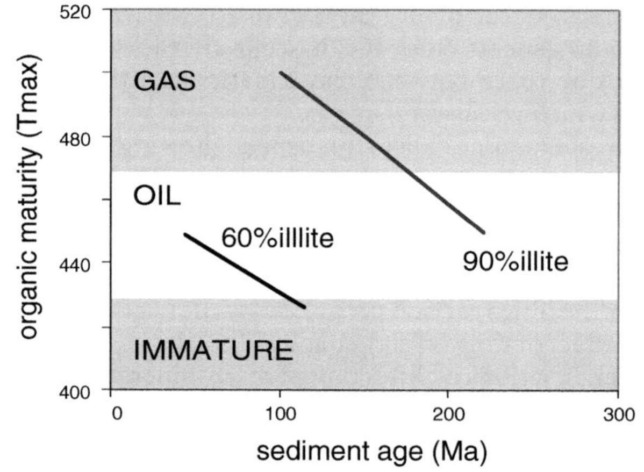

Fig. 4.11. Organic maturity measured as Tmax pyrolysis values compared to illite content in I/S minerals for diagenetic sediments having experienced different thermal gradients. Zones of oil and gas development (organic maturity, Tmax measurements) are presented. Data from Velde and Espitalié (1989)

(1992) with that proposed by Sweeney and Burnham (1990), for example, to compare the two, having data for one or the other of the natural materials found in sedimentary rocks. The stage of organic matter maturation is much more dependent on temperature and can be compared, more or less directly

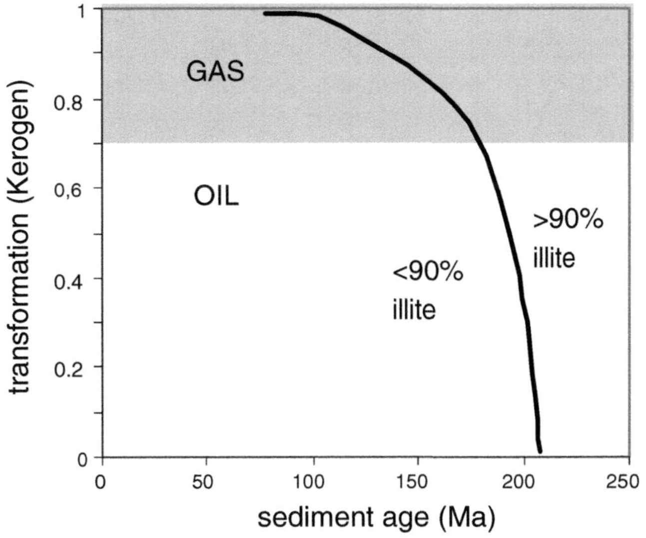

Fig. 4.12. Organic reaction progress (kerogen transformation) plotted against the approximate end-point of clay smectite-to-illite transformation (90% illite). Based on data from Velde and Espitalié (1989)

to temperature while the evolution of clays is much more dependent on time than temperature, at least under conditions common in burial diagenesis.

It should be noted that in both cases, with clay and organic matter, the initial observations are extremely important. Only the same shale lithologies can be used to compare clay mineral changes in the smectite-to-illite transformation. In the case of organic matter, only the same type of organic mater, I, II or III can be compared, as their major constituents respond differently to thermal increase. These precautions are all too often ignored, giving rise to misleading information concerning the interpretation of past thermal histories under conditions of burial in sedimentary basins.

4.1.2.6
Thermal Pulses

The discussion above is confined to thermal regimes of moderate geothermal gradients, 20–60 °C/km. Under such conditions, clays respond in a relatively regular manner for the smectite-to-illite transformation in shales. However, under the conditions of a high thermal regime, it appears that clays do not always respond as quickly as does the organic matter in shales. Velde and Lanson (1993) have explored this possibility by observing a sequence of Permian sediments which has experienced a thermal pulse due to a volcanic intrusion at depth. Since the thermal regime is high, initial comparison can be made between clay-organic matter evolved under conditions of geothermal alteration, regimes of more than 200 °C/km gradients. Agreement between the clay

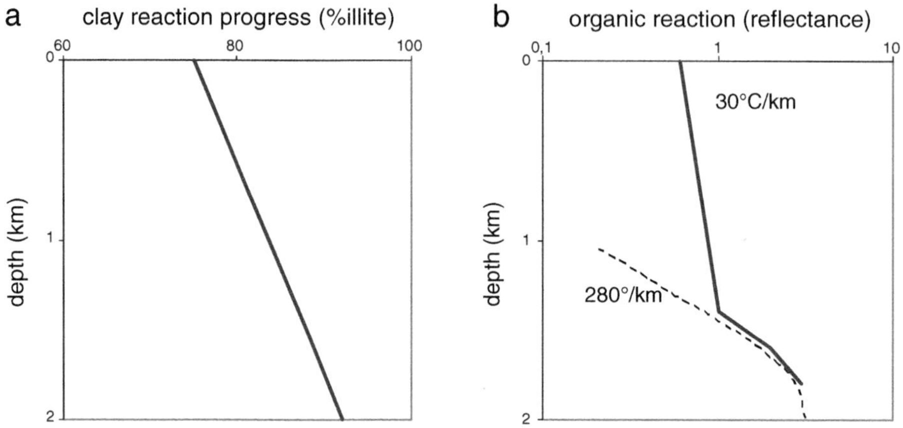

Fig. 4.13a,b. Thermal pulses. **a** Depth-reaction progress relations for clays in Permian sediments of eastern Paris Basin. **b** Depth organic matter composition (vitrinite reflectance) relations in Well Chv showing the strong shift in organic matter evolution in mid-well. *Dashed line* compares observed change in vitrinite reflectance for a Salton Sea well with a geothermal gradient near 280 °C/km

and organic matter transformations observed for Salton Sea (USA) sediments establish a relationship for the clay minerals. Clay composition data (reaction progress measured by illite content in the I/S minerals) is shown for a well from the eastern Paris Basin in Permian sediments. The values of depth and sediment age are shown in the diagram Fig. 4.13. The rate of change for the clays indicates a thermal gradient of about 30 °C/km, compared to other sediments in the Paris Basin.

In this example, there is a strong and abrupt change in the vitrinite reflectance (reaction progress of organic matter) at 1,500 m depth while the clay compositions continue to change regularly, as indicated by the line in Fig. 4.13a. Here the organic matter is much more sensitive to temperature change, and the clays show a slow reaction, due to the strong time factor in their kinetics of transformation. The thermal pulse responsible for this change in organic matter reflectance is on the order of 280 °C/km, if one can compare this data to that for the Salton Sea. It is clear that the time factor in clay kinetics is such that clays will not react, in closed systems of shales, to short thermal events. This leads to a sort of blindness of clays to short thermal pulses which can strongly affect organic matter, while such events go essentially unrecorded by the clays.

4.1.3 Illite Crystallinity and Organic Matter

As mentioned in Sect. 2.2.2, one of the initial hopes for the identification of illite (clay mineral) evolution was a correlation with organic matter. It is now apparent that this is a complicated procedure in the case of diagenetic change, i.e. at temperatures below approximately 200 °C. Kisch (1987) and others have attempted to correlate the change in organic matter with that of the evolution of illite peak width. If one can use the data for diagenesis as a measure of the reactions at higher temperature, this task seems to be very difficult. The evolution of clays and organic matter is such that there is not much correlation. It is true that, as clays approach their ultimate evolution (WCI > PCI), the evolution of organic matter will be very great and both will show relatively small variations. However, careful measurements should show strong differences depending upon the time span of the thermal or metamorphic events. If one considers that orogenic processes take up to a million years to be completed and that temperature maxima are maintained for a reasonable portion of these periods, it is possible that the long periods of time necessary and the high absolute temperatures at which the processes occur may give some coherent results in correlations. However, much of this might merely be due to the relatively small change in illite (WCI/PCI) contents which affect the peak widths in the ultimate stages of clay change (above diagenesis where there is an obvious existence of I/S minerals, < 95% illite). Since the organic matter responds rapidly to temperature, the variation in illite crystallinity index compared to organic maturity will be due to differences in the period over which the heating or tectonic event lasted which will produce more or less of WCI material and hence true illite.

Sandstones If the above-mentioned indications concerning the use of illite or I/S minerals in petroleum prospecting are deceiving, there is a glimmer of hope in the use of illite precipitation in sandstones as a guide for petroleum geologists. As mentioned in Sect. 2.2, direct precipitation of illite in mature, clean sandstone reservoirs is relatively common. The replacement of kaolinite by illite or the direct precipitation of illite indicates fluid flow where the chemical potential of the fluids is in disequilibrium with the reservoir sandstone. The result of illite crystallization is often to block fluid movement by decreasing permeability. This is of course not favourable for the petroleum production. However, the existence of secondary illite does indicate aqueous fluid flow. Such movement is fundamental to the accumulation of petroleum in reservoirs. Thus the appearance of illite means that the layer containing them is poor for extracting petroleum but adjacent layers could be promising as reservoirs. Hence illite (WCI) could be used to indicate fluid flow and petroleum potential. Its presence can also explain production problems in certain reservoirs and give information as to the value of such accumulations.

In these cases the chemical precipitation of illite, due to strong potassium concentrations in solution, can provoke key geological phenomena indicating the accumulation and exploitation of petroleum.

Summary The use of clay mineral evolution (reaction progress in the smectite-to-illite, PCI, transformation) can be useful but its determination must be subject to caution. First and foremost, the clay environment must be stable. The best examples are in shales, in which the system is largely closed to chemical migration and the reaction can proceed solely according to time-temperature influences. Sandstone or bentonites will show the influence of chemical potential which can strongly affect the rate of change of the clays.

Estimates of thermal history can be made in sedimentary basins, but it is best to compare depth sequences as to their clay transformation. If the stratigraphic history is similar, differences in thermal history can be ascertained. It will be difficult to estimate the change in temperature in sequences which have had different burial histories, with for example periods of erosion. In such cases thermal change is often present, for the periods of uplift and faulting, for example.

Comparison of thermal histories of clays and predictions of organic maturation are difficult due to the slow reaction of clays to thermal input. The organic matter responds very rapidly while the clays take much longer time periods to change their maturation state. Thus it will be difficult to compare or predict organic maturation from clay mineral sequences in diagenetic situations. This is unfortunate but true.

The real use of illites in the petroleum industry is more likely to reside in the production side of the problem of extraction. Illites form obstructions to fluid flow when they form by precipitation in porous sandstones. However they can be used as indicators of fluid movement and hence signal the possibility of petroleum migration.

4.1.4
Ore Resources

The aim of this chapter is to give some specific examples of how illite can be used in prospecting rather than an exhaustive presentation of its occurrences in all types of ore deposits. We shall focus on two specific points which illustrate how illite may be used as an indicator for ore prospecting: 1) the presence of illite (sericite) typifies an alteration process (phyllic alteration) in porphyry copper deposits; 2) the polytype change in unconformity-type uranium ore deposits.

4.1.4.1
Porphyry Copper Deposits

The Geological Structure of Porphyry Copper Deposits Different hydrothermal alteration zones are classically described around the porphyry copper deposits (see Fig. 2.31a). Generally one finds a large, extensive volume of rocks altered to the prophylitic facies which define the porphyry alteration zone. Within this volume one finds potassic, phyllitic, and argillic zones (Creasey 1959). The deep roots of alteration zones as they have been observed in the porphyry copper deposits are the location of the origin of interactions of high-temperature (and mostly high-salinity) fluids and intrusive rocks with their immediate enclosing rocks. These interactions have been described as "potassic", "phyllic" and "argillic" alterations. They overprint a large pervasive alteration aureole: the propylitic alteration (Creasey 1959; Lowell and Guilbert 1970).

The typical mineral parageneses are the followings:

- propylitic alteration: chlorite + epidote + albite ± calcite,
- potassic alteration: K-feldspar + biotite + muscovite,
- phyllic alteration: sericite + quartz + pyrite,
- argillic: two facies are distinguished:
  1. the intermediate argillic facies: sericite + I/S MLM + kaolinite ± chlorite
  2. the advanced argillic facies: kaolinite/dickite + pyrophyllite.

The most typical secondary phyllosilicates produced in the phyllic and intermediate alteration zones are small-size white mica type minerals classically called "sericite". Meyer and Hemley (1967) stated that the chemical composition of these small sized white micas varies between that of muscovite or phengite and that of illite or hydromuscovite. More recently, Parry et al. (2002) redefined sericite as being typically illite. Phyllic and argillic alterations are superimposed on earlier propylitic and potassic alterations and are themselves followed by late low-temperature alterations and weathering. The alteration zones or facies can be described as follows:

Propylitic Alteration The propylitic alteration is pervasive and involves a large volume of rocks around the magmatic intrusion. The greater the amount of heat (i.e. the size of the magmatic body), the greater the rock volumes in which temperature is increased by conduction. Thermal models show that heat dissipation is a slow phenomenon, owing to the very low thermal conductivity of rocks (Jaeger 1968; Cathles 1977; Norton and Knight 1977). The fluids entering these rocks through fractures and microfissures (intergranular joints, intramineral fissures) trigger the dissolution of primary minerals and the precipitation of secondary ones. This alteration is similar to metamorphism, where the whole rock participates in a recrystallization of minerals in a new assemblage.

The mineral parageneses typical of the propylitic alteration have been described in the past (Creasey 1959; Lowell and Guilbert 1970; Titley et al. 1986 etc.). They are formed due to local equilibria (microsystems) and they retain the initial rock microstructure (Meunier et al. 1987). Mass transfers take place over short distances of the order of magnitude of the size of primary minerals. The main mineral reactions are interdependent in that one reaction provides the elements in solution needed by another:

$$\text{plagioclase} + K^+ \rightarrow \text{illite} + Al^{3+}$$

$$\text{biotite} + \text{amphibole} + Al^{3+} \rightarrow \text{chlorite} \pm \text{epidote} \pm \text{calcite} \pm \text{sphene} + K^+$$

The CO_2 partial pressure controls the formation of epidote or calcite. Quartz crystallises in microfractures, consuming the excess silica. The quantities of fluids involved are so low that the composition of secondary phases is essentially controlled by that of primary phases (Berger and Velde 1992). This mineral facies and alteration process strongly reminds one of low-grade metamorphism.

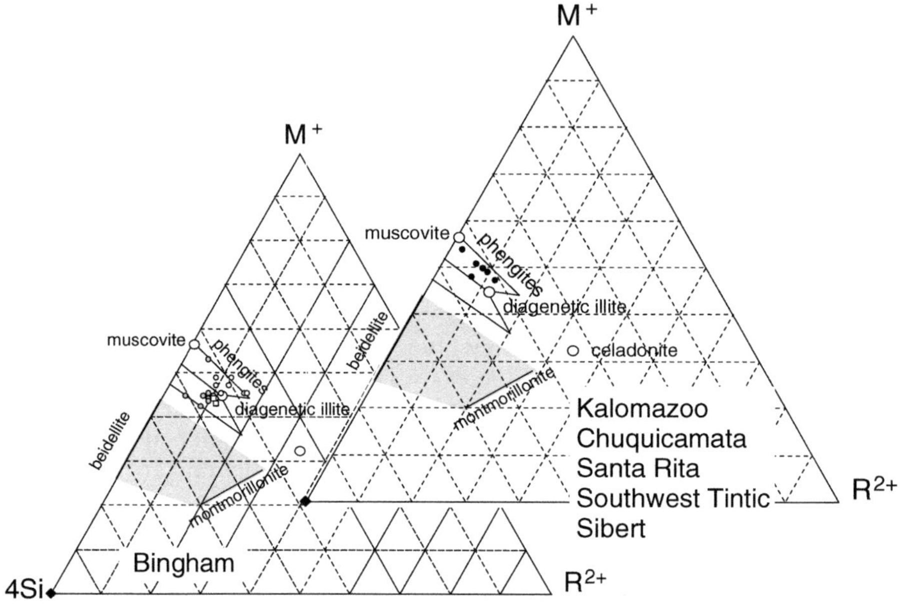

Fig. 4.14. Chemical composition of "sericites" from veins cross- cutting potassic alteration zones (*empty circles*), and a propylitic zone (*dots*). High potassium content (above the intermediate line from diagenetic illite), are probably mixtures of illite and muscovite-celadonite

Phyllic Alteration Phyllic and intermediate argillic alterations involve more narrow zones in the rocks concerned and often overprint the propylitic and potassic facies.The phyllic zone is generally mineralized (pyrite, chalcopyrite) and is composed of a dense vein network. Sericite is observed inside the fracture and adjacent intensely altered wall rocks. The outer zones, located away from the veins, are less altered and contain expandable minerals (smectite and I/S MLM). The chemical composition of sericite depends in part on the that of the rocks in which it is found. Indeed, Parry et al. (2002) showed that the K content is higher when veins crosscut potassic alteration zones compared to those found in propylitic zones (Fig. 4.14). In the case of an overprint in potassic alteration zones, the two polytypes 1M and $2M_1$ coexist; the second being dominant in coarser fractions.

Potassic Alteration This type of alteration shows a strong increase in potassium content in the alteration mineralogy. Potassium feldspar and sericite-mica are present. Potassic alteration appears to be more closely associated spatially with the center of the alteration volume, and directly associated with the intrusive magma which produces the alteration. This is the highest temperature facies of the alteration types.

From a large number of observations, it appears that the phyllic and argillic alteration are especially related to vein type alteration.

Vein Alteration The phyllic and intermediate argillic alterations are controlled by the opening of fractures in which hydrothermal fluids are injected. Their interactions with the surrounding rocks typically form veins sealed by minerals which have precipitated from the solutions and alteration halos composed of remains of primary and secondary species. Even reduced to a few microns in thickness, the halo is most often zoned (Lovering 1949; Bonorino 1959). Each zone is characterised by an assemblage or by a dominating mineral the proportions of which vary with distance from the vein (see Fig. 2.31). These two authors had established early on, that zonations resulted from chemical diffusion processes under isothermal conditions (the heat diffusion rate being higher than the chemical diffusion rate by several orders of magnitude). They considered that the width of the halo was dependent, among other factors, on the system temperature. The origin of zonation was subsequently reconsidered as resulting from crystallisation kinetics (Page and Wenk 1979). These authors based their conclusion on the observation of a sequence of phyllosilicates from the vein to the altered rock: $2M_1$ phengite, 1M then 1Md illite, illite/smectite mixed layers (I/S). Assuming isothermal conditions, they considered that I/S minerals are metastable precursors of illite and phengite.

However, thermal gradients at various scales have been confirmed using different techniques:

- 200 °C over 100 m in an amethyst vein (Horton 1985),
- 60 °C over 6 m in uranium veins (Al Shaara 1986),
- 75 °C over 11 mm in a phengite vein (Turpault et al. 1992b).

Thus, the isothermal models should be reconsidered, the question being: how can a thermal gradient be stabilized for a period of time long enough for the mineral reactions to proceed? This aspect has been investigated by Turpault et al. (1992a,b) in sericite veins crosscutting the La Peyratte granite (Deux Sèvres, France). A thermal gradient was described by observing secondary fluid inclusions in the wall rock quartz crystals. This thermal gradient was stabilised by a pulsed flow regime of hot fluids in the open fractures: the residence time of fluids in the fracture and the expulsion frequency are of the same order of magnitude as those observed in surface geysers. Additionally, the statistical analysis of veins whose width varies from 20 to 300 μm showed that the wall rock alteration obeys the two following processes:

1. formation of mineral reaction "fronts" bounding chlorite, albite or plagioclase dissolution (oligoclase) zones;
2. propagation of fronts such that the various zones do not vary independently (Fig. 4.15).

The propagation of alteration in wall rocks cannot be explained by the dissolution of primary minerals at the contact of fluids alone. Once the chemical equilibrium between hydrothermal solutions and the mineral has been reached, the alteration rate becomes infinitely low. Therefore, in the absence of a thermal gradient which maintains a chemical potential gradient for all

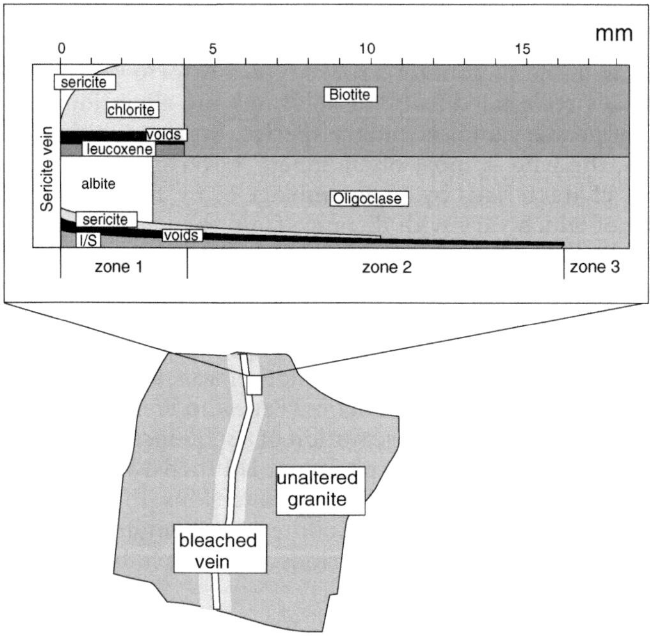

Fig. 4.15. Formation of mineral reaction "fronts" (albite, chlorite and plagioclase dissolution) through dissolution and crystallization processes in hydrothermal veins from the La Peyratte granite (from Turpault et al. 1992a)

elements, mass transfers are believed to be reduced to values that cannot be measured in hydrothermal veins.

The chemical balance of alteration is achieved as follows: 1) transfer of H_2O from the fracture to the wall rock, 2) transfer of Si^{4+}, Al^{3+}, K^+, Ca^{2+}, Na^+ cations from the rock to the fracture and 3) total consumption of cations in the mineral reactions with the exception of Si^{4+} and Ca^{2+}. The mobility of low-solubility cations such as Al^{3+} is maintained by the chemical potential difference of this element between the "source" site, i. e. dissolution front of plagioclases and the "consumption" sites (biotite→chlorite and oligoclase→albite reaction front). The intensity of the plagioclase dissolution is shown by the variation of the size of the dissolution holes (Fig. 4.16) which accounts for the progression of alteration fronts (Meunier 1995).

Alteration involves several mineral reactions which form a "trophic chain" (Fig. 4.17). The ultimate consumer of dissolved material is the phengite which forms in the open space of the fracture. This chain starts operating at the opening of the fracture and stops when one of the constituent reactions stop (Fig. 4.18). In the present case, the crystallisation of phengites ceases when the whole open space is occupied (Berger et al. 1992). The vein is definitively sealed by the precipitation of fluorite±pyrite. The bulk rock chemical balance (fracture+altered wall rocks) shows that only Si^{4+} and Ca^{2+} are exported in the

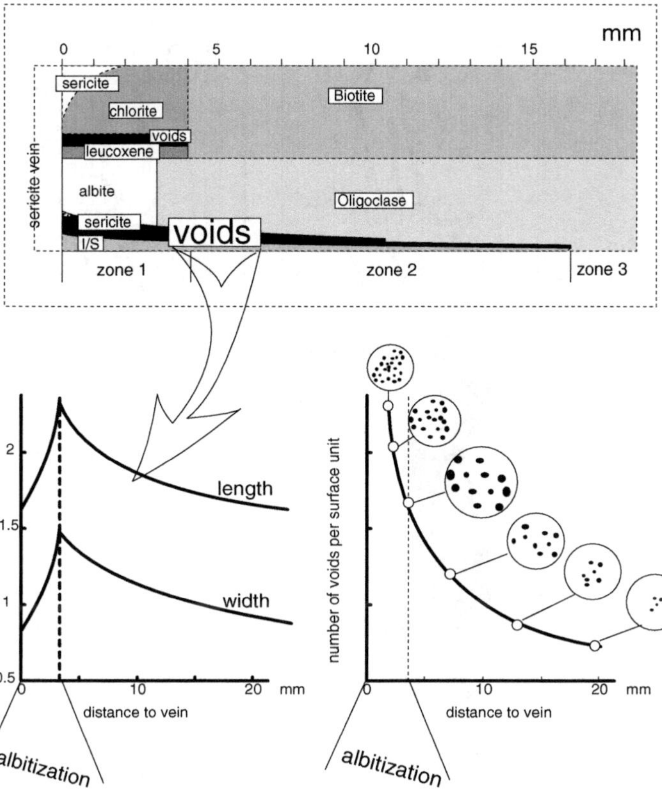

Fig. 4.16. Vein alteration mechanism (example of La Peyratte granite, Turpault et al. 1992a,b). The change in size and shape of oligoclase dissolution voids with distance from the vein

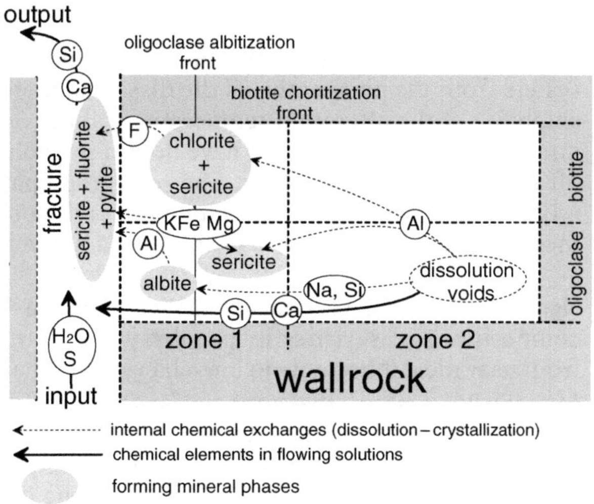

Fig. 4.17. The "trophic" mineral reaction chain which "feeds" the nucleation and growth of sericites

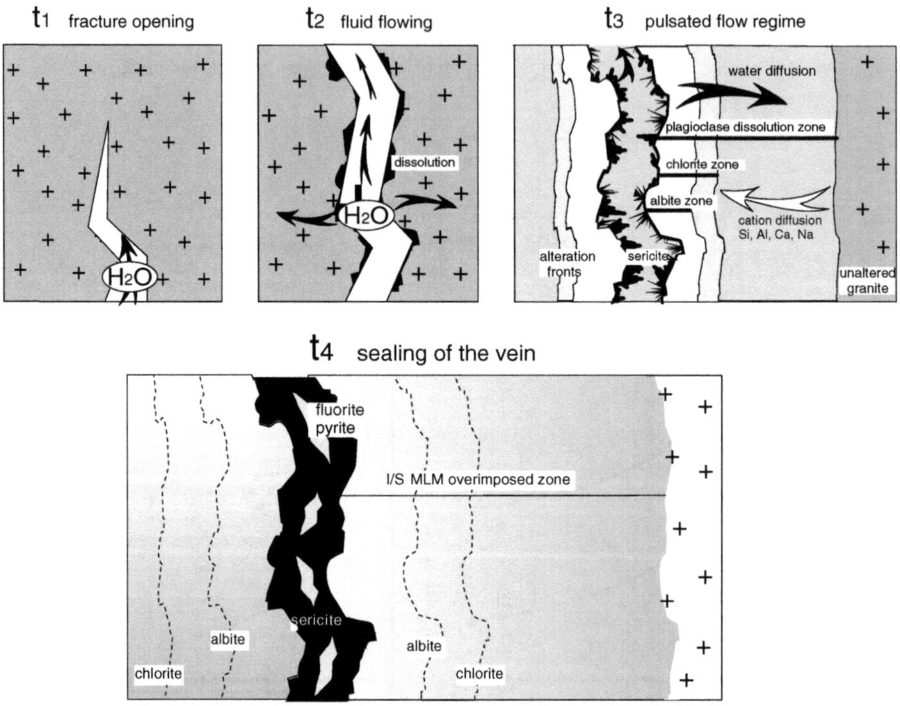

Fig. 4.18. Schematic representation of the hydrothermal processes in vein alteration. t_1: mechanical strains open a fracture that brings granite minerals and hydrothermal fluids into contact; t_2: dissolution of the fracture wall rocks; t_3: wall rock alteration maintained by the establishment of chemical potential gradients. The crystallisation of sericites in the fracture starts at this stage as the last link in the chain of mineral reactions; t_4: alteration propagation by displacement of the mineral reaction fronts

La Peyratte granite. Oddly, the alteration process took place in a nearly closed system from the perspective of the dissolved material. Considering the quartz solubility at the alteration temperature (about 300 °C), one can calculate the number of litres of fluid that have flowed per volume unit of fracture to drain off the silica lost by the rock. The retrograde path (cooling of the system) is recorded in the rock by the precipitation of illite/smectite mixed layers that are increasingly rich in smectite near the vein in the plagioclase dissolution holes.

Significance of Sericite Formation: Application to Ore Prospection The chemical compositions of sericites in porphyry copper deposits were shown to vary from near ideal muscovite to low-charge illite (Parry et al. 2002). Plotted in the M^+–4Si-R^{2+} system, the compositions are systematically under the mica line and thus belong to the illite domain (Fig. 4.14). The compositions of sericite, as determined by electron microprobe, are different when the phyllic alteration overprints a potassic or a propylitic zone. The former are mostly high-charge

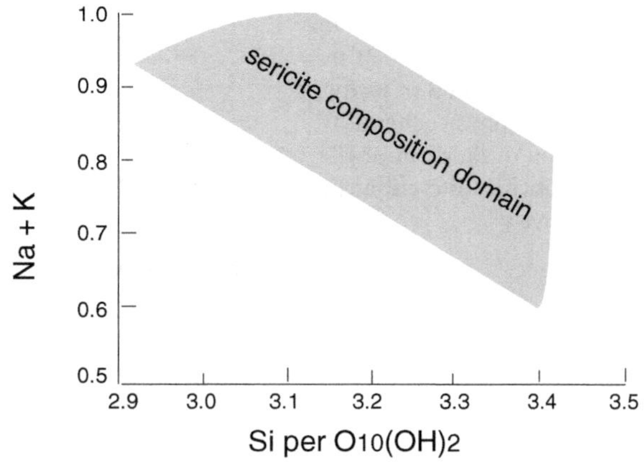

Fig. 4.19. Layer charge vs Si content for sericites in porphyry copper deposits (from Parry et al. 2002). These compositions represent electron microprobe analyses which are not necessarily of single phase material. Most likely the compositions between 0.9 and 1.0 K are mixtures of illite (WCI) and muscovite-phengite minerals. The shaded zone then represents bulk compositions of micaceous materials

$2M_1$ illites (WCI) while the latter are of 1M polytype and frequently contain a few smectite layers (PCI). This is probably related to the differences in the rock chemical compositions between the potassic and the propylitic facies. The dissolution of the typical initial paragenesis enriches the altering solutions in either K^+ or Fe^{2+}+Mg, respectively. Thus, sericite should be richer in Al and K in potassic zones than in propylitic ones. This was shown by Parry et al. (2002) who compared the chemical compositions of sericites from different porphyry copper deposits. They showed that the layer charge decreases with the tetrahedral substitution degree. Indeed, the Si^{4+} and the $K^+ + Na^+$ contents per $O_{10}\ (OH)_2$ are roughly related and range in a continuous domain from 3 to 3.5 and 1 to 0.7, respectively (Fig. 4.19).

Does the sericite composition depend only on the chemical composition of the local rock? This is not as simple as it sounds. Indeed, it is classically thought that the higher the content of the muscovite component, the higher the temperature of illite crystallization. Consequently, the lower the temperature, the lower the K+Na contents. Fluid inclusion studies indicate that the temperatures at which sericite is formed vary between 200 and 350 °C in most of the porphyry copper deposits. The lower temperature conditions are encountered far from the magmatic intrusion in the propylitic zone. Higher temperatures occur inside the potassic alteration zone which is located at the outer interfaces of the intrusion. Thus, the local chemical and temperature conditions act in the same way to favour either K-rich $2M_1$ or K-poor 1M sericites in the potassic or propylitic zones respectively. Since illite has been observed to form from muscovite (Meunier and Velde 1982) one can expect that there is a gap in solid solution between muscovite and illite. However, in prograde alteration (increasing temperature and potassium content) one would expect to find both illite (0.9 K) and muscovite-phengite (1.0 K content). Thus the high potassium minerals are probably an assemblage of illite and muscovite-phengite.

In porphyry copper deposits, the fluids which alter the rocks in the phyllic zones are generated by magmatic processes. They are initially in equilibrium

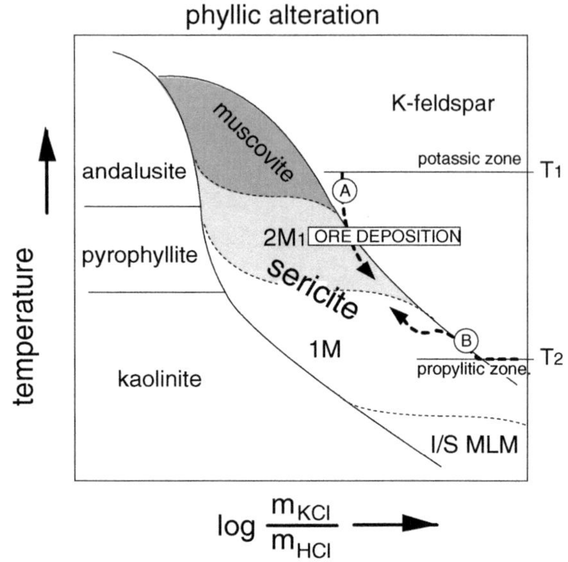

Fig. 4.20. Hypothetical phase diagram modified from Montoya and Hemley (1975) in which the A and B trajectories depict the physicochemical evolution of fluids in phyllic and propylitic alteration zones respectively

with K-feldspars, but, when injected into open fractures, these fluids cool and are no more in equilibrium with K-feldspar. Alteration processes begin: H^+ protons are consumed from the aqueous phase while K^+ ions are extracted from the solids (Beane 1983). The alteration mineral reaction is the following (for simplification, sericite is represented by a muscovite formula):

$$3KAlSi_3O_8 + 2H^+ \rightarrow KAl_3Si_3O_{10}(OH)_2 + 6SiO_2 + 2K^+$$

Using the phase diagram calculated by Montoya and Hemley (1975), the trajectory of fluid composition during the phyllic alteration stage is given by the curve A (Fig. 4.20). It crosscuts the K-feldspar-white mica phase boundary at a temperature at which the potassium activity is sufficiently high to induce the crystallization of the $2M_1$ sericite (Fig. 4.20).

Sericites in propylitized rocks generally appear inside the plagioclases while K-feldspars are not altered. The biotites are transformed into chlorites concomitantly and liberate K^+ ions. The altering fluids are resident in the microporosity of the rocks and do not originate from the intrusion body. They are warmed and heat accelerates chemical transfer from the solution which was previously in equilibrium with the minerals in contact. Alteration processes are then triggered inside microsystems. It is to be noted that the texture of the rock is conserved during alteration. The interdependent mineral reactions can be formulated as follows:

1. Biotite + H^+ $\rightarrow$ chlorite + K^+
2. Plagioclase + H_2O + K^+ $\rightarrow$ sericite + H^+

A simplified version of reaction 2 may be developed as follows:

$$3CaAl_2Si_2O_8 + 3H_2O + 3CO_2 \rightarrow 2KAl_3Si_3O_{10}(OH)_2 + 3CaCO_3 + 2H^+$$

Calcium ions are consumed by epidote instead of calcite crystallization when the CO_2 partial pressure is very low. The trajectory in the phase diagram from Montoya and Hemley (1975) is given by curve B. It shows that neither temperature nor K^+ ion activity are sufficiently high to induce the crystallization of $2M_1$ sericites. Next to this, the formation of 1M sericites is favoured by the diffusion of Mg^{2+} and Fe^{2+} ions in the solutions from biotite to plagioclase microsystems. This also explains the composition differences between sericites formed during phyllic alteration of potassic or propylitic zones as it was observed by Parry et al. (2002) and reported in Fig. 2.31.

From the discussion above, it appears that the "sericite" problem is more complicated than previously considered. However, this complexity may be used as a tool for ore prospecting. Indeed, the sulphide deposits are located near the magmatic body in the vicinity of the potassic-phyllic zone boundary (Lowell and Guilbert 1970). The sericites in these local conditions are of the $2M_1$ polytype. The predominance of the 1M polytype indicates that the phyllic alteration is away from the ore deposit inside the large propylitic zone. Potassic zone sericites are probably mixtures of illite (WCI) and muscovite-phengite. These minerals signal the probable presence of ore deposits.

4.1.4.2
Unconformity-Type Uranium Deposits

The Geological Structures The largest uranium ore deposits are found in the vicinity of the Archean-Proterozoic unconformity of Precambrian cratons. The Archean basement is composed of granodioritic and gneissic rocks while the Proterozoic formations are unmetamorphosed sandstones and shales. The Australian deposits in the McArthur basin and the Paterson Province are located inside the Archean basement just above the unconformity. In the Canadian Shield, the deposits are located either in the basement, at the unconformity or just above inside the overlying sedimentary formations.

The unconformity-type uranium deposits are considered to result from large-scale fluid flow under diagenetic-hydrothermal conditions (Raffensperger and Garven 1995a,b; Komninou and Sverjensky 1996; Kyser et al. 2000). Introduction of fluids occurs around fault zones in the basement-sediment complex. In this type of deposit, clay minerals are concentrated around the ore and form a "alteration halo". Quartz is dissolved at a zonal scale (Fig. 4.21). Most of the mineralogical studies show the presence of di-trioctahedral, tri-trioctahedral chlorites and illite (Nutt 1989; Quirt 2002, among others). The newly formed clay paragenesis overprints those which were formed under diagenetic conditions (mainly kaolinite and dickite).

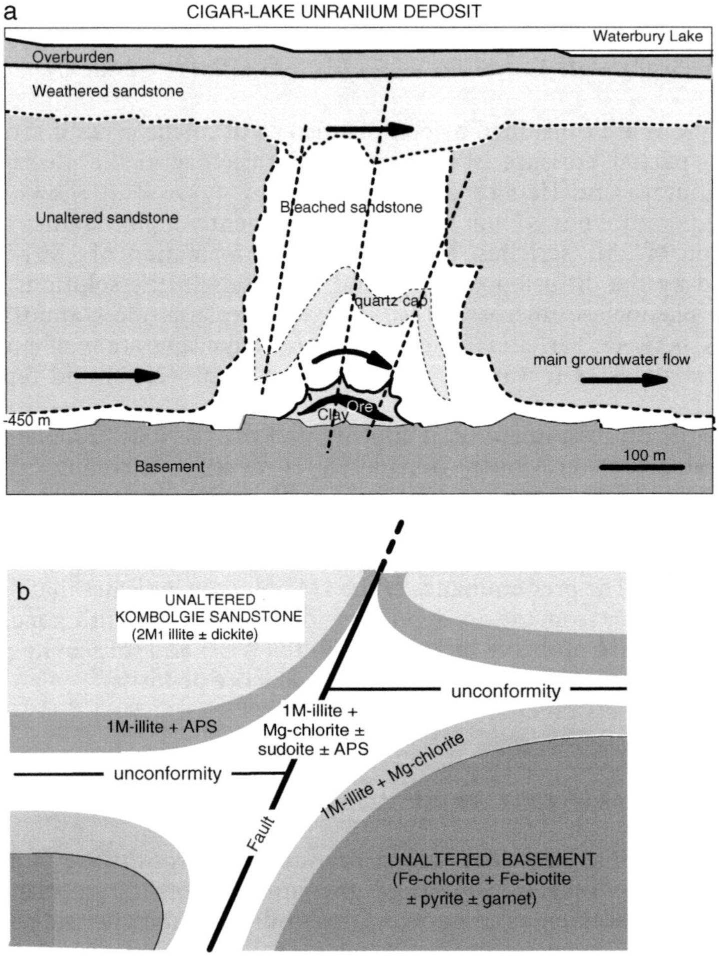

Fig. 4.21a,b. Schematic representation of an unconformity-type uranium ore deposit. **a** The example of the Cigar Lake ore deposit in Athabasca basin (Canada). **b** Clay mineral zoning in alteration haloes around the faults (from Beaufort et al. 2003)

4.1.4.3
Illite Polytype: a Prospecting Tool

Diagenetic illite is observed in sandstones as grain coatings, pore lining and bridging, pseudomorphs of dickite crystals and partial replacement of detrital white micas (Laverret 2002; Beaufort et al. 2003). Depending on the burial conditions in the Canadian and Australian basins, these illites are different (Table 4.2). In barren zones (no uranium mineralization), the platy hexagonal illites have a high-charge (0.9 per $O_{10}(OH)_2$) and are of the $2M_1$ polytype. In altered sandstones, they are mixed with 1M lath-shaped illites (Fig. 4.22).

Table 4.2. Crystallochemical characteristics of illites in Canadian and Australian Proterozoic basins (Laverret 2002; Beaufort et al. 2003)

Crystallochemical properties	Shea Creek (Athabasca, Canada)			Alligator Rivers (Northern Territory, Australia)		
	Diagenetic	Hydro-thermal	Basement	Diagenetic	Hydro-thermal	Basement
Particle size (µm)	2–5 µm	< 1 µm	< 1 µm	< 2	2–5	> 5
Layer charge per $O_{10}(OH)_2$	0.80–0.90	0.75–0.85	0.80–0.90	0.85–0.95		0.90–095
Tetrahedral charge	0.60–0.80	0.70–0.80	0.70–0.80	0.83–0.93	0.50	0.65–0.80
Octahedral Mg content	0.15–0.30	0.10–0.20	0.15–0.30	≤ 0.02	< 0.70	0.15–0.30
Crystal habitus	Laths	Hairy	Laths	Platy hexagonal	Laths	Platy hexagonal
Polytype	1Mc	$1M_t$	$2M_1$–$1M_t$	$2M_1$	$1M_t$	$2M_1$

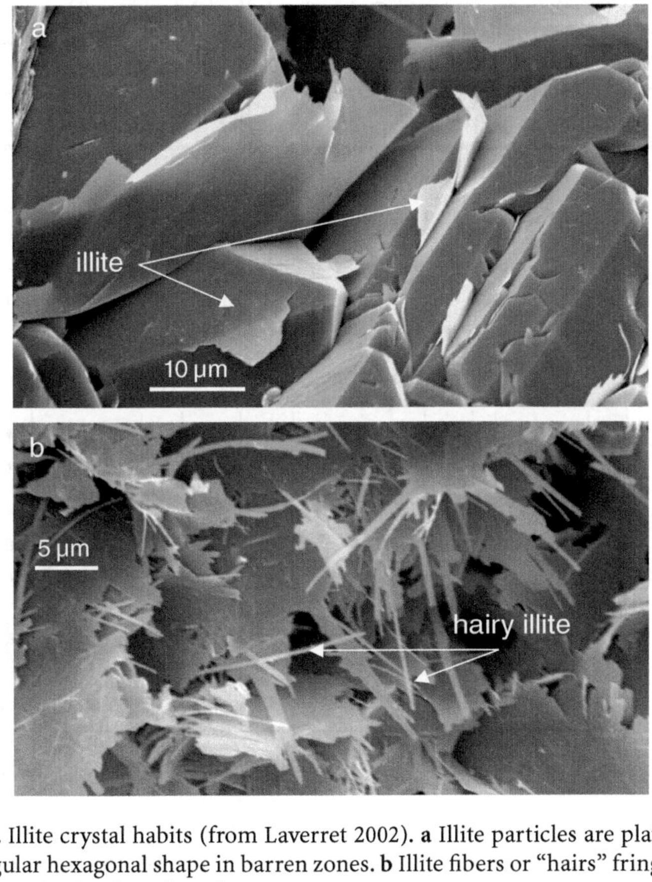

Fig. 4.22a,b. Illite crystal habits (from Laverret 2002). **a** Illite particles are platy with elongated or regular hexagonal shape in barren zones. **b** Illite fibers or "hairs" fringe diagenetic illites or quartz in the altered sandstones

These differences allow one to utilize illite polytype identification as a tool for uranium ore prospecting in these basins. Indeed, the increasing proportions of 1M polytype relative to $2M_1$ indicate that the "hydrothermal-like" alteration has locally affected the rocks because of the presence of faults (Fig. 4.21b). This information was used in interpretations for some exploration drill-holes (Fig. 4.23). The presence of the two illite populations is clearly shown by

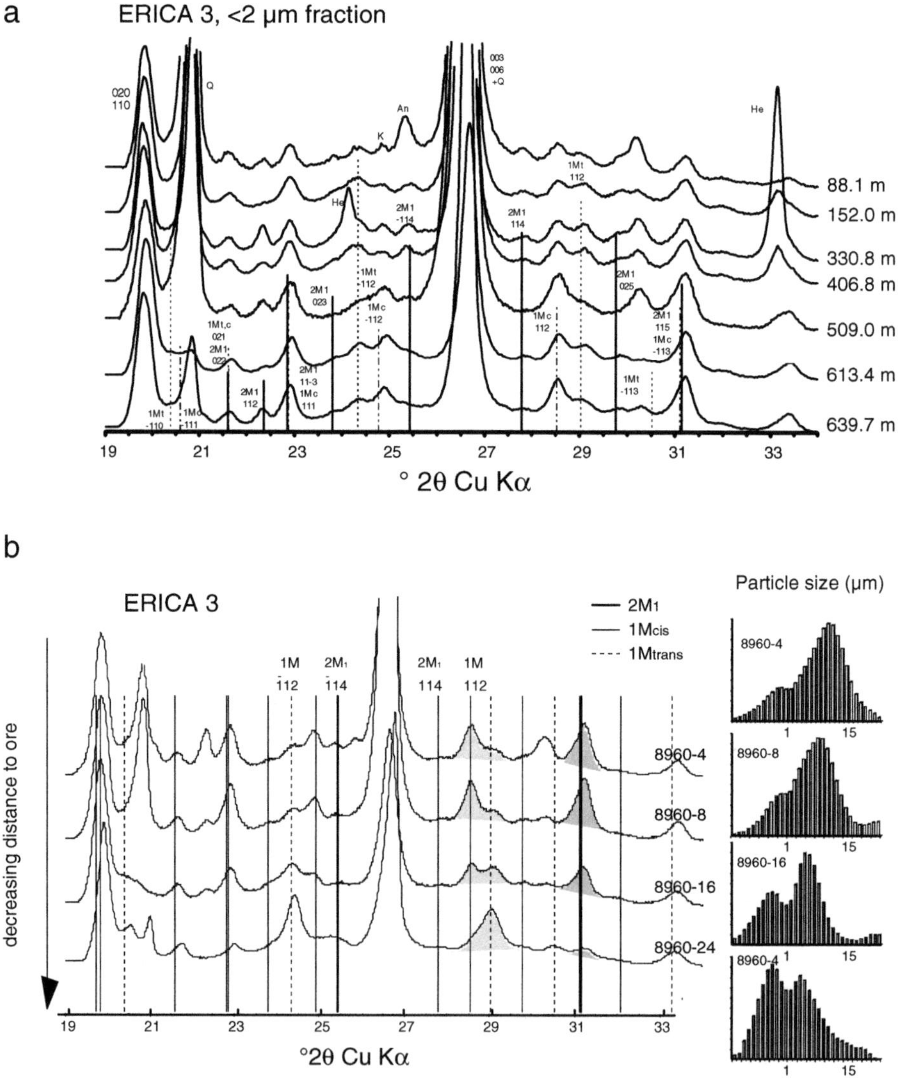

Fig. 4.23a,b. Illite polytype study in the Erica 3 drill-hole, Athabasca basin, Canada (data from Laverret 2002). The relative proportion of 1M illite polytype increases with depth (**a**), that is to say with decreasing distance from the uranium ore deposit (**b**)

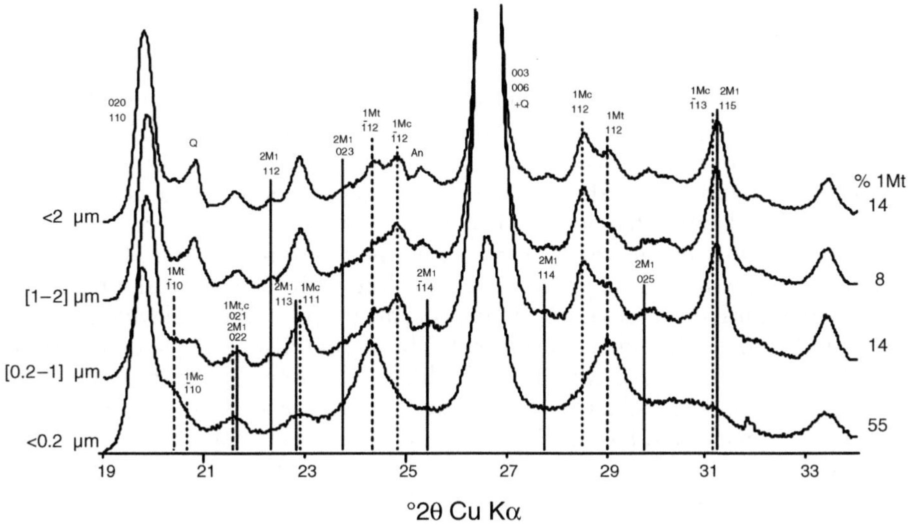

Fig. 4.24. Relation between illite polytype and size fraction in sample SHE 14 – 643.0 m from Athabasca basin, Canada (data from Laverret 2002). The M_t polytype relative proportions is concentrated in the < 0.2 μm fraction

the bimodal distribution of particle size: < 1 μm and 1–4 μm. Their relative proportions change with distance from the ore deposit: the finest one increasing toward the ore deposit (Fig. 4.23b). The intensity of the 1M diffraction bands and particularly that of the $1M_t$ increases concomitantly. Further, Laverret (2002) showed that, for a given sample, the $1M_t$ polytype is concentrated in the < 0.2 μm size fraction (Fig. 4.24). Thus, one can estimate the intensity of the diffuse "hydrothermal-like" event at any place in the basin by simply comparing the intensities of the $1M_c$, $1M_t$ and 2M diffraction bands in several size fractions of the sample.

Summary Laverret (2002) shows that, in the Athabasca basin, the hairy illites which are dominantly of the $1M_t$ polytype are formed specifically in channel ways along which the mineralized fluids flow from their emergence point at the top of the Archean basement. These illites replace the minerals of the kaolin group in the pores of the Proterozoic sandstones. Thus, detecting $1M_t$ illites in these sandstones means that the rocks have been "hydrothermalised". These minerals are a sort of distal signature for a potential mineralization. The more intense the replacement, the highest the hydrothermal fluid quantities, and, consequently, the nearest the ore deposit. Sudoite is associated with $1M_t$ illite in the vicinity of the uranium deposit.

4.2 Environmental Problems

Two subjects are discussed here addressing environmental problems. These are just two of many, but they indicate the extreme importance of understanding clay minerals in problems of surface geology. The great challenge today is for modern civilizations to safely harness the physico-chemical effects of anthropomorphic activity. This activity is almost by definition in the realm of clay mineralogy; that of the Earth's surface. As the beginning of the 21st century is already showing us, the problems are dealing with the chemical interaction between natural clays and man-made concentrations of material which are not at equilibrium with their environment. The strong disequilibria create rapid reactions which can quickly affect the life environment of mankind. In the initial stages of the industrial era, some 150 years ago, it was believed that human activity would have little impact on nature. One had a feeling that the world was a great chemical buffer, which could be depended upon to absorb the changes induced by factories and new farming practice. The emphasis was on increasing goods available to improve human conditions.

Today we know that the world is not an infinite reservoir of correcting influences. The increase in atmospheric CO_2, the loss of high level atmospheric ozone in certain areas, the loss of decent, clean, chemically pure drinking water in many areas of industrially developed countries are all problems which have been discussed in the scientific and popular press. Two major areas of concern are in energy resources and food production: the problems of nuclear waste and fertilizer use. We give some insight into these two problems below.

4.2.1 Illite and Mixed-Layer Minerals in Soils: Questions of Fertility

One crucial problem is how to maintain fertility in soils without polluting the water resources draining from them. Potassium and ammonium are two of the three major fertilizing agents used in modern agriculture. The use of potassium dates from the 19th century and ammonium from the mid-20th century. They are of primary importance in obtaining good crop production. However, too much of a good thing can become a problem. In many areas, nitrogen, coming from the ammonium additions to the soil, finds it way into drinking water resources. This presents a health hazard to developing human bodies. Further, high costs of fertilizer and other chemical products have made it impractical for farmers in industrialized societies to produce at a profit without massive aid from their governments. Essentially, the more farmers produce, the more fertilizer is needed and the higher their costs become. They are victims of their success, as are taxpayers.

In order to break this cycle, it is necessary to return to a chemically less intensive farming protocol. However, it is not realistic to return to practices used in the early 20th century if we expect to eat as much meat and vegetables as we do at the present day. The rest of the world is developing the same tastes, and this gives more importance to finding a solution to a modern, less energy- and chemical-intensive agricultural practice.

Below we investigate the clay mineral dynamics in some agricultural soils which reveal the very important role of illite (source of potassium) in soils and the stability of illite related to agricultural practice.

4.2.1.1
I/S MLM in Soils

In Chap. 2 we have shown some examples of soil clay mineral assemblages in which illite (PCI and WCI) occurs. Illite/smectite mixed layer minerals are common in soils but seem to be present more often in some specific types of soil. Very broadly speaking, I/S minerals are typical of prairie-induced soils. This is in opposition to forest soils where soil vermiculite is more often dominant. Let us quickly review the clay mineral differences in these two major soil environments.

Forest Soil Clay Mineralogy The origin of soil vermiculite is frequently attributed to an alteration of mica or chlorite transformation under acid soil conditions (see Righi and Meunier 1999 for a summary of this type of occurrence). The soil vermiculites are formed, in most cases, by the extraction of potassium from a mica mineral. This is possible through the change in layer charge of the structure. However, the vermiculite that remains has a relatively high layer charge, in the vicinity of 0.7–0.8 charges per $O_{10}(OH)_2$. These soil vermiculites, hydroxy-interlayered vermiculites (HIV) are strong acceptors of ammonium ions in laboratory tests (Evangelou and Lumbanraja 2002). Introduction of potassium tends to collapse layers to the exclusion of ammonium. It is generally assumed that soil vermiculites are replaced by smectite minerals in more mature acidic forest soils (see Righi et al. 1999).

Prairie Clay Mineralogy An extensive study of agricultural top soil clay mineralogy in the central United States indicates the illite/smectite mixed-layer clays are predominant in soils coming from a certain range of climatic conditions which were initially prairie soils. In these soils one most often finds two I/S minerals which occur together, one of higher smectite content (usually greater than 50% smectite) and another more illitic I/S mineral (with less than 30% smectite layers or in fact 70% illite layers). The relative amount of one or the other mixed layer mineral varies with climate and perhaps initial mineral substrate in the parent material of the soil. Both of the mixed-layer minerals are of a disordered interlayering type, unlike those found in diagenetic series of clays (see Sects. 2.2.2 and 3.4.1). Chronosequences of clay mineral development in prairie soils (Velde et al. 2003) indicate that the relative amounts of illite and smectite in both I/S clay minerals can change with time and soil maturity depending upon the context of the organic matter present. The type and relative amounts of the mixed-layer minerals is dynamically affected by soil chemistry.

Hence we have on the one hand soil vermiculite, dominant in the upper portions of forest soils, which is derived from micas for the most part, and illite/smectite mixed-layer minerals in prairie soils, derived from various sources. In both cases illite or potassic clay minerals are of greatest importance.

A study on the effect of vegetation on the production of clay mineral types can elucidate some of these aspects of mineralogy (Velde and Moore, unpublished). In this study a forest area and an adjacent cultivated forest soil were observed in Northern Illinois. The source material for the soils is the Peoria loess, a smecite-rich silty material. At lower levels in the profile of the cultivated forest soil one finds the typical two I/S mineral assemblage with a minor chlorite-vermiculite component (Fig. 4.25). This soil, cultivated since the mid-nineteenth century looses the chlorite-vermiculite component and the illitic I/S mineral. Overall, illite content (peak areas) decreases, as do both PCI and WCI. In the mid- 1950s a small portion of the forest soil was put into native grasses. After 50 years of such vegetative action there is a significant increase in the illitic I/S, and illite (PCI and WCI) content of clays in the lower portion

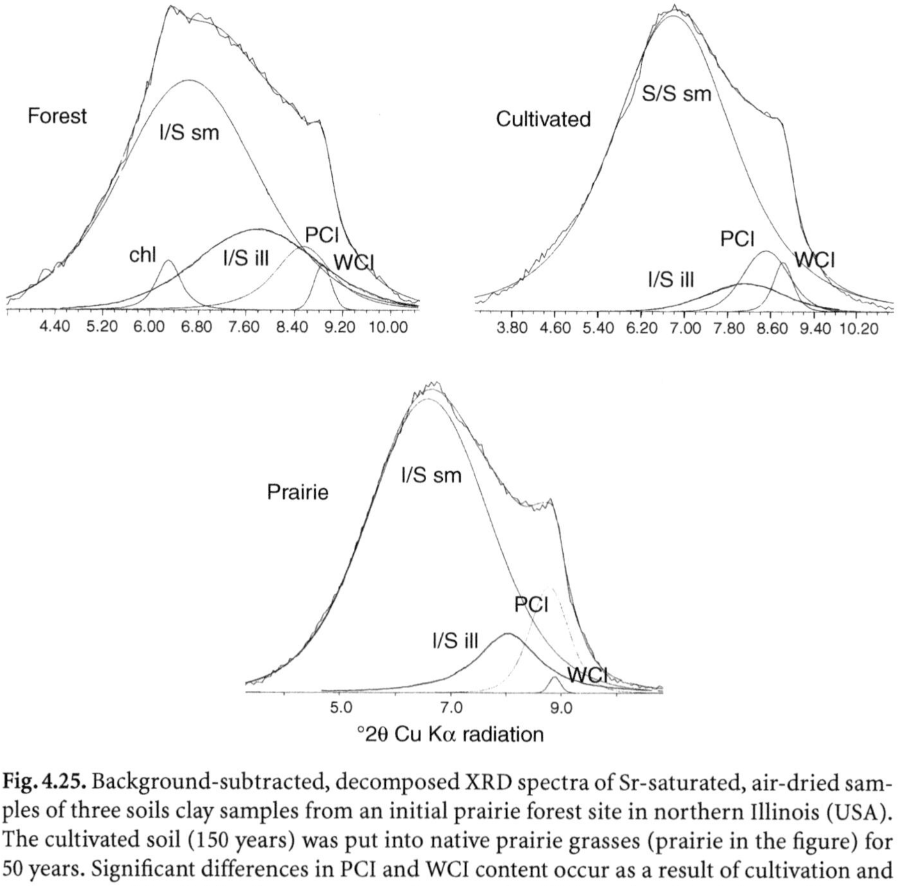

Fig. 4.25. Background-subtracted, decomposed XRD spectra of Sr-saturated, air-dried samples of three soils clay samples from an initial prairie forest site in northern Illinois (USA). The cultivated soil (150 years) was put into native prairie grasses (prairie in the figure) for 50 years. Significant differences in PCI and WCI content occur as a result of cultivation and return to prairie. Cultivation appears to decrease the illitic I/S band (I/S_{il}) as well as PCI and WCI. Prairie grasses regenerate the I/S illite and PCI content of the soils

of the profile near 1 m depth. In fact PCI becomes much more important than WCI. From these observations it is clear that the effect of the prairie is to increase illite content, that the agricultural practices common in the Midwest for 150 years decreased the illite content of the soil until it was exhausted. There was no vermiculite regeneration. Thus it appears that the prairie regime favors illite and the forest soil favors some vermiculite formation, even though little mica or chlorite is initially present in the initial substrate of the these forest soils.

The long-term fertility of forest and prairie soils are quite different. The American experience (USA) indicates that the long-range fertility of forest soils is relatively short. One can consider the East Coast USA where much forest farmland is now abandoned, as well as in the eastern, old Midwest of Ohio and Pennsylvania where much farmland is now put into pasture or forest. The formerly rich cotton fields of Georgia are now largely piney forest. The areas or rich farmland in the 18th and early 19th centuries are no longer fertile enough to be consistently farmed. However, the prairie soils in the Midwest seem to maintain much of their initial fertility, up until now at any rate. The farming experience of the Midwest is shorter, only 130–150 years. In many areas of the USA fertility is maintained at the cost of intensive use of fertilizers. However, in the long run, the prairie soils maintain and can regain organic matter fertility within short time periods while the forest soils seem to be more difficult to regenerate. Given the rather different nature of the major clay minerals in forest and prairie soils, one might look to the clays in order to gain insight into the variations in fertility.

The following discussion attempts to shed some light on the problems of soil fertility and the role of illite and I/S minerals in this fertility.

4.2.1.2
Role of Potassium in Agriculture

Potassium was one of the first artificial fertilizer materials to be widely used in Europe, originating from the KCl salt deposits near Strassfurth in Germany in the 1860s (Larbaletrier 1891, p 56). These cheaper sources of potassium replaced the use of more expensive wood ash imported from Russia or America. The great importance of this element for crop growth has been recognized since. Potassium is especially important in the periods of fast plant growth. It is taken up differently by various crops, corn, rice and potatoes being particularly important sinks for this element (Mussa 1887). In general, most reasonably fertile agricultural soils have sufficient potassium contents to sustain crop growth, but the problem is often how to release this element to the plants (Getman and Ladd 1925, p 117). In most soils, the most readily available source of potassium is in the clay or silt fraction, where micas, illite and I/S minerals are abundant (Lagatu and Sicard 1901). From these few citations in the literature, it is clear that much has been known for a long time pertaining to the role of clays in soil fertility. The problem is to understand how the clays release or capture the plant nutrients. We will focus on the illite content of soils here as a key to fertility because it is an obvious source of potassium.

4.2.2
Some Effects of Agricultural Practice and their Bearing upon the Loss of Illite Content in Soils

4.2.2.1
Rice Culture on Red Soils in China

Li et al. (2003) present data from a sequence of flood irrigation rice paddies in China developed on red soils which show a dramatic change in the clay mineralogy. The most important change is in the illite-mica (PCI plus WCI) content which is shown in Fig. 4.26. Initial clay mineralogy is an illite, chlorite and a mica-chlorite mixed-layer mineral assemblage. Iron oxide is very prominent in the soil, giving it a strong red pigment coloring. The portion of illite, determined by peak ratio compared to the other clay minerals present decreases dramatically with rice culture as seen in the figure. At the same time, there is an increase in magnesium chlorite in the soil clays. Within 30 years most of the potassium-bearing clays are lost from the cultivated horizon. This

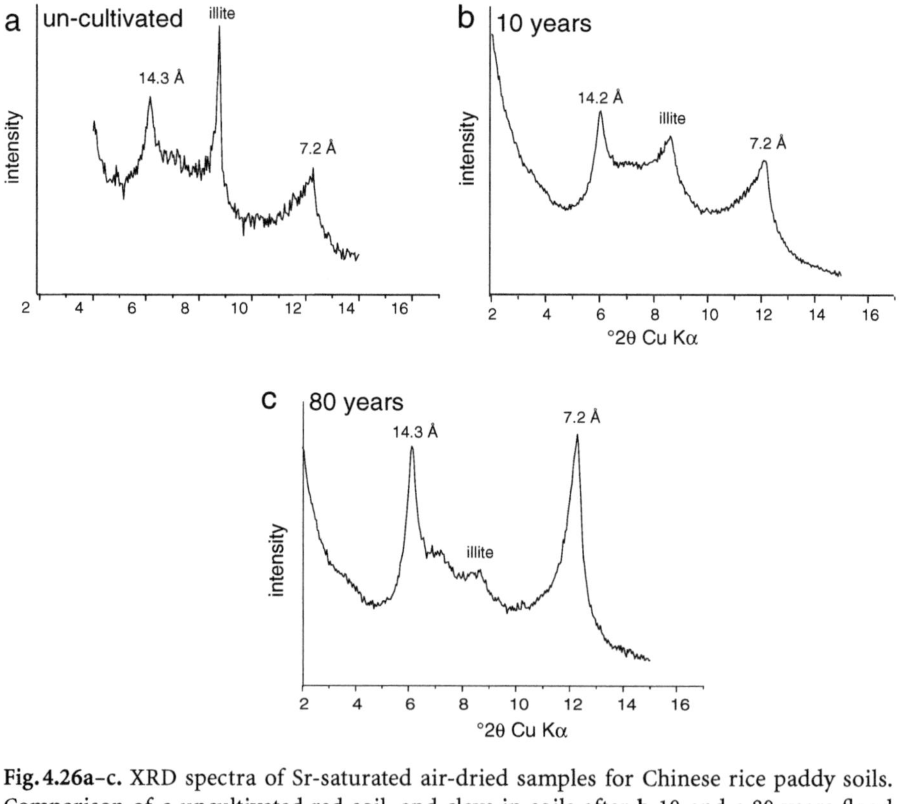

Fig. 4.26a–c. XRD spectra of Sr-saturated air-dried samples for Chinese rice paddy soils. Comparison of **a** uncultivated red soil, and clays in soils after **b** 10 and **c** 80 years flood irrigated culture. Illite (PCI and WCI) decreases greatly and chlorite (14.3 Å) increases in relative proportion

is shown by a strong decrease in potassium content of the soils. The result is that during the rice cultivation on these soils, 2 to 3 times the normal amount of potassium fertilizer is needed to maintain a average rice production. Here it is clear that there is a relationship between crop productivity and the presence of potassic clay minerals.

The explanation of illite loss is probably complex, in that the rice culture under flood irrigation induces a strongly reducing environment in the cultivated soil for a large part of the year. One observes a strong change in soil color going from bright red to dark grey after 80 years of culture. Iron content is lost from the soil with this color change. A new composition, chlorite, magnesian, forms during these mineral transformations. Hence illite is lost and replaced by an aluminous magnesian silicate. The chemical potentials producing these changes are induced by the change in oxidation state of the total environment.

The result is a loss of fertility due to the loss of illite from the soil mineral assemblage.

4.2.2.2
Corn in the Midwest

Velde and Peck (2002) have studied the effects of continuous corn planting on Illinois (USA) Mollisols. The study was performed in an experimental plot of the University of Illinois where several tens of meter square plots were cultivated in a systematic manner since 1913. Crop rotation was the main theme of the initial study but the generalized use of NKP (nitrogen, potassium, phosphorous) fertilizers was taken into consideration in the 1950s. Clay mineralogy is dominated, as would be expected in a prairie environment, by the two disordered I/S minerals, one of high smectite content (> 50%) and another of high illite content (> 60%). Some illite (PCI and WCI) is present in the initial clay assemblage. The effect of continuous corn cultivation is to decrease the illite

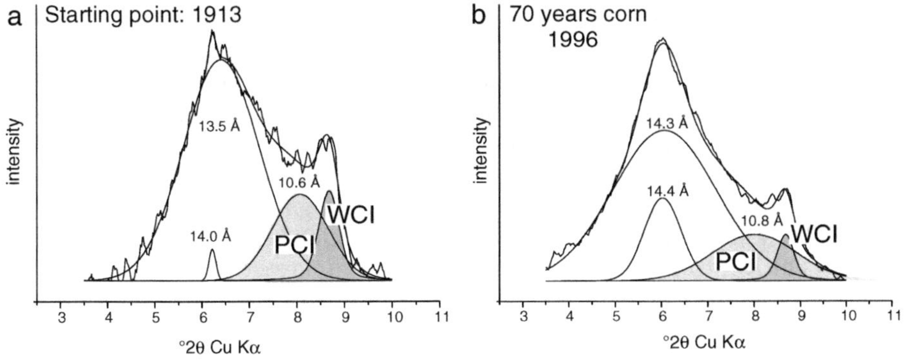

Fig. 4.27a,b. Background-subtracted, decomposed XRD spectra of Sr-saturated air-dried samples of Illinois Mollisol agricultural clays in initial state (**a**) and after 70 years continuous corn planting (**b**). Note loss of illite (PCI and WCI) due to corn plant activity in the soils

content (PCI-WCI) and the quantity of illitic I/S minerals (Fig. 4.27). Also, the illite content of the smectitic I/S mineral is reduced as a result of corn cultivation for 70 years, witnessed by a shift to higher spacings (smectite layers) in the mixed-layer mineral. A linear correlation between smectite content of the I/S mineral and corn production per hectare can be established. However, in crop-rotated plots (corn-oats-hay) the clay minerals seem to be unaffected by these years of cultivation. Further, in plots fertilized (NKP) since 1955, the clay mineral assemblage is essentially that of the initial materials and the crop-rotated plots. This study indicates that the extraction of potassium by corn depletes and destabilizes the mica or illite content of the clay assemblages, either in the form of individual illite particles or potassic interlayers in mixed layer clays. However in this case, where I/S mixed-layer minerals are present, illite or potassium-bearing clays can be regenerated by chemical treatment unlike the case of the Chinese red soil rice cropping. The Chinese soils do not contain an I/S mixed layer mineral.

4.2.2.3
Stabilizing Illite in Soils – Effect of Manure and Humic Matter

Manure amendment has been an age-old practice used to maintain and enhance fertility in soils (Pattullo, 1758 for example). It has also been known that liming (addition of CaO, in some form) has a beneficial effect in combination with manure. One major, and evident action of lime is to reduce the acidity. Since acid soils are not conducive to the formation of I/S minerals, one assumes that this use had some underlying, but unsuspected to the operators, effect in stabilizing I/S minerals in the soil. Getman and Ladd (1925) insist on the capacity of manure to render the potassium in soils available to plants.

Several recent studies have been performed which indicate the importance of organic matter on the stability and formation of potassic clay minerals.

Uncultivated systems Pernes-Debuyser et al. (in press) give data concerning a 70-year experiment investigating the effects of fertilizer on soils. The initial starting material is a cultivated forest soil in the Paris area of France. The period of cultivation is unknown, but is probably in the range of thousands of years, though perhaps not in continuity. Various mineral fertilizers were used in the experiment in different plots, and plant life was removed by plowing, and cultivation throughout the year. No effects on clay minerals were observed under the different chemical fertilizers (pH ranging from 3.5 to 8.5) except in one, KCl. Figure 4.28 shows the initial state of the clay assemblage and that after 70 years of KCl treatment. There is a significant increase in the PCI and WCI content of the clay assemblage. However, analysis of the XRD spectra of the smectitic I/S mineral indicates that the I/S mineral gains illite content, about 20% or more with KCl treatment but slightly more when manure is used (30% increase in illite content of the I/S mineral). Here we see that the KCl increases the illite content but the manure increases the illite layer content of the I/S minerals.

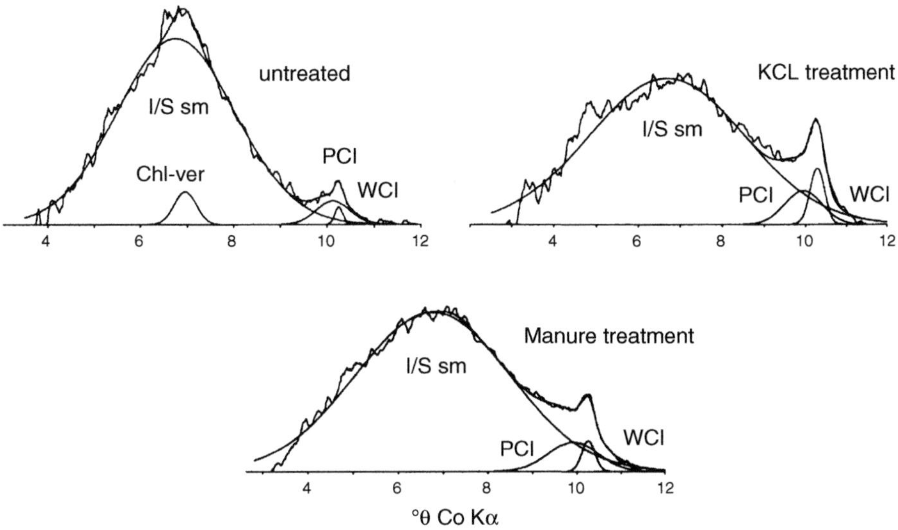

Fig. 4.28a–c. Background-subtracted, decomposed XRD spectra of Sr-saturated air-dried samples of uncultivated soils having experienced different fertilizer treatments. **a** Untreated soil. **b** KCl treatment. **c** Manure treatment. KCl shows increase in WCI and manure treatment shows increase in PCI

Vertisol Agricultural Site Huang and Velde (unpublished, work funded by the Soil Science Institute, Chinese Academy of Sciences, Nanjing) have found some interesting results for Vertisol plots in Anhui Province, China. Here the soils have been agricultural for an unspecified time. For the last 40 years three major treatments were performed: unmanaged, NKP fertilizer and manure fertilization. In these soils the classical I/S mixed-layer minerals of high and low illite content are present but dominated by the smectitic I/S mineral (Fig. 4.29). Chemical fertilizer changes the clay assemblage somewhat, slightly increasing the illite content. This is in contrast to the unplanted plots in Versailles where illite content increased when no crops were present. However, the major change occurs in the manure-managed plot where there is a dramatic increase in WCI content. It is clear from this experiment that the effect of manure is to increase the illite content in highly smectitic soil clay assemblages.

Natural Grass Lands In another study, one observes the effects of a buildup of organic matter in a developing prairie soil. No manure or potassium source is brought into play in the soil. Velde et al. (2003) show the significant impact of prairie humic matter on the formation of potassic clays. Figure 4.30 shows the change in clay mineralogy in poldered sediments put into pasture since 1158. The initial sediment contains the typical two I/S mineral assemblage with some illite. The humic zones in the profiles produce a more illitic I/S mineral and some WCI compared to PCI. By contrast, in the lower, non-humic zones the I/S minerals become more smectitic in nature. Here the

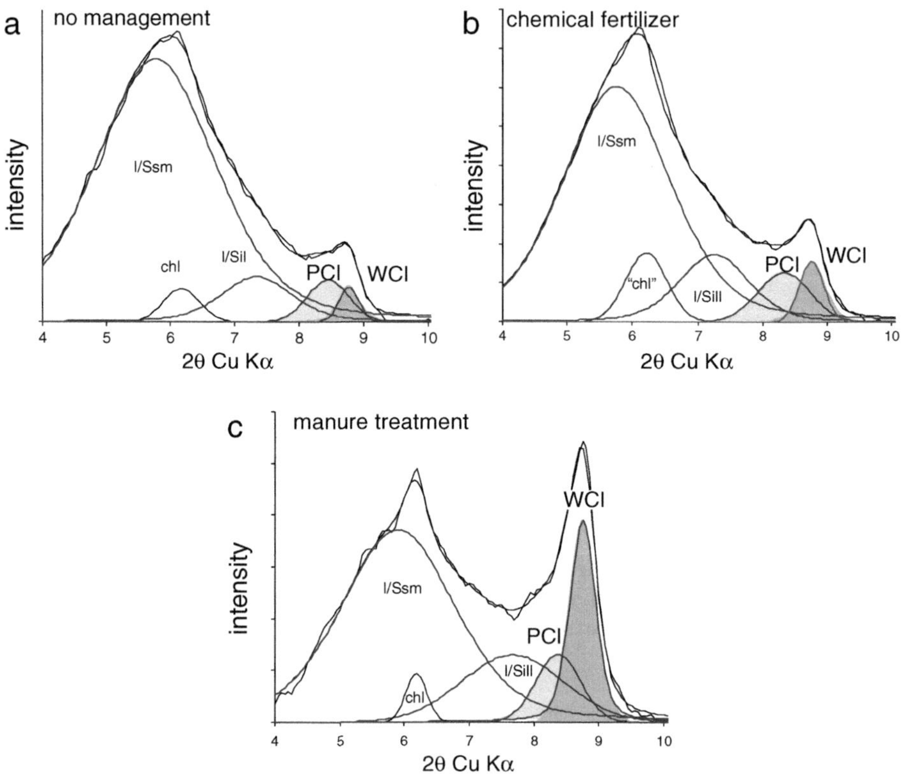

Fig. 4.29a–c. Background-subtracted, decomposed XRD spectra of Sr-saturated air-dried samples of Chinese Vertisol samples from experimental plots. **a** No management, traditional planting with no amendments. **b** Use of NKP fertilizers. **c** Use of farmyard manure. Not very strong increase of WCI in the last figure

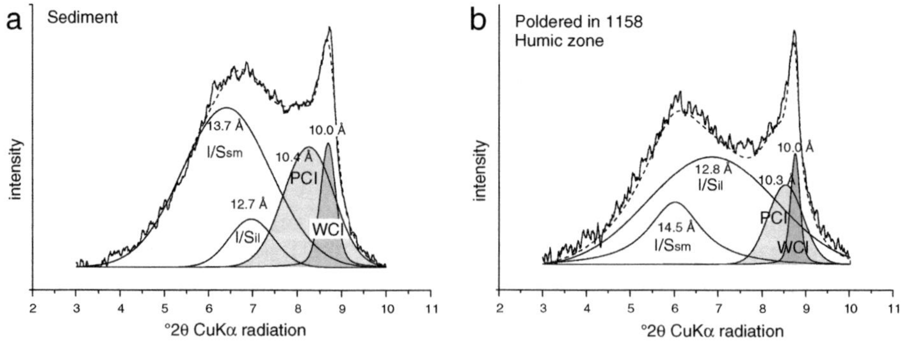

Fig. 4.30a,b. Background-subtracted, decomposed XRD diagram of Sr-saturated, air-dried samples. **a** Poldered salt marsh sediments. **b** Humic horizon of sediments after 850 years of prairie soil development. Note the relative increase in illite (PCI and WCI) content

action of grass humus leads to increase the illite content of the mineral assemblage.

In the studies discussed above, it is clear that the action of grass humic matter, either natural or in the form of evolved material in manure which is a combination of animal excrement and different grass plant stems, increases or stabilizes the illite or potassic mineral content of soil clay mineral assemblages. This occurs in the presence or in the absence of growing plants. Hence, the age-old practice of manuring crop land seems to have the result of stabilizing the potassium in soil clays. Since the reason for this practice is to increase fertility, one must conclude that it in fact does, and the way is by producing or protecting illitic layers in clays.

Summary In these case studies we see that intensive cultivation tends to deplete the potassium minerals (PCI, WCI, and illite layers in I/S minerals) but that application of organic matter in the form of manure can greatly enhance illite content. Prairie humic matter increases the illite content of the various soil clay minerals as well. There is an obvious but unexplained relationship between conservation and generation of illitic material and the action of organic matter. This is clearly a subject for research searching for sustainable agricultural practice.

4.2.2.4
Clay Behavior in the Presence of Organic Matter: Laboratory Tests

If certain agricultural practices appear to stabilize illite, can one be sure that this is in fact what is called illite now in this book (WCI) and if so how does it change from illite to smectite to illite again? Some initial answers are proposed below.

Methods and Observations The usual treatment of soil clays in routine pedological analysis is to oxidize the organic matter, extract the free iron and then look at the clays using X-ray diffraction. The studies cited above in this section have not performed according to this protocol. Instead, they conducted the analysis using untreated material to characterize the clays. With this unorthodox procedure, it was possible to make the classical treatments in a second step, and then to compare the untreated and oxidized (organic-poor) clays. Another classical treatment is the saturation of the clay particles with potassium. This operation is used in order to distinguish the high-charge smectite layers from low charge ones, when they are present. This treatment can be used to estimate the approximate amount of vermiculite-type layers present in an I/S mineral.

The studies of soil clays cited above have shown that both the change in organic content (elimination of a large portion) and potassium saturation affect the relative amounts of non-expanding layers in the clays. The importance of these observations is twofold,

1. change in clay mineralogy upon potassium treatment indicates the relative under saturation of the clays with respect to this fundamental nutrient in the soil complex.

2. elimination of the organic matter can give some idea of its role in the cation exchange capacity and structure of the clays. If the clay structures become "illitic" upon loss of organic matter, they will have a lower exchange capacity. With organic matter present they will have a higher exchange capacity.

Results Velde (2001) surveyed agricultural-horizon clay minerals in a variety of grassland sites in the central United States. About 20 samples of the untreated, potassium-saturated and oxidized clay samples were compared. In these samples, the effect of oxidization (loss of much of the organic matter) was to systematically change the peak position of the smectitic I/S mineral to higher values, indicating a higher smectite content for this I/S mineral. The illitic I/S mineral behaved in roughly the same way. Potassium treatment increased the illite content (decrease in smectite content) and the peak surface of the WCI peak and to a lesser extent that of the PCI peak. The effect of potassium is to close some layers to form illite-like material and as a result the remaining I/S material becomes less illitic and more smectitic. Approximately the same type of behavior was noted in a study of cultivated soils in the Morrow Plots, University of Illinois experimental farm (Velde and Peck 2002). All of these soils are of the prairie type, having experienced cultivation for approximately a century and a half.

We can appreciate the effects of these two treatments better when observing the poldered prairie soils reported by Velde et al. (2003). In Fig. 4.31a,b we have XRD data for samples from two parts of the oldest soil profile (poldered in 1158) with an upper humic horizon and a lower, apparently nonhumic horizon. Potassium saturation of untreated (containing initial organic matter) samples in both portions of the profile enhances the relative intensity of the PCI peak significantly. Overall, the sum of the illitic components (PCI plus WCI) increases relative to the I/S minerals, according to peak area. In both samples, the smectitic I/S peak shifts to higher values upon potassium saturation, indicating more smectite content in the smectitic I/S mineral. The illitic I/S remains at about the same peak position (illite content) but its relative intensity is significantly enhanced. Thus it appears that a portion of the initially smectitic I/S becomes illitic I/S whereas the vermiculitic smectite layers close upon potassium saturation. Some of the smectitic I/S minerals seem to have a relatively high number of high charge layers which are closed to the illite spacing under potassium saturation, producing PCI. The overall effect of potassium saturation is to close many of the smectite layers in the illitic I/S minerals to form PCI and to close a significant portion of the smectite layers in the smectitic I/S minerals. These clays then show a behavior indicating undersaturation with respect to potassium in their natural state.

Oxidation of the organic matter (or most of the aliphatic C–O linkage molecules) produces an effect similar to that of the potassium saturation. However, the illite components are more greatly enhanced, especially the WCI, than was the case of potassium saturation. If the use of potassium to close high-charge layers is not surprising, the effect of the destruction of organic matter closing expanding layers is much less expected. In order to explain these observations, one can propose that the organic matter initially present

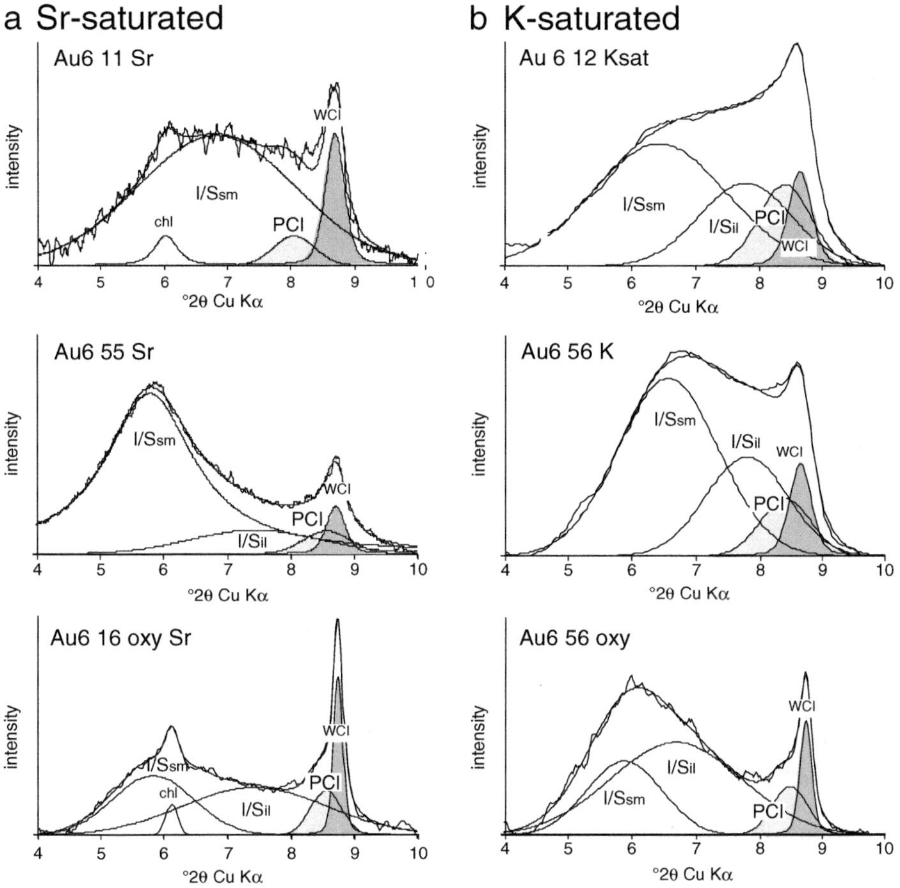

Fig. 4.31a,b. Background-subtracted, decomposed XRD spectra of soil clay minerals after various laboratory treatments. All samples in the air-dried state. **a** Sr-saturated, **b** Potassium saturated, oxidized sample (unpublished data)

propped open high-charge smectite layers in the I/S minerals. Its oxidation (destruction) allows the layers to collapse.

However, reintroduction of organic matter in the form of polar glycol molecules does not reopen the closed layers to any great extent. Thus the organic material in these samples does not prop open the layers when introduced in abundance. The layers are irreversibly closed after loss of organic matter. Saturation of the oxidized clay sample with potassium changes the illite content of the illitic I/S somewhat, but does not significantly affect the apparent illite content (Fig. 4.32). Thus there is a problem in interpreting the results of potassium and oxidation of organic matter in soil clays.

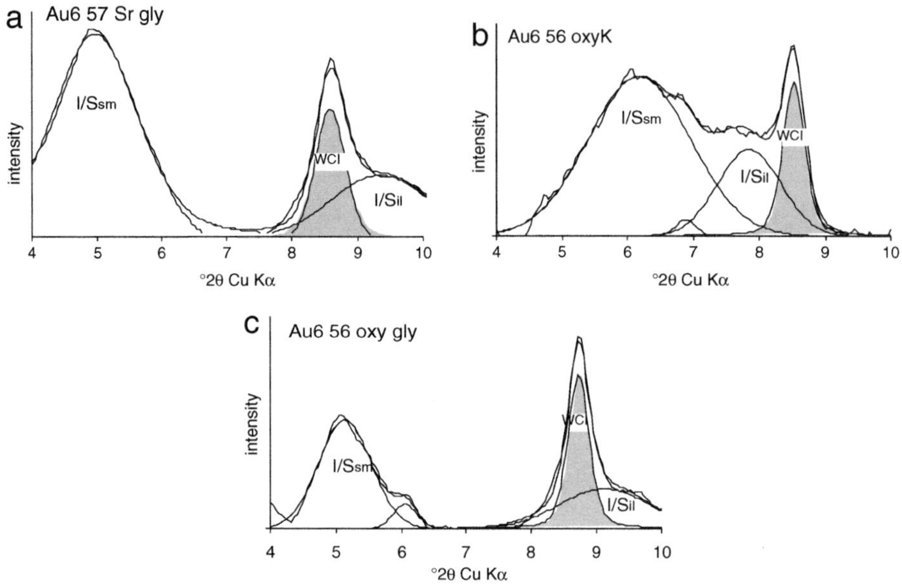

Fig. 4.32a,b. Background-subtracted, decomposed XRD spectra of a soil clay mineral after various laboratory treatments, **a** Sr-saturated and glycol-saturated, **b** oxidized and potassium-saturated, **c** oxidized and glycol-saturated (unpublished data)

4.2.2.5
Potassium and Ammonium

From the observations on prairie soil clay minerals under different laboratory treatments, we can conclude that the effect of eliminating a portion of the organic matter is to create much more discrete illite and illite layers in I/S minerals. The enhancement of the WCI peak as well as the PCI peak is consistent in most of the samples observed. This being the case, since there is no potassium added to the systems when organic matter is oxidized with H_2O_2, where only a loss of C–O linked organic matter occurs, the collapse of the layers indicates that some ions must replace the organic matter and hold the high charge layers to form a 10 Å structure. One common ionic molecule found in soils is ammonium. The attraction of ammonium for the high-charge interlayer sites in vermiculites is well known. In general, ammonium is fixed by the fine fraction in agricultural soils. Thus, ammonium is a likely candidate for the formation of illite layers. During the hydrogen peroxide oxidation procedure, much of the carbon in aliphatic chains is most likely destroyed to form CO_2 gas. However, it is probable that some of the organic nitrogen, in the form of aromatic compounds, which are resistant to oxidation, will remain in the soil clay complex after treatment. This material could well evolve into ammonium ions during the treatment process. Since it is established that ammonium has a strong tendency to fill in the interlayer sites of high-charge smectites (vermiculite), it might be possible that the observed decrease in swelling capacity (smectite content) upon oxidation of the organic matter could be due to the substitution of ammonium in the high-charge layer sites to form an illite.

Ammonium illite is known to form under high-temperature conditions (coal layers Daniels and Altaner 1993) and intermediate potassium-ammonium illites have been synthesized at high temperatures (Sucha et al. 1998). Recent laboratory experiments on interlayer ion exchange (Tatard, unpublished) indicate that only about 3% of the potassium can be exchanged from a natural illite (Illite du Puy) by ammonium at 90 °C after 3 days equilibration with a 0.3M ammonium chloride solution. Thus the possibility of introducing ammonium into the interlayer site of high-charge (illite) phyllosilicate minerals replacing potassium is not implausible, but the quantity is small at low temperatures over short periods of time. It is more likely that the ammonium enters into the high-charge smectite (vermiculite) sites where it remains strongly but not irreversibly fixed. It is very likely that the clay mineral changes observed by XRD are in fact due to responses to high-charge smectite or vermiculite layers in the structures. If the illitic I/S minerals are inhomogeneous in illite and vermiculite content, saturation either with potassium or ammonium could well close the few remaining high-charge layers to form discrete illite particles.

These illite layers, formed with vermiculite charge, can well be the labile portion of clays under the influences of agricultural practice. If there is a strong potential for potassium extraction, as in the case of corn agriculture, the illite (vermiculite charge) layers will loose potassium and behave as smectite layers, expanding and with high CEC (for ions other than potassium, of course). When ammonium fertilizer is introduced into the soil, the vermiculite (high-charge smectite) layers can absorb the NH_4 and behave as illite. However, this ammonium is also not strongly fixed and can be eventually exchanged when it is depleted in the soil fluid media by plant growth. Thus the vermiculite layers in the I/S and PCI phases will act as an exchange reservoir for the two major plant nutrients, potassium and ammonium.

4.2.2.6
Summary

From the reports discussed above it appears that high-charge smectite layers in I/S minerals in soil clays can be illitized through the introduction of large quantities of potassium or ammonium into the soil clay environment. However, most of the I/S minerals seem to be unsaturated with respect to both ammonium and potassium in their natural soil state. Hence when fertilizers are introduced into the soil, the clays can absorb large amounts of these plant nutrients. However, this is easily reversed by agricultural practices. Such an effect was noted long ago by Larbaletrier (1891) and undoubtedly well understood since by farmers and agronomists. The present problem is to better understand these functions in order to enhance the natural storage capacity of the clays and to release these nutrients from the I/S interlayer minerals.

Such research should constitute part of the future programs of agronomy institutes. The need to reduce the usage of fertilizer, especially ammonium is very evident today. In many areas of advanced, industrialized countries, farm lands have polluted the water resources through the use of large quantities of ammonium fertilizer. It is clear that such practices must stop. However, it is not reasonable to do this unless a viable alternative to high fertilizer use is proposed. One of the keys to future sustainable agriculture is a detailed understanding of the mechanisms by which both potassium and ammonium are held and released from soil clay minerals. It is also clear that such an understanding can only originate from an understanding of the complex but fundamental interactions of organic matter and soil clay minerals. If we can understand why horse manure induces such high fertility in soils, we will have made a great step in approaching sustainable agriculture.

Sustainable agriculture means a rational use of ammonium and the elimination of excess nitrates from drinking water and other water resources. Recent events have shown that the water supply in populated areas is to a large extent limited. Poisoning it with excess nitrates due to irresponsible usage in agriculture is to be condemned, only if a viable alternative exists. The key to this is in understanding the equilibria between ammonium and potassium in illitic clay layers in soil clays.

4.2.3
Nuclear Waste Barriers – Strategy and Illite Mineralogy

Much brain energy and money have been invested into the problems of nuclear waste confinement. The high-energy (and longest-lived) waste presents some specific engineering of geological problems which have been investigated in many ways. The general agreement to date is that high-energy nuclear waste should be buried at an intermediate depth (less than one kilometer) so that it will always be possible to access it. The internal, close-field strategies vary from one country or research group to another, but medium-field strategy generally proposes the use of compacted bentonite (smectite) as a barrier to ionic migration. The assumption is that some of the waste can be dissolved in aqueous solution and transported in the ionic state by convection or possibly diffusion. The concept is to ion-exchange the dangerous radionucleide and daughter element radioactive ions from solutions onto smectites. These minerals are selected for their relatively high ionic exchange capacity and their viscosity. The viscosity is assumed when the initially dried bentonite of the barrier is hydrated by the convecting fluids. Smectite will expand, become fluid and penetrate fractures, holes and so forth, thus forming a strong physical barrier of low permeability.

Thus the use of smectite as a medium field barrier is twofold in design: first it expands and blocks, physically, the movement of fluids by its great impermeability and second it absorbs the unwanted ions from the percolating solutions.

One problem in the use of smectite is its inherent thermal instability, due to the smectite-to-illite transformation. The further the reaction proceeds, the lower the cation-fixing capacity of the clay. A second inherent problem is that the clay is not a permanent site for unwanted cations but one of temporary residence. This is inherent to the concept of cation exchange. When the clay contains much less unwanted cation due to its initial composition, it will exchange for that in solution which has a relatively high concentration compared to the clay. However, when the solutions contain less of the unwanted material, the equilibrium of cation exchange will force the clay to release the cations to the solution. Therefore, the clays will stop strong ionic migration in the initial stages of fluid flow contact but will either become saturated with the ion and allow further flow or will release the ions when the concentration in solution decreases. As long as water flows through the clay barrier, the unwanted cations will be dispersed beyond the barrier itself.

It is clear that it would be preferable to find a material which would fix the unwanted cations in a permanent, thermally stable material. Below we outline an alternative barrier design based upon observations of natural systems and certain properties of illite.

4.2.3.1
Natural Analogues of Nuclear Waste Assessment: the Oklo Natural Reactors

The Oklo natural reactors were discovered during uranium prospecting in Gabon in the 1970s because of anomalous ^{235}U isotope abundances. These reactors formed in several mineralized zones (mainly uraninite and pitchblende) which are enclosed in sandstones and conglomerates at their unconformity contact with the Archean basement (Fig. 4.33). The mineralized bodies are several meters long and wide and few tens of centimeters thick. These rocks, covered by black shales, were deposited in a basin about 1.74±0.20 Ga ago, forming the base of the Francevillian sedimentary formations. The uranium probably originated in nearby Archean igneous rocks. It was dissolved during weathering that is to say alteration by oxidative surface waters. The atmospheric oxygen pressure was sufficiently high 2 Ga ago. At that time, the relative ^{235}U isotope abundance was approximately 3% which is comparable to the concentrations used in man-made reactors. The conditions for criticality were attained because of high uranium concentrations, high concentrations of fissile ^{235}U isotope, and presence of water and organic matter as neutron absorbers. It is likely that criticality was not continuous because heat diffusion, reducing the water density, decreased its effectiveness as a moderator and reflector. The duration of the shutdown periods depended on the water supply which locally regulated the temperature and pressure conditions.

Clay minerals are present in three geological facies (Table 4.3):

1. in the undisturbed diagenetic rocks (sandstones and shales);
2. around the ore bodies, where they form an isolating "cap";
3. inside the ore bodies (Fig. 4.33).

Illite and chlorite are the principal species represented in the three facies. Sudoite (lithium, aluminous chlorite) has been detected through the chemical compositions of chlorite (Gauthier-Lafaye et al. 1989; Pourcelot and Gauthier-Lafaye 1999). The maximum temperature reached during diagenesis before

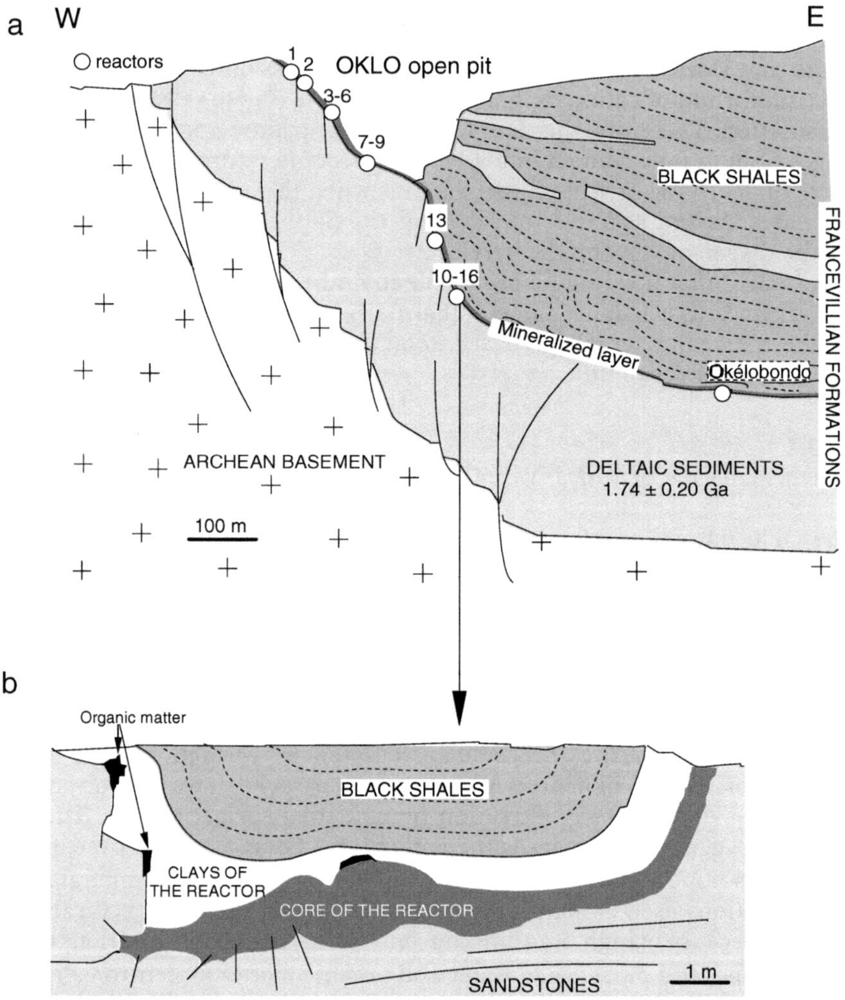

Fig. 4.33a,b. The Oklo natural reactors (from Pourcelot and Gauthier-Lafaye 1999). **a** Simplified cross section of the Oklo-Okélobondo area. **b** An example of the reactor structure (reactor 10)

Table 4.3. Clay minerals in the Oklo reactor FA formations from (Gauthier-Lafaye et al. 1989; Pourcelot and Gauthier-Lafaye 1999)

Rock facies	Clay mineral assemblages	Illite polytype
Black shales and sandstones	Illite+chlorites+quartz	1M to $2M_1$
Barren sandstone	Illite+Fe-chlorite+quartz	$1M_d$
Mineralized sandstone	Illite+Fe,Mg-chlorite	1M
Edges of the reactor	Illite+Mg-chlorite+sudoite	$1M+2M_1$
Core of the reactor	Illite+Mg-chlorite	1M

the uranium deposit formation was measured to be 180 ± 20 °C using fluid inclusion analyses and ^{18}O isotopic composition of clay minerals.

The discovery of ^{235}U enrichment of clays at the contact of the uranium ore showed that fluids have percolated through the reactor during its activity. They removed a fraction of the plutonium which was afterwards incorporated in the crystal structure or on the outer surfaces of the Al-Fe chlorites (Bros et al. 1992).

In these rocks, the crystallinity index of illite increases with depth to values typical of the anchizone-epizone transition indicating a high proportion of WCI. Illites in the undisturbed black shales and sandstones of the FA formation are predominantly of the 1M polytype. They are uniquely of the $2M_1$ polytype in the pore fillings of sandstones at the base of the formation (Gauthier-Lafaye and Weber 1989). The predominant polytype is 1M in the reactor and its surroundings except in heated edges of the reactor where the $2M_1$ polytype was observed. Illite crystals are identified in thin sections as highly birefringent particles contrasting with the low birefringent chlorites.

Surprisingly, the chemical compositions, obtained by electron microprobe, of so-called "illites" are extremely variable (Savary 1995) and scatter out of the classical illite composition domain toward that of smectites (Fig. 4.34). They overlap the composition domain of illite-smectite mixed layers. In absence of any XRD evidence for I/S MLM, this high composition variability must be related to illite being damaged by the ore radioactivity. This effect is so dramatic for chlorites that they do not form, or they disappear in the ore

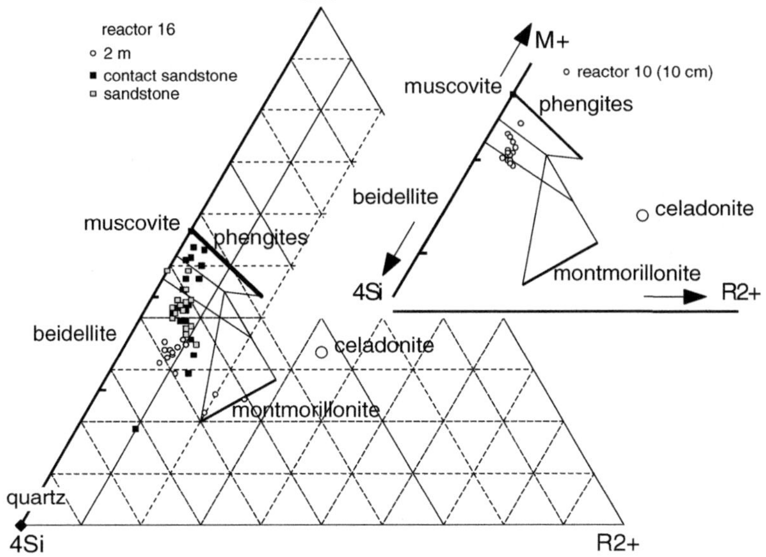

Fig. 4.34. Chemical composition of "illite" particles in the vicinity of reactors 10 and 16 (data from Savary 1995)

deposit itself. They formed after the reactor activity during the cooling period (Pourcelot and Gauthier-Lafaye 1999).

During the activity and cooling periods, many trace elements were mobilized and flowed away from the reactor zone:

- the volatile fission products such as noble gases,
- alkaline earth elements such as Ba and Sr.

However, the clay minerals (illite, chlorite and sudoite) retain some fission elements, mainly of the REE group and exceptionally ^{235}U, which is the decay product of ^{239}Pu (Bros et al. 1993). The question is how to measure the containment efficiency of the natural clay cap. Pourcelot and Gauthier-Lafaye (1999) studied the dispersion of the fission Sm around the reactors in different geological positions: below and above the weathering boundary present today. The variation of the $^{149}Sm/^{147}Sm$ isotopic ratio of the clay materials shows that Sm migrates around the reactor when the clay cap is altered by meteoric waters while it is strictly confined in the unweathered materials. The most important weathering effect on the clay cap is the vermiculitisation of the chlorites. Surprisingly, in spite of the fact that the CEC of the secondary mineral is much more important than that of parent chlorites, the fission decay elements are not retained. On the contrary, the effect of weathering is the loss of these elements (LREE, Ru, PD, Mo, Ag and Cd).

In conclusion, the illite-chlorite assemblage appears to hold, in a rather permanent way, many radionucleides. These elements are incorporated into the minerals under high-temperature conditions. In spite of the fact that the clay minerals forming the cap above the uranium ore have a very small cation exchange capacity, the radioactive decay elements are efficiently confined within the reactor area. Whatever the position of the reactors in the Oklo-Okélobondo area, water circulates through them in any case during the low temperature post-reactor history. The clay cap in this full scale, natural analog is not an impermeable barrier as would have been hoped to be the case for engineered bentonite barriers envisaged in the nuclear waste storage programs. The replacement of chlorite by vermiculite during weathering decreases the containment function of the clay cap.

The relations discussed above show that the efficiency of a clay barrier is not related to the exchange capacity of the minerals which compose it.

4.2.3.2
Cation Retention Properties of Illite

It has been known since the 1960s that in spite of the lower cation exchange capacity of illite compared to vermiculite or smectite, illite adsorbs more Cs from dilute solutions (Tamura and Jacobs 1960). Sawhney (1970) found that illite has a larger selectivity than vermiculite for K^+ and Cs^+ ions at low concentrations because of the presence of frayed-edge sites. The higher selectivity of frayed-edge sites compared to interlayer sites has been shown by alteration experiments on micas (LeRoux and Rich 1969). The selectivity

Table 4.4. Selectivity coefficient, free energies of cation exchange for Cs–M^+ and Cs–M^{2+} and relative CEC for the three types of sorption sites on illite (from Brouwer et al. 1983)

Exchange Equilibrium	Site I (frayed-edges) LnK	ΔG kJ per eq.	Site II LnK	ΔG kJ per eq.	Site III LnK
Ca^{2+}–Cs^+	30.6	−38.5	13.8	−18.5	4.6
Na^+–Cs^+	13	−32.5	6.8	−17.5	3.6
K^+–Cs^+	9.4	−23.4	3.2	−8.0	1.5
% of total sorbed Cs	71.7–98.7		3.5–1.0		24.8–0.3
Site capacity (%CEC)	0.25–0.55		2.5–3.3		96–97

increases on removal of K^+ and then decreases. This results in a stronger polar bonding of the interlayer cations.

^{137}Cs is important because it is one of the most mobile radioactive decay elements which may reach the biosphere around a nuclear waste storage area. A large number of studies have been performed to understand how Cs can be sorbed on minerals and specifically on clay minerals (Cornell 1993). Cesium adsorption on illite has been interpreted as a multi-site ion exchange system (Maes et al. 1985; Comans et al. 1991). According to the value of the selectivity coefficient for Cs^+ against Ca^{2+} cations ($K_{Ca^{2+}}^{Cs^+}$ or K_c), three adsorption sites are proposed (Table 4.4):

1. interlayer highly selective edge sites ($\ln K_{Ca^{2+}}^{Cs^+} = 30$) or "frayed"-edge sites,
2. surface sites. intermediate and low selective sites ($\ln K_{Ca^{2+}}^{Cs^+} = 5$),
3. internal interlayer sites.

Cesium cannot penetrate inside the interlayer space, it is fixed on the layer charge accessible from the edges and surfaces of the crystals. When entering from the edges inside the interlayer sites, cesium loses its hydration shell and is strongly bonded to the structural oxygens of the ditrigonal cavities. Thus, its fixation causes the collapse of the layers to 10.8 Å. These sites are located on the outer basal surfaces.

What are the "frayed-edge sites" in illite crystals? It is commonly recognized that they are formed during hydrothermal alteration or weathering with a partial removal of K^+ ions (Fig. 4.35a). This model was established since the 1960s in many studies devoted to the transformation of mica into vermiculite. Natural and experimental alterations showed that micas are transformed inward from their edges. Is this the case for illite? If so, we must consider that the CEC of illite depends on their degree of alteration. As this is obviously not the case, we must consider that the illite structure itself favours the formation of adsorption sites in which cations are exchangeable. The main difference between micas and illites is the presence of a "pyrophyllite" component which lowers the average layer charge. It could be assumed that the "frayed-edge" sites are linked to pyrophyllite zones as indicated in Fig. 4.35b. If this is true, cesium

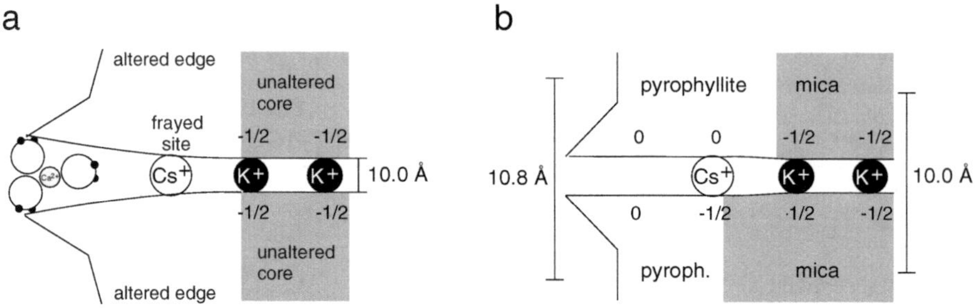

Fig. 4.35a,b. The highly selective sites or "frayed sites" in illite crystals. **a** The origin of the frayed site is an alteration developing inward. **b** A possible alternative: the frayed sites are bordering the "pyrophyllite" zones in illite crystals

ions, for example, could be used to study the distribution of the "pyrophyllite zones" in the crystal structure of illite.

The retention capacity of the frayed sites on illites for Cs ions has been checked by desorption experiments. Whatever their origin (edge alteration or pyrophyllite channel), Cs ions are strongly bound but not irreversibly fixed. The rate of the Cs adsorption is initially fast when all the frayed sites are available for the cation exchange. Later it slows down because the remaining sites are less selective. In a Ca environment, cesium sorption on illite proceeds significantly faster and reaches much higher K_d values than in environments dominated by K^+ ions.

4.2.3.3
Illite or Smectite for Medium Field Retention Materials?

The Facts Bentonites (smectites) are usually considered to be the best candidates for engineered clay barriers in waste storage projects. Of course, several factors other than mineralogical properties must be taken into account for the choice of the best material. The physical properties of barrier material such as the workability can be limiting factors. Assuming that illite and smectite will have similar physical properties during barrier construction, we will focus solely on the mineralogical point of view here. Why is smectite thought to be better than illite for radioactive decay retention? The first argument was the higher CEC or absorption capacity for ions in solution. The second was the rheological properties of wet smectite muds which should be able to fill the fractures in a rock when an initially dehydrated clay is hydrated due to near field leakage. A foreseen drawback to the use of smectite in barrier construction was the smectite-to-illite transformation which may occur in the nuclear storage sites through an increase in temperature due to radioactive decay. This mineral reaction was thought to alter the durability of the engineered barriers and to destroy a part of the cation exchange capacity.

Let us consider the cation exchange capacity argument of smectite for barrier construction. Given that the CEC of smectite is initially higher than that of illite (120 meq compared to 5 for illite), we must consider the exchangeability of various cations on the two materials. In this case we can consider the case of Cs. It is shown above that illite fixes Cs more strongly than smectite, and hence it would be a more efficient long-term barrier material comparing the two minerals. Normal smectite is a cation exchanger, ie. it holds ions in proportion to their chemical activity in solution and to the relative chemical affinity of different ions competing for absorption sites. However, there is always an exchange between ions absorbed on solids and solutions according to ionic concentrations in both phases. Initial sorption of ions on clays occurs when the solutions contain more ions than the clay. However, when solution concentration decreases, the higher concentration of ions on clays comes to equilibrium with the solutions and ions are desorbed from it. A cation exchanger does just that, and as a result the smectite barrier has only a temporary effect. Therefore, the engineered smectite barrier will not prevent the migration of radioactive elements toward the biosphere, it will only delay it. From that point of view, illite is undoubtedly a better candidate than smectite as it fixes Cs ions much more firmly on the clay substrate.

What do we learn from the natural Oklo nuclear reactor analogues? In these systems, the radioactive elements migrate from the reacting mass into the surrounding clay mass which becomes an illite-chlorite assemblage. The natural analogue Oklo reactors show that radioactive decay species other than ^{137}Cs (samarium, neodynium, rubidium and strontium) have been strongly retained in the clays capping the ore deposit. This suggests that plutonium has been fixed on the surfaces of chlorite crystals (Bros et al. 1992). The ions do not migrate further. They are only released through the action of weathering which produces smectitic minerals from the chlorite. It is clear that illite-chlorite is a stable holding medium for radioactive ions as long as these minerals remain stable. If any lesson can be learned from these natural occurrences, it seems to be that a porous mixture of illite and chlorite could be an efficient filter around radioactive waste storage sites.

New Suggestions: an Illite-Chlorite Filter Instead of a Smectite Barrier The usual engineered barrier concept for the safety of nuclear waste storage is based on two objectives:

1. to lower the permeability as far as possible,
2. to absorb the radioactive cations (mainly ^{137}Cs).

These ions can clearly be bound by illite-chlorite assemblages. The usage of an illite-chlorite material would be our proposal for high-energy radioactive burial barriers.

Summary In this section on environmental problems and illite mineralogy, we have two rather different problems where illite stability is called into consideration. In the case of nuclear waste management, illite is useful because it is stable under the conditions of its employment, burial at 500–1,000 m depth under conditions of high temperatures, approximately 200 °C. Here the absorption properties of illite for Cs will make it a useful component of the containment barrier. These properties are more useful than those of the more unstable smectite.

In the case of of soil fertility problems, it appears that the illite layers or minerals are a key to maintaining or retaining potassium or ammonium in the soil substrate. Illite responds rapidly to new chemical situations engendered by agricultural practice. Proper use of illite in soils should lead to a more rational use of soil resources for modern agriculture.

In these two examples, the understanding of illite mineralogy leads to a more rational use of resources and a more safe environment.

Glossary

Stable Phase, Metastable Phase, and Phase Metastably Present

A *stable phase* is one which has a field in P, T, x space where it will remain present for an infinite period of time. This phase will crystallize when P, T, x conditions are such that other phases become unstable. It will not persist outside of the P, T, x conditions of its stability and will transform to another or other phases. Hence, a stable phase can be created or destroyed by changing the chemical and physical constraints of its stable system.

A *metastable phase* is one which exists outside of its field of stability (P, T, x variables). It will transform into another stable phase or phases when sufficient energy is available to effect the transformation. Some phases are assumed to be present even though they have no field of stability. These are true metastable phases. Such phases cannot be formed by reversing the variables (changing the parameters of field stability) once the phase has disappeared from a system.

A phase present in P, T, x space outside of its field of stability is *metastably present.* Such a phase will disappear when enough energy is available to overcome barriers to its destabilization in the field of variables where it is metastably present. It will form within the boundaries of its stability field.

Illite, Mica, PCI, WCI (Terms Used to Designate X-Ray Diffraction Peaks)

Chemical

Illite is assumed to be a low potassiium ($\leq$ 0.9 K per $O_{10}(OH)_2$), dioctahedral aluminous mica (i.e. 10 Å structure) mineral. The mica is assumed to be of the muscovite-phengite series, basically aluminous and dioctahedral, with a nearly 1.0 potassium ions in the interlayer. These chemical characteristics are not known for most materials observed by X-ray diffraction.

XRD

PCI and WCI indicate poorly and well-crystallized illite. These designations are based upon decomposition characteristics of the peaks near 10 Å (10–10.5 Å). PCI is a wide band (1.2–0.5 °2θ Cu Kα), which by its width and position indicates small diffraction domains ($N < 9$) of an essentially 10-Å-mineral. Glycol saturation does not displace the peak position and hence the proportion of expandable layers is less than 5%. WCI designates a narrow peak (0.4–0.15 °2θ Cu Kα) at 10 Å indicating a mica dimension. The narrow peak indicates a larger number of coherently diffracting layers ($N > 14$). These designations are not based upon chemical data which would determine the interlayer ion saturation. Hence, WCI could be a mixture of true illite (low potassiuim content) and a mica of high-temperature origin.

Vermiculite, Smectite

The term vermiculite has two historical origins, and quite different mineralogical meanings. The mineral vermiculite is a high-charge, dioctahedral smectite (with appropiate swelling characteristics), usually derived from hydrothermal alteration of biotite-bearing eruptive rocks . It origin is high temperature (i. e. outside the range of soils, sediments and sedimentary rocks). Its crystal size is greater than 2 µm for the most part and hence, technically, it is not a clay mineral, according to the grain size criterion for this mineral group.

Vermiculite of low-temperature (soils for the most part) origin and has a non-smectite behavior, i. e. poor and irregular swelling. It is generally considered to be a smectite with hydroxy Al and Fe interlayer ions. Other names have been used to designate soil vermiculite (HIVn HIS where HI = hydroxy-interlayer). Usually potassium saturation collapses some of the soil vermiculite layers to 10 Å. Heating to above 200 °C changes the structure to a position between 10 and 15 Å.

Recent use of the term vermiculite has been extended to certain types of the smectitic interlayer units in illite/expanding interlayer clay mineral structures. Here, the grain size is by definition small, the swelling properties are dependent upon the charge and the charge site of the 10 Å layer part of the structure, as is the case for a smectite. These vermiculite layers are of high charge or with a specific charge site such that the hydration state of monovalent cations is less than that of the other exchange cations (Ca, Mg or Sr for example). This gives a lower, but not 10 Å spacing to the vermiculite layers.

In this book, we will use the term *vermiculite* for some layers in mixed-layer minerals to signify a high-charge smectite layer. Vermiculite found in soils (HIV, HIS and other poorly expanding and contracting phases of greater than 10 Å basal spacing) will be referred to as *soil vermiculite*.

Identification of Expandable Layer Types Using XRD

Montmorillonitic Layer

Fully expandable layer (2 water layers or 2 ethylene-glycol layers at 80% RH, whatever the interlayer cation, except for K^+ and NH_4^+). Most of the charge originates from the octahedral sheet. Total layer charge varies from 0.3 to 0.6 per $O_{10}(OH)_2$.

Vermiculitic Layer

High-charge layer which is partly expandable (1 water layer or 1 ethylene glycol layer at 80% RH) if the interlayer is not K- or NH_4-saturated. If so, the interlayer is collapsed to $\approx$ 10 Å. In this book the layer charge is assumed to vary from 0.6 to that of illite: 0.9 per $O_{10}(OH)_2$, and to originate mostly from the tetrahedral sheets surrounding the interlayer space. Glycol saturation produces a one layer, 13.5-Å spacing.

Basic Definitions

Illite or Smectite Layer

There are two ways to define an illite or smectite layer:

- a 2/1 unit with an interlayer sheet whose charge is distributed symmetrically on each side of the interlayer space, with the crystallographic origin being located in the middle of the octahedral sheet.
- a "fundamental particle" whose symmetry plane is in the middle of the interlayer cation sheet

MacEwan Crystallite

The McEwan model provides a one-dimensional description of I-S along the c^* direction. Layers of illite and smectite are pictured as intimately interlayered, stacked either as a random, as an ordered (even partially), or as a segregated sequence. Such a sequence of strictly parallel layers of different nature and/or thickness, acting as a coherent unit, scatters X-rays (Moore and Reynolds 1989). The number of diffracting layers (N) indicates the coherent scattering domain size (CSDS). After K-saturation and heating (Drits et al. 1997c), the CSDS may be computed from the peak profile using a modified Scherrer or Bertaut-Warren-Averbach procedure (Drits et al. 1997c, 1998b). This parameter may be derived by fitting the complete XRD profile by a trial-and-error approach (Drits and Tchoubar 1990).

Fundamental Particles (FP)

Particles which have interfaces that are capable of absorbing water and organic molecules and, as a consequence, behave as a smectite interlayer if two such particles are stacked. A smectite FP is 10 Å thick whereas thicker particles are illite FPs (Nadeau et al. 1984b). Fundamental particles can be considered as sub-units of a MacEwan crystallite (Eberl and Srodon 1988; Reynolds 1992).

Crystal

Three-dimensional unit of strictly parallel layers that scatters X-rays coherently along the three axes. The periodicity along the three axes may be disrupted by crystal defects (turbostratism included). Three different layer types can be distinguished in I/S crystals: smectite, vermiculite, and illite. The crystal thickness (T_{cry}) is identical to the N value used by Reynolds (1985) for the CSDS.

Particle

Superunit composed of several crystals connected by coalescence or formed by epitaxial growth process. Crystal defects may interrupt the periodicity along any crystallographic axis.

Stacking Sequences in Illite-Smectite Mixed-Layered Minerals

This chapter is inspired from the book "Clays" (Meunier, in press).

Stacking Sequence

This sequence is statistically determined as a function of the ordering type which is characterized by the Reichweite (R) parameter (Jagodzinski 1949). This parameter indicates how far a layer may influence the occurrence of another layer type in the stacking sequence. However, the R parameter alone does not permit a description of the whole sequence, which is determined by the relative proportions of the different layer types and by the junction probabilities used to describe the influence of a sequence fragment on the nature of the next layer. It should be noted that the occurrence of a regular ISIS sequence does not imply an R1 stacking, with or without maximum possible degree of order, as such a sequence could be possible in a R0 structure (e.g.).

Conditions of Interstratification

The most commonly described two-component mixed-layer minerals are illite and dioctahedral smectite, kaolinite and dioctahedral smectite, chlorite and saponite. The condition that apparently best explains their frequency is the slight difference between the a and b dimensions of the two types of layers. Mixed-layer minerals formed by the stacking of trioctahedral and dioctahedral layers are unquestionably rare. Recent studies show that, even though rarely described in the literature, naturally occurring three-component mixed-layer minerals may be more abundant than commonly thought (Drits et al. 1997).

A mixed-layer mineral is identified when its components, their proportions, and the degree of order of their stacking sequence have been determined. Let us consider a mixed-layer mineral composed of two components A-B occurring in varying relative proportions W_A and W_B. It will be fully described if succession probabilities of A and B layers ("nearest-neighbour") are known: P_{AA}, P_{AB}, P_{BA} and P_{BB}. Generally, these six parameters are linked by four independent relationships:

$$W_A + W_B = 1\,, \qquad P_{AA} + P_{AB} = 1\,,$$
$$P_{BA} + P_{BB} = 1\,, \qquad W_A P_{AB} = W_B P_{BA}\,.$$

So there are six variables and four non-redundant equations which permit their calculation if two are fixed. Usually, the composition as well as one of the junction probabilities are fixed ($W_A = 0.4$ and $P_{BB} = 0.8$, for instance). The development of probability calculations can be found in classical books (Brindley and Brown 1980 for instance).

Random Stacking Sequence (R0)

In a random stacking sequence, an A layer may be followed by a A or a B layer without any forbidden sequence (Fig. A.1a). The succession probability of A and B layers depends only on the relative proportions W_A and $W_B = 1 - W_A$. Therefore, the probability for A to follow B is given by $P_{AB} = W_B$; we know

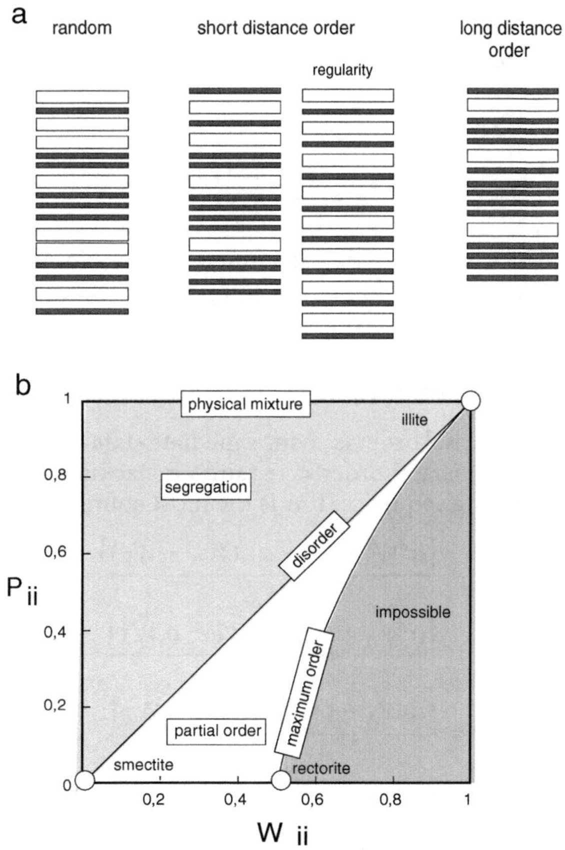

Fig. A.1a,b. Interstratification. **a** Schematic diagram of randomly ordered, ordered and regularly ordered interstratifications of two types of A and B layers such as illite (10 Å) and smectite (17 Å). **b** Relationship between succession probability of two illite layers (P_{AA}) and illite content (W_A) establishing mixing, segregation and interstratification domains (domain in *white*)

that $P_{AB} + P_{AA} = 1$, so $P_{AA} = W_A$. The variation of P_{AA} as a function of W_A of the random stacking sequence is represented by a straight line which slope is equal to 1 (Fig. A.1b)

Maximum Degree of Order (R1)

If $W_B \leq 0.5$, the maximum degree of order is reached if the probability of finding a B-B pair is zero (Fig. A.1a). Let's consider a 10-layer stacking sequence containing 40% of B; a sequence such as B-A-B-A-A-B-A-B-A-A may exist, but a sequence such as B-A-A-A-B-B-A-B-A-A is forbidden. As W_B decreases, the probability of formation of B-A-B-type sequences becomes zero. A long-distance order may be established with at least two, then three consecutive A after each B. Although the actual influence of the B layer on the subsequent layers is not established in terms of chemical or physical interactions, A-rich crystal structures can still be described using these long-distance ordering modes: order for a triplet R=2; order for a quadruplet R=3 etc.

In R1 stacking sequences, if A is the most abundant type of layer ($W_A \geq 0.5$), the maximum order implies that no BB layer pair can exist ($P_{BB} = 0$). Therefore, the following relations may be inferred:

$$P_{BB} = 0$$

$$P_{BA} = 1$$

$$P_{AB} = \frac{W_B}{W_A}$$

$$P_{AA} = 1 - P_{AB} = \frac{(W_A - W_B)}{W_A}$$

The variation of P_{AA} as a function of W_A for the maximum order is given by the curve originating in the composition of rectorite in the case of illite/smectite mixed layers (Fig. A.1b).

Partial Order (R1.)

The partial order is an intermediate state between random order and maximum order: partial order = α random stacking sequence $+(1-\alpha)$ maximum order stacking sequence. If A is the most abundant type of layer, then:

$$P_{AA} = \frac{\left[\alpha\, W_A^2 + (1-\alpha)\,(W_A - W_B)\right]}{W_A}$$

$$P_{AB} = \frac{\left[\alpha\, W_A + (1-\alpha)\, W_B - \alpha\, W_A^2\right]}{W_A}$$

$$P_{BA} = \frac{\left[\alpha\, W_A + (1-\alpha) W_B - \alpha\, W_A^2\right]}{W_B}$$

$$P_{BB} = \frac{\left[\alpha\, W_A^2 - \alpha\, W_A + \alpha\, W_B\right]}{W_B}$$

The α parameter varies between 0 and 1, which requires that P_{AA} vary between the values calculated for the random stacking sequence and the maximum order stacking sequence. The partial order domain is limited by the maximum order curve and the random order straight line (Fig. A.1b).

Segregation (R1)

This type of stacking sequence is intermediate between random state and physical mixing (Fig. 1.13b). Consequently P_{AA} must be fixed as a function of W_A in order to be located within the domain of segregation:

$$P_{AA} = \alpha + (1 - \alpha) W_A$$
$$P_{AB} = (1 - \alpha) W_B$$
$$P_{BA} = (1 - \alpha) W_A$$
$$P_{BB} = \alpha + (1 - \alpha) W_B$$

Probabilities do not describe the heterogeneity of natural clay minerals in which at least two parameters may vary: the number of layers in a stacking sequence, and the proportion of each type of layer within this stacking sequence. Variations of the second parameter are described by Markovian probabilities applied to quasi-homogeneous structures (same number of layers in the stacking sequence for all crystals). In this manner, proportions of the various possible types of stacking sequence can be calculated in a population of crystals exhibiting the same number of layers and the same degree of order: randomly ordered (Fig. A.2a), and ordered (Fig. A.2b). This probabilistic theory is described in detail in a book by Drits and Tchoubar (1990). It is used as a basis by modelling software to calculate diffraction patterns, for instance NEWMOD (Reynolds 1985) or MLM2C (Plançon and Drits 2000).

Méring (1949) has proposed an elegant method for readily identifying randomly ordered or ordered mixed-layer minerals by X-ray diffraction. Indeed, he has shown that, when the components of the mixed-layer mineral show neighbouring peaks, the latter interfere, forming a wider diffraction band with an intermediate angular position. This position varies with the respective amounts of the two components. Therefore, except in case of perfect regularity, mixed-layer minerals can be identified by non-rational series of peaks. Only regularly ordered mixed-layer minerals show rational series as do pure minerals. Let's consider illite-smectite mixed layers. The rational 00*l* series characterise the following components:

- ethylene glycol-saturated smectite: 17 Å, 8.5 Å, 5.67 Å, 3.4 Å ...
- illite: 10 Å, 5 Å, 3.33 Å ...
- rectorite (50%illite, 50% smectite): 27 Å, 13.5 Å, 9.00 Å, 6.75 Å, 5.40 Å, 4.50 Å ...

Randomly ordered mixed-layer minerals are characterised by a non-rational 00*l* series of peaks at 17 Å, 10 to 8.50 Å, 5.67 to 5.00 Å, and 3.40 to 3.33 Å.

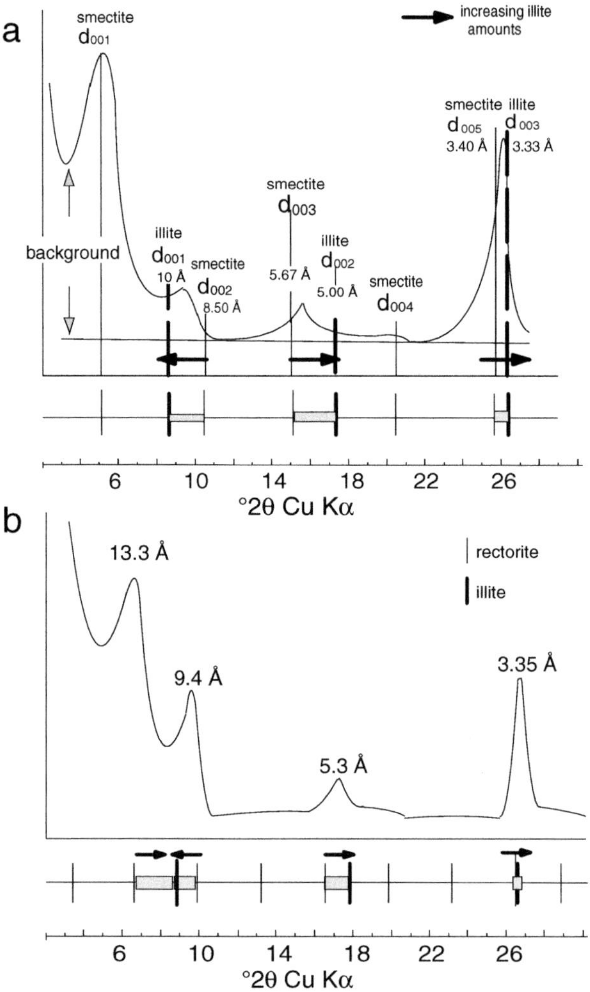

Fig. A.2a,b. Application of Méring's law (1949). **a** Diffraction pattern of randomly ordered illite-smectite mixed layer. **b** Diffraction pattern of an ordered mixed-layer mineral

The greater the illite content, the more the 2nd-, 3rd- and 4th order diffraction bands shift towards the typical positions of illite at 10, 5 and 3.33 Å, respectively (Fig. A.2c). Irregularly ordered mixed-layer minerals with more than 50% illite may be considered as randomly ordered rectorite and illite mixed layers (Drits et al. 1994). Application of Méring's law permits a determination of the non-rational series of their characteristic peaks (Fig. A.2d). Based on this principle, several methods for determining the illite content of illite-smectite mixed layers have been proposed (Srodon 1980, 1981, 1984; Watanabe 1988). The same reasoning applies to other types of mixed layers (chlorite-smectite, kaolinite-smectite ...) by adopting the rational 00*l* series characteristic of each component.

Three-Component Mixed-Layer Minerals and Non-Markovian Probabilities

Recent work dedicated to illitisation in diagenetic series (Drits et al. 1997) show that mixed-layer clays frequently cannot be properly described in a 2-component structure such as illite+smectite (2 ethylene glycol layers). The position and width of X-ray diffraction peaks are only reproduced well if a third component is introduced: vermiculite (1 ethylene glycol layer). Layer stacking sequences are governed by the same probabilistic laws (Plançon and Drits 2000). Although far beyond the scope of this book, these observations can by no means be ignored, even if identification of 3-component mixed-layer minerals is relatively complex for the present time. An expert system and a MLM3C calculation software for three-component diffraction patterns have been developed by these authors.

The calculation of the theoretical diffraction pattern that best fits the experimental diffraction pattern modifies its interpretation to a significant extent. Accordingly, Claret (2001) has shown that the diagenetic series of the eastern part of the Paris Basin and of the Gulf Coast (USA), considered as a classical transition from randomly ordered (I/S R=0) to ordered (I/S R=1) illite-smectite mixed layers, may be regarded as a mixture of smectite, I/S R=0 with a high illite content (65–70%) and illite. Furthermore, he has shown that the superstructures visible in samples collected from the basis of the sedimentary series of the Gulf Coast cannot be reproduced by calculations using Markovian probabilities.

Cation Exchange Capacity (CEC)

External CEC and Internal CEC

Clays have the property of fixing reversibly some cations dissolved in the surrounding solutions. The cation exchange capacity (CEC) corresponds to the number of negative charges likely to fix cations in this manner. It is expressed in centimols per kg ($\mathrm{cmol\,kg^{-1}}$), which is a translation in the international system of units of the milliequivalents per 100 g (meq) which have been used traditionally for decades. Cations can only be exchanged if they are weakly bonded to the external or internal surfaces (interlayer spaces) of crystals.

The external CEC depends on the number of bonding sites of cations on the external crystal surfaces. These negatively charged sites can correspond to charges resulting from the tetrahedral or octahedral substitutions of those sheets forming the (001) faces, or to defects emerging on these faces. The interrupted bonds of the *hk*0 faces can be added to this. Therefore, the external CEC is a direct function of the crystal size: for a given volume or mass, the greater the external surfaces, the smaller the crystal size. Consequently, the measurement of the external CEC gives information on the mean crystal sizes.

The properties of external sites depend on pH; this is why they are called *variable charges* of the clay material where H^+ compensates negative CEC sites.

The internal CEC reflects the charge deficiency of 2:1 layers in the case of vermiculites and smectites. Consequently, the internal CEC depends on the *permanent charges* of clay species. One might think that the higher the structural charges, the greater the CEC. This would mean that the CEC of micas should be greater than that of smectites or vermiculites. In reality, it is the opposite because when structural charges are too high, cations are irreversibly fixed in the interlayer spaces.

Example: theoretical calculation of the CEC of a montmorillonite The exchange capacity is defined as the amount of cations retained by all the negative charges (permanent charges) in 100 g of clay at pH 7. It is expressed in milliequivalents (meq) per 100 g of clay. The milliequivalent is equal to (charge/mass) × 1,000; it is equal to one centimole of unit charge per kilogram of dry matter ($\mathrm{cmol\,kg^{-1}}$). The exchange capacity is calculated following the relation:

$$\mathrm{CEC} = (\mathrm{charge/mass}) \times 1{,}000 \times 100$$

Let's consider a montmorillonite half unit cell whose formula is: $Si_4\,O_{10}\,Al_{1.7}\,Mg_{0.3}\,(OH)_2\,Na_{0,3}$. The mass of the half unit cell is 367 g; the charge is 0.33 so

$$\mathrm{CEC} = (0.33/367) \times 10^5 = 89.9\,\mathrm{meq/100\,g}$$

This value of 89.9 meq/100 g corresponds to the exchange capacity of the interlayer sites. The CEC related to the external surfaces of crystals (or quasi-crystals) must be added. The overall value for smectites varies from 100 to 120 meq/100 g.

Selectivity Coefficient K_s

Eberl (1980): "Cation selectivity and cation fixation in clay both result from the interplay of two competing forces: 1) the force of attraction of a cation for its hydration shell; 2) the force of attraction of a cation for clay surfaces. Selectivity arises because these forces differ for different cations. Fixation occurs when the second force exceeds the first".

A clay, which has negatively charged exchangeable sites (X_2) that are saturated by Mg^{2+}, is dispersed in a $CaCl_2$ solution and energetically stirred. Once in the neighbourhood of these sites (diffusion), the Ca^{2+} ions replace the Mg^{2+} ions, which enter the solution. Although the exchange is not a classical chemical reaction, because only low-energy bonds are involved, it can be written following the same formalism (McBride 1994):

$$CaCl_{2(sol)} + MgX_{2(clay)} \Leftrightarrow MgCl_{2(sol)} + CaX_{2(clay)}$$

The exchange equilibrium constant K_{eq} is considered as the equilibrium constant of a classical reaction and is expressed from the activities of reactants and products:

$$K_{eq} = \frac{(MgCl_2)(CaX_2)}{(CaCl_2)(MgX_2)}$$

The activity of solids is equal to 1 by convention,

$$K_{eq} = \frac{(MgCl_2)}{(CaCl_2)}$$

Thus defined, the constant K_{eq} imposes the following rule: as long as the solid can exchange Mg^{2+} ions, the activities of those ions in solution are constant. This is not the case in real exchanges because the activity of solids is not equal to 1. Indeed, MgX_2 and CaX_2 values vary as functions of the proportion of exchangeable sites occupied by Mg^{2+} and Ca^{2+}. The true variable is the ratio of the concentrations of adsorbed ions $[CaX_2]$ and $[MgX_2]$:

$$M_{Ca} = \frac{[CaX_2]}{[CaX_2] + [MgX_2]}$$

$$M_{Mg} = \frac{[MgX_2]}{[CaX_2] + [MgX_2]}$$

Therefore, since the equilibrium constant K_{eq} does not describe the exchange phenomenon, it is replaced by a constant K_s – or selectivity coefficient – which is expressed as follows:

$$K_s = \frac{(MgCl_2)M_{Ca}}{(CaCl_2)M_{Mg}}$$

Deviation from Ideality

Ion exchange modifies the composition of clay which varies between calcic and magnesian end-members. If this variation is considered similar to a solid solution, the selectivity coefficient is likely to deviate from ideality by the addition of mixing energy. The more similar the charges and diameters of ions, the lower this energy; the more different, the higher this energy. This means that the concentrations $[CaX_2]$ and $[MgX_2]$ need to be "corrected" by selectivity factors: $(CaX_2) = f_{Ca}\ M_{Ca}$and $(MgX_2) = f_{Mg}M_{Mg}$. The exchange equilibrium constant then turns into:

$$K_E = \frac{(MgCl_2) \times f_{Ca}M_{Ca}}{(CaCl_2) \times f_{Mg}M_{Mg}} = K_S\frac{f_{Ca}}{f_{Mg}}$$

where K_S is the selectivity coefficient.

The example of the exchange of Rb^+ cations in a Na-saturated montmorillonite shows that K_S is greater when fewer exchangeable sites have fixed this

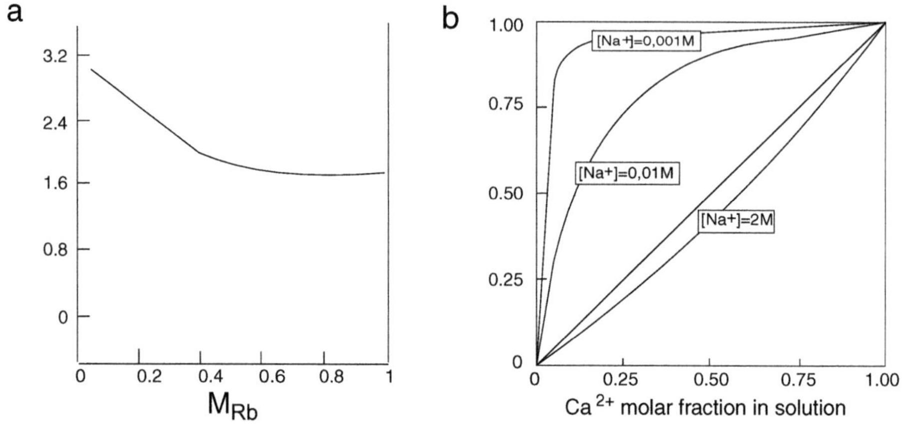

Fig. A.3a,b. Variation of the selectivity coefficient K_S. **a** Rb^+- Na-montmorillonite exchange. K_S varies as a function of the respective proportions of the exchangeable sites occupied by Rb^+ and Na^+ ions (M_{Rb}). **b**) Relationship between Ca^{2+} contents in the exchanger (montmorillonite) and the solution. The more diluted the electrolyte, the greater the selectivity (K_s) for Ca^{2+}

cation (Fig. A.3a). For monovalent cations, the selectivity order is as follows: $Cs^+ > K^+ > Na^+ > Li^+$; for bivalent cations: $Ba^{2+} > Sr^{2+} > Ca^{2+} > Mg^{2+}$. This order is defined by the size (diameter) of the association cation-hydration sphere: the cation with the smaller diameter displaces the one with the greater diameter. Selectivity varies with the electrolyte concentration: the more diluted the solution, the greater the selectivity (Fig. A.3b).

If the exchange is performed at constant temperature, the selectivity coefficient (K_S) determines a cation exchange isotherm. Exchange isotherms are modified when pH conditions vary. Indeed, the behaviour of the H^+ ion is similar to that of the other cations with which it competes to bond to the variable or permanent charges. This necessitates that the measurement of isotherms is performed under controlled pH conditions. What does the selectivity coefficient mean at the crystallite level? Obviously, it corresponds to exchange energies that differ according to sites. If one considers smectites, for which variable charges are considered negligible, the exchange sites are located in the interlayer zone. Talibudeen and Goulding (1983) have shown with microcalorimetric analyses the occurrence of six groups of sites with an enthalpy of exchange (exothermal reaction) that varies from 5.7 to 10.9 kJ/eq. For low-charge smectites, the enthalpy of most of the exchange sites varies from 5.7 to 7.5 kJ/eq. They have a few sites of higher enthalpy. These results have been confirmed by the study of Cs for Ca exchanges (Maes et al. 1985). Although not confirmed yet, it seems obvious that the variation in the exchange energy is related to the charge heterogeneities at the surface of 2:1 layers. These heterogeneities could result from the way in which ionic substitutions are distributed between tetrahedral and octahedral sheets.

Layer Charge and CEC

The charge (tetrahedral+octahedral) of the layers making up the crystal structures of smectites can vary from approximately 0.30 to 0.65 per Si_4O_{10}. Such a difference (factor of two) entails changes in the chemical and physical properties of crystallites. Indeed, cations are weakly bonded in the interlayer space of low-charge layers; they are totally exchangeable and polar molecules such as water, glycol or glycerol can enter this space. But, in contrast, the much more strongly bonded cations in the interlayer space of high-charge layers are not all exchangeable. The K^+ ions in particular adopt a configuration similar to the one they have in the structure of micas or celadonites. In this case, the layers lose their expansion capacity by absorption of polar molecules.

The presence of high-charge or low-charge layers can be identified readily using a method based upon the differences in chemical properties. Suspended clays are saturated with potassium by agitation in a 1 M KCl solution (pH 7). After rinsing, they are deposited on a glass slide (Malla and Douglas 1987). The resulting oriented sample is heated at 110 °C for 12 h so that all the molecular water is removed from the interlayer space. The K^+ ions are then irreversibly bonded in the high-charge sites and prevent polar molecules from entering. The structure of smectite acts like that of a mica (10 Å). Since low-charge sites remain accessible to polar molecules, the expansion takes place after glycerol saturation (18 Å). Therefore, a calcium-saturated smectite exhibiting a homogeneous swelling of all the interlayer spaces – whether of high or low charge – behaves like a complex mixed-layer mineral consisting of layers with 2, 1 or 0 glycol layers when saturated with potassium.

This simple method can be improved by saturating anew the K-smectite with Ca^{2+}ions. This permits identification of the various types of expandable layers:

- low-charge smectite: Ca- or K-saturated → 2 layers of ethylene glycol (EG),
- intermediate-charge smectite: Ca-saturated → 2 EG; K-saturated → 1 EG,
- "vermiculite"-like: Ca-saturated → 1 EG; K-saturated K → 0 EG.

Nevertheless, the method is not accurate enough to measure the value of these charges. This measurement is performed using a more complex method of difficult implementation: the absorption of alkylammonium ions. This method, devised by Lagaly and Weiss (1969), is thoroughly described by Mermut (1994). An empirical simplified method has been proposed by Olis et al. (1990).

References

Aja SU, Rosenberg PE, Kittrick JA (1991) Illite equilibria in solution: I. Phase relationships in the system K_2O-Al_2O_3-SiO_2-H_2O between 25 and 250 °C. Geochim Cosmochim Acta 55:3431–3435

Al Shaara M (1986) Etude géochimique et métallogénique des minéralisations (U-Ba) du nord du massif des Palanges (Aveyron, France). Thèse Univ Paris VI, p 174

Altaner SP, Hower J, Whitney G, Aronson JL (1984) Model for K-bentonite formation: evidence from zoned K-bentonites in the disturbed belt, Montana. Geology 12:412–415

Altaner SP (1989) Calculation of K diffusional rates in bentonite beds. Geochimica Cosmochimica Acta 53:923–931

Altaner SP, Ylagan RF (1997) Comparison of structural models of mixed layer illite/smectite and reaction mechanisms of smectite illitization. Clays Clay Minerals 45:517–533

Altaner SP, Weiss CA, Kirkpatrick RJ (1988) Evidence from ^{29}Si NMR for the structure of mixed-layer illite-smectite clay minerals. Nature v 331:699–702

Aronson JL, Hower J (1976) Mechanism of burial metamorphism of argillaceous sediment: 2. Radiogenic argon evidence. Geol Soc Amer Bull 87:738–744

Awwiller DN (1993) Illite/smectite formation and potassium mass transfer during burial diagenesis of mudrocks: a study from the Texas Gulf Coast Paleocene-Eocene. J Sed Petrol 63:501–512

Bailey SW (1980) Structures of layer silicates. In: Brindley GW, Brown G (eds) Crystal structures of clay minerals and their X-ray identification. Mineralogical Society Monograph 5:1–123

Baronnet A (1975) Growth spirals and complex polytypism in micas. I. Polytypic structure generation. Acta Cryst A31:345–355

Baronnet A (1976) Polytypisme et polymorphisme dans les micas. Contribution à l'étude du rôle de la croissance cristalline. Thèse Univ Aix-Marseille III, p 256

Baronnet A (1982) Ostwald ripening in solution. The case study of calcite and mica. Estudios Geologicos v 38:185–198

Baronnet A (1991) Mûrissement d'Ostwald des minéraux: rappel des conditions limitantes. Bulletin de Liaison de la Société Française de Minéralogie et Cristallographie v 2:28–28

Bauer A (1997) Etude du comportement des smectites et de la kaolinte dans des solutions potassiques (0.1–4 M). Thèse Univ Paris 6, France

Bauer A, Velde B, Gaupp R (2000) Experimental constraints on illite crystal morphology. Clay Miner 35:587–597

Beane RE (1983) Hydrothermal alteration in silicate rocks. Southwestern North America. In: Titley SR (ed) Advances in geology of the porphyry copper deposits. Southwestern North America. The University Arizona Press, Tucson pp 117–137

Beaufort D (1987) Interstratified chlorite/smectite ("metamorphic vermiculite") in the upper precambrian greywackes of Rouez, Sarthe, France. In Proc Inter Clay Conf, Denver In: Schultz LG, Olphen H Van, Mumpton FA (eds) 1985, pp 59–65

Beaufort D, Baronnet A, Lanson B, Meunier A (1997) Corrensite: a single phase or a mixed layered phyllosilicate of the saponite-chlorite conversion series? The case study of the Sancerre-Couy deep drill-hole (France). Amer Miner 82:109–124

Beaufort D, Berger G, Lacharpagne JC, Meunier A (2001) An experimental alteration of montmorillonite to a di+trioctahedral smectite assemblage at 100 and 200 °C. Clay Miner 36:211–225

Beaufort D, Cassagnabère A, Petit S, Lanson B, Berger G, Lacharpagne JC, Johansen H (1998) Kaolinite-to-dickite conversion series in sandstone reservoirs. Clay Miner 33:297–316

Beaufort D, Papapanagiotou P, Patrier P, Fujimoto K, Kasai K (1995) High temperature smectites in active geothermal field. In: Kharaka YK, Chudaev OV (eds) Proc 8th Inter Symp Water-Rock Interact, pp 493–496

Beaufort D, Patrier P, Laverret E, Bruneton P, Mondy J (2003) Clay alterations associated with Proterozoic unconformity-type uranium deposits in the East alligator Rivers Uranium Field (Northern Territory, Australia). Econ Geol (in press)

Beaufort D, Papapanagiotou P, Patrier P, Traineau H (1995) Les interstratifiés I-S et C-S dans les champs géothermiques actifs: sont-ils comparables à ceux des séries diagénétiques? Bulletin des Centres de Recherche et d'Exploration-Production Elf Aquitaine v 19:267–294

Beaufort D, Papapanagiotou P, Patrier P, Fujimoto K, Kasai K (1995) High temperature smectites in active geothermal systems. In: Kharaka YK, Chudaev OV (eds) Proceedings of the 8th International Symposium on Water-rock interaction, pp 493–496

Berger G, Velde B (1992) Chemical parameters controlling the propylitic and argillic alteration process. Eur J Mineral 4:1439–1454

Berger G, Lacharpagne JC, Velde B, Beaufort D, Lanson B (1995) Mécanismes etc contraintes cinétiques des réactions d'illitisation d'argiles sédimentaires, déduits de modélisations d'interaction eau-roche. Bull Centre de Recherche Exploration et Production 19:225–234

Berger G, Lacharpagne JC, Velde B, Beaufort D, Lanson B (1997) Kinetic constraints on illitization reactions and the effects of organic diagenesis in sandstone/shale sequences. Applied Geochim 12:23–35

Berger G, Turpault MP, Meunier A (1992) Dissolution-precipitation processes induced by hot water in a fractured granite. Part 2: Modelling of water-rock interaction. Eur J Mineral 4:1477–1488

Berg-Madsen V (1983) High-alumina glaucony from the Middle Cambrian of Öland and Bornholm, southern Baltoscandia. J Sedim Petrol 53:875–893

Besson G, Drits VA (1997a) Refined relationships between chemical composition of dioctahedral fine-grained mica minerals and their infrared spectra within the OH stretching region. Part I: Identification of the OH stretching bands. Clays Clay Miner 45:158–169

Besson G, Drits VA (1997b) Refined relationships between chemical composition of dioctahedral fine-grained mica minerals and their infrared spectra within the OH stretching region. Part II: The main factor affecting OH vibration and quantitative analysis. Clays Clay Miner 45:170–183

Besson G, Drits VA, Daynyak LG, Smolliar BB (1987) Analysis of cation distribution in dioctahedral micaceous minerals on the basis of IR spectroscopy data. Clay Miner 22:465–478

Bethke CM, Altaner SP (1986) Layer-by-layer mechanisms of smectite illitization and application to a new rate law. Clays Clay Miner 34:136–145

Biscaye PE (1965) Mineralogy and sediment of recent deep-sea clay in the Atlantic and adjacent seas and oceans. Bulletin Geological Society America 76:803–832

Bishop BP, Bird DK (1987) Variation in sericite compositions from fracture zones within the Coso Hot Springs geothermal system. Geochim Cosmochim Acta 51:1245–1256

Bjorlykke K, Aagaard P (1992) Clay minerals in Noth Sea sandstones. In: Houseknecht DW, Pittman ED (eds) Origin, Diagenesis, and Petrophysics of Clay Minerals in Sandstones. SEPM Special Publication 47:65–80

Bjorlykke K, Aagaard P, Dypvik H, Hastings DS, Harper AS (1986) Diagenesis and reservoir properties of Jurassic sandstones from the Haltenbanken area, offshore Mid-Norway. In: Spencer AM (ed) Habitat of hydrocarbons on the Norwegian Continental Shelf. Graham & Trotman London pp 275–286

Boistelle R (1982) Mineral crystallization from solution. Estudios Geol 38:135–153

Boles JR, Franks SG (1979) Clay diagenesis in Wilcox sandstones of southwestern Texas: Implications of smectite diagenesis on sandstone sedimentation. Journal Sedimentary Petrology 49:55–70

Bonhomme MG, Dos Santos RP, Renac C (1995) La datation potassium-argon des minéraux argileux. Etat des connaissances. Bull Centre Rech Explor-Prod Elf Aquitaine 19:197–223

Bonorino FG (1959) Hydrothermal alteration in the front range mineral Belt, Montana. Bull Geol Soc Amer 70:53–90

Bouchet A, Lajudie A, Rassineux F, Meunier A, Atabek R (1992) Mineralogy and kinetics of alteration of a mixed-layer kaolinite/smectite in nuclear waste disposal simulation experiment (Stripa site, Sweden). In: Meunier A (ed) Clays and hydrosilicate gels in nuclear fields, pp 113–123

Brindley GW (1980) Order-Disorder in clay mineral structures. In: Brindley GW, Brown G (eds) Crystal structures of clay minerals and their X-ray identification. Mineralogical Society Monograph 5:125–195

Brindley GW, Brown G (1980) Crystal structures of clay minerals and their X-ray identification. Mineralogical Society Monograph No. 5, p 485

Bros R, Turpin L, Gauthier-Lafaye F, Holliger P, Stille P (1993) Occurrence of naturally enriched 235U: implications for plutonium behaviour in natural environments. Geochim Cosmochim Acta 57:1351–1356

Brouwer E, Bayens B, Maes A, Cremers A (1983) Cesium and Rubidium ion equilibria in illite clay. J Phys Chem 87:1213–1219

Brown G, Norrish ML (1952) Hydrous micas. Miner Mag 29:929–932

Bruce C (1984) Smectite dehydration – Is relation to structural development and hydrocarbon accumulation in the Northern Gulf of Mexico Basin. Ame Assoc Petrol Geol Bull 68:673–683

Brusewitz A-M (1986) Chemical and physical properties of Paleozoic bentonites from Kinnekulle, Sweden. Clays Clay Minerals 34:442–454

Bühmann C, Schoeman JL (1995) A mineralogical characterization of Vertisols from northern regions of the Republic of South Africa. Geoderma 66:239–257

Burst JF (1959) Post diagenetic clay mineral-environmental relationships in the Gulf Coast Eocene in clays and clay minerals. Clays Clay Miner 6:327–341

Burst JF (1969) Diagenesis of Gulf Coast clayey sediments and its possible relation to petroleum migration; Bulletin Association Petroleum Geologists 53:73–93

Byström A-M (1954) 'Mixed layer' minerals in Ordovician bentonite beds a Kinnekulle Sweden. Nature 173:783
Carr A (2000) Suppression and retardation of vitrinite reflectance, Part I. Formation and significance for hydrocarbon exploration. Jour Petrol Geol 23:313–343
Cassagnabère A, Iden IK, Johansen H, Lacharpagne JC, Beaufort D (1999) Kaolinite and dickite in Froy and Rind sandstone hydrocarbon reservoirs (Brent formation from Norwegian continental shelf). 11th Intern Clay Conf Proc, Ottawa, pp 97–102
Cathles LM (1977) An analysis of the cooling of intrusives by ground-water convection which includes boiling. Econ Geol 72:804–826
Chermak JA (1989) The kinetics and thermodynamics of clay mineral reactions. PhD thesis, Virginia Polytechnic Institute, Blacksburg, Va, USA
Cicel B, Machajdik D (1981) Potassium- and ammonium-treated montmorillonites. I. Interstratified structures with ethylene glycol and water. Clays Clay Miner 29:40–46
Claret F (2001) Caractérisation structurale des transitions minéralogiques dans les formations argileuses: Contrôle et imlications géochimiques des processus d'illitisation. Cas particulier d'une perturbation alcaline dans le Callovo-Oxfordien. Laboratoire souterrain Meuse, Haute Marne Thèse Univ Joseph Fourier, Grenoble 1, p 174
Clauer N, Chaudhuri S (1995) Clays in crustal environments. Isotope dating and tracing. Springer, Berlin Heidelberg New York, pp 359
Clauer N, Rinckenbach T, Weber F, Sommer F, Chauduri, S, O'Niel J (1999) Diagenetic evolution of clay minerals in oil-bearing Neogene sandstones and associated shales, Mahakam delta basin, Kalimantan, Indonesia. Ame Assoc Petrol Geol Bull 83:62–87
Comans RNJ, Haller M, De Preter P (1991) Sorption of cesium on illite: non-equilibrium behaviour and reversibility. Geochim Cosmochim Acta 55:433–440
Compton JS (1991) Origin and diagenesis of clay minerals in the Monterey Formation, Santa Maria basin area, California. Clays Clay Minerals 39:449–446
Cook MG, Rich CI (1962) Weathering of sodium-potassium mica in soils of the Virginia Piedmont. Soil Sci Soc Amer Proc 26:591–595
Corcoran D, Clayton G (1999) Interpretation of vitrinite reflectance profiles in the Central Irish Sea area: Implications for the timing of organic maturation. Jour Petrol Geol 22:261–286
Cornell RM (1993) Adsorption of cesium on minerals: a review. J Radioanal Nuclear Chem 171:483–500
Cornell RM (1993) Adsorption of cesium on minerals: a review. Journ Radioanal Nucl Chem Articles 171:483–500
Cox A, Dalrymple GB (1967) Statistical analysis of geomagnetic reversal data and the precision of potassium-argon dating. J Geophys Res 72:2603–2614
Creasey SC (1959) Some phase relations in hydrothermally altered rocks of porphyry copper deposits. Econ Geol 54:351–373
Cuadros J, Altaner SP (1998) Characterization of mixed-layer illite-smectite from bentonites using microscopic, chemical, and X-ray methods: Constraints on the smectite-to-illite transformation mechanism. Amer Miner 83:762–774
Cuadros J, Altaner SP (1998) Compositional and structural features of the octahedral sheet in mixed-layer illite-smectite from bentonites. Eur J Miner 10:111–124
Cuadros J, Linares J (1995) Some evidence supporting the existence of polar layers in mixed-layer illite/smectite. Clays Clay Miner 43:467–473
Dabkowska-Naskret H, Dligosz J (1996) Occurrence and characteristics of layer silicates in alluvial soils from the lower Wisla river valley. Applied Clay Science 11:77–83

Daniels E, Altaner S (1993) Inorganic nitrogen in anthracite from eastern Pennsylvania. USA Jour Coal Geology 22:21–35

Darby D, Wilkinson M, Fallick AE, Haszeldine RS (1997) Illite dates record deep fluid movements in petroleum basins. Petroleum Geoscience 3:133–140

Dong H, Peacor DR (1996) TEM observation of coherent stacking relations in smectite, I/S and illite of shales: evidence for McEwan crystallites and dominance of 2M1 polytype. Clays and Clay Minerals v 44:257–275

Drits VA, Weber F, Salyn AL, Tsipursky SI (1993) X-ray identification of one-layer illite varieties: application to the study of illites around uranium deposits of Canada. Clays Clay Miner 41:389–398

Drits VA (1985) Mixed-layer minerals: Diffraction methods and structural features. In: Schultz LG, Olphen H Van, Mumpton FA (eds) Proc Int Clay Conf, Denver, pp 33–45

Drits VA, Lindgreen H, Sakharov BA, Salyn AS (1997) Sequence structure transformation of illite-smectite-vermiculite during diagenesis of Upper Jurassic shales, North Sea. Clay Miner 33:351–371

Drits VA, Varaxina TV, Sakharov BA, Plançon A (1994) A simple technique for identification of one-dimensional powder X-ray diffraction patterns for mixed-layer illite-smectites and other interstratified minerals. Clays Clay Miner 42:382–390

Drits VA, Tchoubar C (1990) X-ray diffraction by disordered lamellar strucutures. Theory and applications to microdivided silicates and carbons. Springer, Berlin Heidelberg New York, p 371

Drits VA, Salyn AL, Sucha V (1996) Structural transformations of interstratified illite-smectites from Dolna Ves hydrothermal deposits: dynamics and mechanisms. Clays and Clay Minerals v 2:181–190

Dudoignon P (1983) Altérations hydrothermal et supergènes des granites. Etude des gisements de Montebras (Creuse), de Sourches (Deux-Sèvres) et des arènes granitiques (Massif de Parthenay). Thèse Doctorat 3eme cycle, Univ Poitiers, p 124

Dunoyer de Segonzac G (1969) Les minérax argileux dans la diagènese. Passage au métamorphisme. Mémoires Service Carte Géologique Alsace-Lorraine, Strasbourg 29:230

Dunoyer de Segonzac G (1970) The transformation of clay minerals during diagenesis and low grade metamorphism. Sedimentology 15:281–346

Eberl DD (1978) The reaction of montmorillonite to mixed layer clay: the effect of interlayer alkali and alkaline-earth cations. Geochim Cosmochim Acta 42:1–7

Eberl DD (1980) Alkali cation selectivity and fixation by clay minerals. Clays Clay Miner 28:161–172

Eberl DD, Srodon J (1984) Illite. In: Ribbe PH (ed) Micas. Reviews in Mineralogy 13:495–544

Eberl DD, Srodon J (1988) Ostwald ripenning and interparticle-diffraction effects for illite crystals. Amer Miner 73:1335–1345

Eberl DD, Srodon J, Northrop HR (1986) Potassium fixation in smectite by wetting and drying. In: Davis JA, Hayes KF (eds) Geochemical processes at mineral surfaces. Amer Chem Soc Symp Ser 323:296–326

Eberl DD, Srodon J, Lee M, Nadeau PH, Northrop HR (1987) Sericite from the Siverton caldera, Colorado: Correlation among structure, composition, origin and particle thickness. Amer Miner 72:914–934

Eberl DD, Whitney G, Khoury H (1978) Hydrothermal reactivity of smectite. Amer Miner 63:401–409

Eberl DD, Hower J (1976) Kinetics of illite formation. Geol Soc Amer Bull 87:1326–1330

Eberl DD, Blum A (1993) Illite crystallite thickness by X-ray diffraction. In: RC Reynolds, JR Walker (eds) Computer applications to X-Ray Powder Diffraction Analysis of Clay Minerals. Clay Mineral Society workshop lectures v 5:124–153

Eberl DD, Srodon J, Kralik M, Taylor BE, Peterman ZE (1990) Ostwald ripening of clays and metamorphic minerals. Science v 248:474–477

Eberl DD, Srodon J, Mingchou Lee Nadeau PH, Northrop HR (1987) Sericite from the Silverton Caldera, Colorado: correlations between structure, composition, origin, and particle thickness. American Mineralogist v 72:914–934

Elliot C, Aronson JL (1987) Alleghanian episode of K-bentonite illitization in the southern Appalachian basin. Geology 15:735–739

Elliot WC, Matisoff G (1996) Evaluation of kinetic models for the smectite to illite transformation. Clays Clay Miner 44:77–87

Elliott WC, Aronson JL, Matisoff G, Gautier DL (1991) Kinetics of the smectite to illite transformation in the Denver basin: clay mineralogy, K-Ar data and mathematical modelling. Bull Amer Assoc Petrol Geol 75:436–462

Elliott WC, Edenfield AM, Wamples JM, Matisoff G, Long PE (1999) The kinetics of the smectite to illite transformation in Cretaceous bentonites, Cerro Negro, New Mexico. Clays Clay Miner 47:286–296

Ellis S, Mellor A (1992) Soils and Environment, Routledge, London pp 364

Erhenberg SN, Nadeau PH (1989) Formation of diagenetic illite in sandstones in the Garn formation, Haltenbanken area, mid-Norwegian continental shelf. Clay Miner 24:233–253

Erhenberg SN (1991) Kaolinized, potassium-leached zones at the contacts of the Garn Formation, Haltenbanken, mid-Norwegian continental shelf. Marine Petrol Geol 8:250–269

Eslinger E, Highsmith P, Albers D, DeMayo B (1979) Rolc of iron reduction in the conversion of smectite to illite in bentonites in the disturbed belt, Montana. Clayc Clay Miner 27:327–338

Espitalié J (1986) Use of Tmax as a maturity index for different types of organic matter: comparison with reflectance. In: Burrus (ed) Thermal Modelling in Sedimentary Basins. Technip Paris, p 475–496

Esposito KJ, Whitney G (1995) Thermal effects of thin igneous intrusions on diagenetic reactions in a tertiary basin of southwestern Washington. US Geol Surv Bull 2085-c, p 40

Evangelou VP, Lumbanraja J (2002) Ammonium-potassium-calcium exchange on vermiculite and hydroxy-aluminium vermiculite. Soil Sci Soc Am J 66:445–455

Fanning DS, Keramidas VZ (1977) Micas. In: Dixon JB, Weed SB (eds) Mineral in Soil Environments. Soil Sci Soc Amer, Madison USA pp 195–258

Feth TH, Bobertson CE, Polzer WE (1964) Sources of mineral constituents in waters from granitic rocks, Sierra Nevada, California and Nevada. US Geol Surv Water Supply Paper 1535, p 70

Foscolos AE, Kodama H (1974) Diagenesis of clay minerals from lower Cretaceous shales of North Eastern British Columbia. Clays Clay Miner 22:319–335

Freed RL, Peacor DR (1989) Geopressured shales and sealing effect of smectite to illite transition. Amer Assoc Petrol Geol Bull 73:1223–1232

Frey M (1970) The step from diagenesis to metamorphism in pelitic rocks during Alpine orogenesis. Sedimentology 15:261–279

Gabis V (1963) Etude minéralogique et géochimique de la série sédimentaire oligocène du Velay. Bull Soc Fr Minér Crist 86:315–334

Garrels RM (1984) Montmorillonite/illite stability diagrams. Clays Clay Miner 32:161–166

Gaudette HE, Eades JL, Grim RE (1966) The nature of illite. In: Bradley WF, Bailey SW (eds) Proc 13th Nat Conf Clays Clay Miner, pp 33–48
Gauthier-Lafaye F, Weber F (1989) The Francevillian (Lower Proterozoic) uranium ore deposits of Gabon Econ Geol 84:2267–2285
Gauthier-Lafaye F, Weber F, Ohmoto H (1989) Natural fission reactors of Oklo. Econ Geol 84:2286–2295
Getman AK, Ladd CE (1925) Crop Production and Soil Management. Wiley, New York, pp 515
Gharrabi M, Velde B, Sagon JP (1998) The transformation of illite to muscovite in pelitic rocks: constraints from X-ray diffraction. Clays Clay Miner 46:79–88
Gharrabi M, Velde B, Sagon JP (1998) The transformation of illite to muscovite in pelitic rocks: constraints from X-ray diffraction. Clays Clay Miner 46:78–88
Giorgetti G, Mata MP, Peacor DR (2000) TEM study of the mechanism of transformation of detrital kaolinite and muscovite to illite/smectite in sediments of the Salton Sea Geothermal Field. Eur J Miner 12:923–934
Glasmann JR, Lundegard PD, Clark RA, Penny BK, Collins ID (1989) Geochemical evidence for the history of diagenesis and fluid migration: Brent sandstone, Heather Field, North Sea. Clay Miner 24:255–284
Grathoff GH, Moore DM (1996) Illite polytype quantification using Wildfire calculated X-ray diffraction patterns. Clays Clay Miner 44:835–842
Grauby O, Petit S, Decarreau A, Baronnet A (1993) The beidellite-saponite series: an experimental approach. Eur J Miner 5:623–635
Grim RE (1953) Clay mineralogy. McGraw-Hill Book, New York, pp 384
Grim RE, Bray RM, Bradley WF (1937) The Mica in Argillaceous sediments. Amer Miner 22:813–829
Grim RE, Kulbicki G (1961) Montmorillonite high temperature reactions and classification. American Mineralogist 46:1329–1369
Güven N, Huang WL (1991) Effects of octahedral Mg^{2+} and Fe^{3+} substitutions on hydrothermal illitization reactions. Clays Clay Miner 39:387–399
Güven N (2001) Mica structure and fibrous growth of illite. Clays Clay Miner 49:189–196
Harper CT (1970) Graphical solutions to the problem of $^{40}Ar^*$ loss from metamorphic minerals. Ecologae Geol Helv 63:1500–1507
Harris WG, Zelazny LW (1985) Criteria assessment for micaceous and illitic classes in Soil Taxonomy. In: Kittrick JA (ed) Mineral Classification of Soils. Soil Sci Soc Amer Special Publication no 16, pp 147–160
Harvey CC, Browne PRL (1991) Mixed-layer clay geothermometry in the Wairakei geothermal field, New Zealand. Clays Clay Miner 39:614–621
Haszeldine S, Brint JF, Fallick AE, Hamilton PJ, Brown S (1992) K-Ar dating of illites in Brent Group reservoirs. In: Morton AC, Haszeldine RS, Giles MR, Brown S (eds) Geology of the Brent Group. Special publication 61, Geological Society, London, pp 377–400
Hathaway J (1972) Regional clay mineral facies in estuaries and continental margin of the United States East Coast. In: Nelson BW (ed) Environmental Framework of Coastal Plain Estuaries. Geological Society America Memoire 133:616
Hemley JJ, Jones WR (1964) Chemical aspects of hydrothermal alteration with emphasis on hydrogen metasomatism. Econ Geol 59:538–569
Hemley JJ (1959) Some mineral equilibria in the system K_2O-Al_2O_3-SiO_2-H_2O. Amer J Sci 257:241–270
Héroux Y, Changnon A, Bertrand R (1979) Compilation and correlation of major thermal maturation indicators. Amer Assoc Petrol Geol Bull 63:2128–2144

Hervig RL, Peacock SM (1989) Water and trace elements in coexisting muscovite and biotite from metamorphic rocks. EOS 70:490

Higashi S (1982) Tobelite, a new ammonium dioctahedral mica. Mineral Journal 11:784–904

Hillier S, Matays J, Matter A, Vasseur G (1995) Illite/smectite diagenesis and its variable correlation with vitrinite reflectance in the Pannonian Basin. Clays Clay Miner 43:174–183

Hoffman J, Hower J (1979)Clay mineral assemblages as low grade metamorphic geothermometers: Application to the thrust faulted disturbed belt, Montana, Aspects of Diagenesis. Society Economic Paleontological Mining Geologists Sp Pub 26:55–81

Hofmann AW, Mahoney JW Jr, Giletti BJ (1974) K-Ar and Rb-Sr data on detrital and post-depositional history of Pennsylvanian clay from Ohio and Pennsylvania. Geol Soc Amer Bull 85:639–644

Holtzapffel T, Chamley H (1986) Les smectites lattées du domaine Atlantique depuis le Jurassique supérieur: gisement et signification. Clay Miner 21:133–148

Horton DG (1985) Mixed layer illite/smectite as a paleotemperature indicator in the Amethyst vein system. Creede district, Colorado, USA. Contrib Mineral Petrol 91:171–179

Howard JJ (1981) Lithium and potassium saturation of illite/smectite clays from interlaminated shales and sandstones. Clays Clay Miner 29:136–142

Howard JJ, Roy DM (1985) Development of layer charge and kinetics of experimental smectite alteration. Clays and Clay Minerals v 33:81–88

Hower J, Mowatt TC (1966) The mineralogy of illites and mixed-layer illite/montmorillonites. Amer Miner 51:825–854

Hower J, Eslinger EV, Hower ME, Perry EA (1976) Mechanism of burial and metamorphism of argillaceous sediments: 1. Mineralogical and chemical evidence. Geol Soc Amer Bull 87:725–737

Huang WL, Longo JM, Pevear DR (1993) An experimentally derived kinetic model for smectite-to-illite conversion and its use as a geothermometer. Clays Clay Miner 41:162–177

Huang* B, Velde B (unpublished) Effect of fertilizer management on agricultural Vertisols in Anhui Province (*Institute of Soil Science, Nanjing)

Huang W-L (1993) The formation of illitic clays from kaolinite in KOH solution from 225 °C to 350 °C. Clays Clay Minerals 41:645–654

Huff WD Türkmenoglu AG (1981) Chemical characteristics of Ordovician K-bentonites along the Cincinnati arch. Clays Clay Miner 29:113–123

Hughes RE, Moore DM, Glass HD (1994) Qualitative and quantitative analysis of clays in soils. In: Amonette, JE, Zelazney, LW (ed) Quantative Methods in Soil Mineralogy. SSSA Miscellaneous Publications Madison, WI, pp 330–359

Hurst A (1985) Diagenetic chlorite formation in some Mesozoic shales from the Sleipner area of the North Sea. Clay Miner 20:69–79

Inoue A, Kitagawa R (1994) Morphological characteristics of illitic clay minerals from a hydrothermal system. Amer Miner 79:700–711

Inoue A, Utada M, Wakita K (1992) Smectite-to-illite conversion in natural hydrothermal systems. Applied Clay Sci 7:131–145

Inoue A, Velde B, Meunier A, Touchard G (1988) Mechanism of illite formation during smectite-to-illite conversion in a hydrothermal system. Amer Miner 73:1325–1334

Inoue A (1995) Formation of clay minerals in hydrothermal environments, pp 268–329. In: Velde B (ed) Origin and Mineralogy of Clays. Springer, Berlin Heidelberg New York, pp 334

Inoue A, Kohyama N, Kitagawa R, Watanabe T (1987) Chemical and morphological evidence for the conversion of smectite to illite. Clays Clay Minerals v 35:111–120
Inoue A, Watanabe T, Kohyama N, Brusewitz A-M (1990) Characterization of illitization of smectite in bentonite beds at Kinnekulle, Sweden. Clays Clay Minerals 34:241–249
Inoue A (1983) Potassium fixation of clay minerals during hydrothermal alteration. Clays and Clay Minerals 31:81–91
Inoue A, Bouchet A, Velde B, Meunier A (1989) A convenient technic for estimating smectite layer percentage in randomly interstratified illite/smectite minerals. Clays and Clay Minerals v 37:227–234
Inoue A, Minato H, Utada M (1978) Mineralogical properties and occurrence of illite/montmorillonite mixed-layer minerals formed from Miocene volcanic glass in Waga Omono district. Clay Sci 5:1234–136
Jaboyedoff M, Cosca MA (1999) Dating incipient metamorphism using $^{40}Ar/^{39}Ar$ geochronology and XRD modeling: a case study from the Swiss Alps. Contrib Mineral Petrol 135:93–113
Jaboyedoff M (1999) Transformation des interstratifiés illite-smectite vers l'illite et la phengite: un exemple dans la série carbonatée du domaine Briançonnais des Alpes suisses romandes. Thèse Univ Lausanne, p 452
Jaboyedoff M, Kübler B, Thélin P (1999) An empirical equation for weakly swelling mixed-layer minerals, especially illite-smectite. Clay Miner 34:443–466
Jaboyedoff M, Kübler B, Sartori L, Thélin P (2000) Basis for the meaningful illite crystallinity measurements: an example from the Swiss Prealps. Schweiz Mineral Petrogr Mitt 80:75–83
Jackson ML (1964) Chemical composition of the soil. In: Bear FE (ed) Chemistry of the Soil Reinhold, NY, pp 71–141
Jaeger JC (1964)Thermal effects of intrusions. Reviews of Geophysics 2:443–466
Jennings S, Thompson GR (1987) Diagenesis in the Pio-Pleistocene sediments of the Colorado River delta, southern California. J Sedim Petrol 56:89–98
Jiang WT, Peacor DR, Merriman RJ, Roberts B (1990) Transmission and analytical electron microscopic study of mixed-layer illite/smectite formed as an apparent replacement product of diagenetic illite. Clays and Clay Minerals v 38:449–468
Joesten R (1983) Grain growth and grain boundary diffusion in quartz from Christmas Mountains (Texas) contact aureole. Amer J Sci 283:233–254
Johnsson PA, Holchella MF Jr, Parks GA, Blum AE, Sposito G (1992) Direct observation of muscovite basal-plane dissolution and secondary phase formation: an XPS, LEED, and SFM study. In: Kharaka YK, Maest AS (eds) Water-rock Interactions. Balkema, pp 159–162
Kim Y, Kirkpatrick RJ, Cygan RT (1996) ^{133}Cs NMR study of cesium on the surfaces of kaolinite and illite. Geochim Cosmochim Acta 60:4059–4074
Kisch H (1987) Correlation between indicators of very low-grade metamorphism. In: Frey M (ed) Low Temperature Metamorphism. Blackie London, pp 552
Kisch HJ (1983) Mineralogy and petrology of burial diagenesis (burial metamorphism) and incipient metamorphism in clastic rocks. In: Larsen G, Chilingar GV (eds) Diagenesis in sediments and sedimentary rocks. Elsevier, Amsterdam 2:289–493
Kisch H (1990) Calibration of the anchizone: A critical comparison of illite "crystallinity" scales used for definitions. J Metamorphic Petrol 8:3–46

Kitagawa R (1995) Coarsening process of a hydrothermal sericite sample using surface microtopography and transmission electron microscopy techniques. In: Churchman GJ, Fitzpatrick RW, Eggleton RA (eds) Clays controlling the environment. Proc Int Clay Conf, Adelaide, pp 249–252
Kitagawa Y, Watanabe Y (1970) Preparation of dioctahedral vermiculite from muscovite. Clay Science 4:31–36
Kittrick JA (1984) Solubility measurements of phases in three illites. Clays Clay Miner 32:115–124
Koch CB, Bentzen MD, Larsen EW, Borggard OK (1992) Clay mineralogy from two Ultisols from central Kalimantan, Indonesia. Soil Science 154:158–167
Kodama H, Dean RS (1980) Illite from Eldorado, Saskatchewan. Can Miner 18:109–118
Kodama H, Ross GJ (1972) Structural changes accompanying potassium exchange in a clay-size muscovite. Proc Int Clay Conf, Madrid, pp 481–492
Komninou A, Sverjensky DA (1993) Geochemical modeling of the formation of an unconformity-type uranium deposit. Econ Geol 91:590–606
Kübler B, Jaboyedoff M (2000) Illite crystallinity. C R Acad Sci Paris 331:75–89
Kübler B (1967) La cristallinité de l'illite et les zones tout-à-fait supérieures du métamorphisme. In: Schaer JP (ed) Colloque sur les étages tectoniques, à la Baconnière, Neuchâtel, pp 105–122
Kübler B (1968) Evaluation quantative du metamorphisme par la cristallinité de l'illite. Bull Centre Rech Pau-SNAP2 2:385–397
Kulik DM, Aja SU, Sinitsyn VA, Wood SA (2000) Acid-base surface chemistry and sorption of some lanthanides on K^+-saturated Marblehead illite: II. A multisite-surface complexation modeling. Geochim Cosmochim Acta 64:195–213
Kyser K, Hiatt E, Renac C, Durocher K, Holk G, Deckart K (2000) Diagenetic fluids in paleo- and meso-Proterozoic sedimentary basins and their implications for long protracted fluid histories. In: Kyser K (ed) Fluids and basin evolution. Mineralogical Association of Canada, Short Courses Series 28:473–506
Lagaly G, Weiss A (1969) Determination of the layer charge in mica-type layer silicates. In: Heller, L (ed) Proceedings, International Clay Conference, Tokyo, Japan 1:61–80
Lagatu H, Sicard L (1901) L'Analyse des Terres Masson, Paris, pp 303
Lanson B, Meunier A (1995) La transformation des interstratifiés ordonnés ($S\geq1$) illite-smectite en illite dans les séries diagénétiques. Etat des connaissances et perspective. Bull Centres Rech Explor-Prod elf aquitaine 19:149–165
Lanson B (1997) Decomposition of experimental X-ray diffraction patterns (profile fitting): a convenient way to study clay minerals. Clays Clay Miner 45:132–146
Lanson B, Beaufort D, Berger G, Baradat J, Lacharpagne JC (1996) Late-stage diagenesis of clay minerals in porous rocks: Lower Permian Rotliegendes reservoir off-shore of The Netherlands. J Sedim Res 66:501–518
Lanson B, Beaufort D, Berger G, Bauer A, Cassagnabère A, Meunier A (2002) Authigenic kaolin and illitic minerals during burial diagenesis of sandstones: a review. Clay Miner 37:1–22
Lanson B, Champion D (1991) The I-S to illite reaction in the late stages of diagenesis. American Journal Science 291:473–506
Lanson B, Velde B (1992) Decomposition of X-ray diffraction patterns: A convenient way to describe complex diagenetic smectite-to-illite evolution. Clays and Clay Miner 40:629–643
Lanson B, Velde B (1993) Comparison of I/S transformation and maturity of organic matter at elevated temperatures. Clays Clay Miner 41:178–183

Lanson B, Velde B, Meunier A (1998) Late-stage diagenesis of illitic clay minerals as seen by decomposition of X-ray diffraction patterns: Contrasted behaviors of sedimentary basins with different burial histories. Clays Clay Minerals 46:69–78
Larbaletrier A (1891) Les Engrais. J-B Baillière et Fils, Paris, pp 352
Lasaga A (1981) Rate laws of chemical reactions. In: Lasaga A, Kirkpatrick (eds) Kinetics of geochemical processes. Reviews in Mineralogy 8:1–68
Laverret E (2002) Evolutions temporelles et spatiales des altérations argileuses des gisements d'uranium sous discordance, secteur de Shea Creek (Bassin de l'Athabasca, Canada). Thèse Univ Poitiers p 192
LeRoux J, Rich CI (1969) Ion selectivity of micas as influenced by degree of potassium depletion. Soil Sci Soc Am Proc 33:684–690
Lonker SW, Fitz Gerald JD (1990) Formation of coexisting 1M and 2M polytypes in illite from an active hydrothermal system. Amer Miner 75:1282–1289
Louks RR (1991) The bound interlayer water of potassic white micas: muscovite-hydromuscovite-hydropyrophyllite solution. Amer Miner 76:1563–1579
Lovering TS (1949) Rock alteration as a guide to ore – East Tintic district, Utah. Econ Geol Mon 1 p 65
Lowell JD, Guilbert JM (1970) Lateral and vertical alteration and mineralization zoning in porphyry ore deposits. Econ Geol 65:373–408
Machadjik D, Cicel B (1981) Potassium- and ammonium-treated montmorillonites. II. Calculation of characteristic layer charges. Clays Clay Miner 29:47–52
Maes A, Verheyden D, Cremers A (1985) Formation of highly selective cesium-exchange sites in montmorillonites. Clays Clay Miner 33:251–257
Majhoory RA (1975) Clay mineralogy, physical and chemical properties of some soils in arid regions of Iran. Soil Sci Soc Amer Proc 39:1157–1164
Malla PB, Douglas LA (1987) Identification of expanding layer silicates: layer charge vs. expanding properties. Proc Inter Clay Conf, Denver 1985:277–283
Mamy J, Gaultier JP (1975) Etude de l'évolution de l'ordre cristallin dans la montmorillonite en relation avec la diminution d'échangeabilité du potassium. Proc Intern Clay Conf Mexico City pp 149–155
Mamy J, Gaultier JP (1975) Evolution de l'ordre cristallin dans la montmorillonite en relation avec la diminution d'échangeabilité du potassium. In: Bailey, SW (ed) Proc Int Clay Conf, Mexico City, pp 149–155
Mathieu Y, Velde B (1989) Identification of thermal anomalies using clay mineral composition, Clay Miner 24:591–602
Matthews J, Velde B, Johansson H (1994) Significance of K-Ar ages of authigenic illitic clay minerals in sandstones and shales from the North Sea. Clay Minerals 29:379–389
McBride MB (1994) Environmental chemistry of soils. Oxford University Press, New York, p 406
McDaniel PA, Nielson GA (1985) Illuvial versus inherited clays in a Cryoboralf of the Boulder Batholith, Montant. Soil Sci Soc Ame Journ 49:156–178
McDowell SD, Elders WA (1980) Authigenic layer silicate minerals in bore-hole Elmer 1, Salton Sea geothermal field, California, USA. Contrib Mineral Petrol 74:293–310
McDowell SD, Elders WA (1983) Allogenic layer silicate minerals in borehole Elmore #1. Salton Sea Geothermal Field, California. Amer Miner 68:1146–1159
Méring J (1949) L'interférence des rayons X dans les systèmes à stratification désordonnée. Acta Crystallogr 2:371–377

Mermut AR (1994) Problems associated with layer charge characterization of 2:1 phyllosilicates. In: Mermut AR (ed) Layer charge characteristics of 2:1 silicate clay minerals, pp 106–122
Merriman R, Frey M (1999) Patterns of very low-grade metamorphism in metapelitic rocks. In: Frey M, Robinson D (eds) Low-grade metamorphism. Blackwell Science, pp 61–107
Meunier A (2004) Clays. Springer 2004 (in press)
Meunier A, Velde B (1979) Weathering mineral facies in altered granites: the importance of local small-scale equilibria. Min Mag 43:261–268
Meunier A (1980) Les mécanismes de l'altération des granites et le rôle des microsystèmes. Etude des arènes du massif granitique de Parthenay. Mémoires Société Géologique de France, 140, p 80
Meunier A (1995) Hydrothermal alteration by veins. In: Velde B (ed) Origin and mineralogy of clays. Clays and the environment. Springer, Berlin Heidelberg New York, pp 247–267
Meunier A, Beaufort D, Parneix JC (1987) Dépôts minéraux et altérations liées aux microfracturations des roches: un moyen pour caractériser les circulations hydrothermales. Bull Soc Géol France, 8, t III, no 5:163–171
Meunier A, Lanson B, Beaufort D (2000) Vermiculitization of smectite interfaces and illite layer growth as a possible dual model for I-S illitization in diagenetic environments: a synthesis. Clay Miner 35:573–586
Meunier A (1995) Hydrothermal alteration by veins, pp 247–267. In: Velde B (ed) Origin and Mineralogy of Clays. Springer, Berlin Heidelberg New York, pp 334
Meunier A, Velde B (1989) Solid solutions in I/S mixed layer minerals and illite. American Miner. 74:1106–1112
Meunier A, Proust D, Beaufort D, Lajudie A, Petit JC (1992) Heterogeneous reactions of dioctahedral illite-smectite and kaolinite-smectite mixed-layers: applications to clay materials for engineered barriers. Applied Geochemistry Suppl Issue v 1:143–150
Meunier A, Velde B, Griffault L (1998) The reactivity of bentonites: a review. An application to clay barrier stability for nuclear waste storage. Clay Minerals 33:187–196
Meyer C, Hemley JJ (1967) Wall rock alteration. In: Barnes HL (ed) Geochemistry of hydrothermal ore deposits. Holt, Rinchart & Winston, New York, pp 166–235
Moe JA, Ryan PC, Elliot WC, Reynolds RC (1996) Petrology, chemistry and clay mineralogy of a K-bentonite in the Proterozoic Belt Subgroup of western Montana. J Sed Res 66:95–99
Montoya JW, Hemley JJ (1975) Activity relations and stabilities in alkali feldspar and mica alteration reactions. Econ Geol 70:577–582
Moore D, Velde B, Mueller C (unpublished) Rapid changes in illitic clay minerals in three cores of forest soils, each with different anthropogenic histories
Mosser C, Gall JC, Tardy Y (1972) Géochimie des illites du grès à Voltzia du Buntsandstein supérieur des Vosges du Nord, France. Chem Geol 9:157–177
Mukhamet-Galeyev AP, Pokrovskiy VA, Zotov AV, Ivanov IP, Samotoin N (1985) Kinetics and mechanism of hydrothermal crystallization of $2M_1$ muscovite: an experimental study. Intern Geol Rev 27:1352–1364
Mussa L (1887) Pratique des engrais Chimiques Lib Agricole de la Maison Rustique, Paris, p 134
Nadeau PH, Bain DC (1986) Composition of some smectites and diagenetic illitic clays and implications for their origin. Clays Clay Miner 34:455–464
Nadeau PH, Wilson MJ, McHardy WJ, Tait JM (1985) The conversion of the smectite to illite during diagenesis. Evidence from some illitic clays from bentonites and sandstones. Miner Mag 49:393–400

Nadeau PH, Wilson J, McHardy WJ, Tait JM (1984a) Interparticle diffraction: a new concept for interstratified clays. Clay Minerals v 19:757–769

Nadeau PH, Wilson J, McHardy WJ, Tait JM (1984b) Interstratified clays as fundamental particles. Science 225:923–925

Nagy KL (1994) Application of morphological data obtained using scanning force microscopy to quantification of fibrous illite growth rates. In: Nagy KL, Blum AE (eds) Scanning probe microscopy of clay minerals. CMS Workshop Lectures 7:204–239

Nagy KL, Chiarello RP, Sturchio NC, Cygan RT (1996) Clay formation by epitaxial nucleation and growth on mica. Geol Soc Amer Abstr Prog 28:A147 (abstract)

Nettleton WD, Nelson RE, Flach KW (1973) Formation of mica in surface horizons of dryland soils. Soil Sci Soc Amer Proc 37:473–478

Newman ACD (1987) Chemistry of clays and clay minerals. Mineralogical Society Monograph 6, Mineralogical Society editor, p 480

Nicot E (1987) Etude minéralogique de néogenèses phylliteuses dans quelques grès terrigènes. Thèse Univ Paris VI, pp 384

Niederbudde EA (1972) Changes in K/Ca properties of clay in loess-derived soils in soil formation. In Proc 9th Colloquium (Landshut, W Ger), Int Pot Inst, Berne, Switzerland, pp 103–107

Niederbudde EA, Schwarzmann A, Schwertmann U (1969) Tonmineralbedingter K-haushalt einer gedüngten parabraunerden and Würm-Geschiebemergel. Z Pflanzenernaehr Dueng Bodenkd 124:212–224

Norrish K, Pickering JG (1983) Clay minerals. In: CSIRO (ed) Soils, an Australian viewpoint. Division of Soils, CSIRO, Academic Pres, London, pp 281–308

Norrish K (1972) Factors in the weathering of mica to vermiculite. Proc 4th Intern Clay Conf Madrid, pp 417–432

Norton D, Knight J (1977) Transport phenomena in hydrothermal systems: cooling plutons. Amer J Sci 277:937–981

Nutt CJ (1989) Chloritization and associated alteration at Jabiluka unconformity-type deposit, Northern Territory, Australia. Canadian Mineral 27:41–58

Olis AC, Malla PB, Douglas LA (1990) The rapid estimation of the layer charges of 2:1 expanding clays from a single alkylammonium ion expansion. Clay Minerals 25:39–50

Page R, Wenk HR (1979) Phyllosilicate alteration of plagioclase studied by transmission electron microscopy. Geology 7:393–397

Papapanagiotou P, Beaufort D, Patrier P, Traineau H (1992) Clay mineralogy of the <0.2 µm rock fraction in the MI-1 drill hole of the geothermal field of Milos (Greece). Bull Geol Soc Greece, XXVIII/2:575–586

Parry WT, Jasumback M, Wilson PN (2002) Clay mineralogy of phyllic and intermediate argillic alteration at Bingham, Utah. Econ Geol 97:221–239

Patrier P, Beaufort D, Laverret E, Bruneton P (2003) High-grade diagenetic dickite and illite-2M1 from the middle Proterozoic Kombolgie Formation (Northern Territory, Australia). Clays Clay Miner 51:102–116

Patrier P, Beaufort D, Mas A, Traineau H (2003) Superficial clay assemblages associated with hydrothermal acitivity of Bouillante (Guadeloupe). J Volcan Geotherm Res 126:143–146

Patrier P, Papapanagiotou P, Beaufort D, Traineau H, Bril H, Rojas J (1996) Role of permeability versus temperature in the distribution of fine (<0.2 µM) clay fraction in the Chipilapa geothermal system (El Salvador, Central America). J Volcan Geotherm Res 72:101–120

Patrier P, Traineau H, Papapanagiotou F, Turgné E, Beaufort D (1998) I-S series in geothermal fields: comparison with diagenetic I-S series. In: Arehart GB, Hulston JR (eds) Water-Rock Interactions. Balkema, Rotterdam, pp 683–686

Pattullo H (1758) Essai sur l'Amelioration des Terres. Durand, Paris, pp 284

Peacor DR, Bauluz B, Dong H, Tillick D, Yan Y (2002) Transmission and analytical electron microscopy evidence for high Mg contents of 1M illite: absence of 1M polytypism in normal prograde diagenetic sequences of pelitic rocks. Clays Clay Miner 50:757–765

Pernes-Debuyser A, Pernes M, Velde B, Tessier D (In: Clays Clay Mineral in press) The specific role of potassium on soil mineralogy evolution in the 42 Plots experiment, Versailles, France

Pernes-Debuyser A, Pernes M, Velde B, Tessier D (2003) Soil clay mineralogy evolution in the INRA 42 plot experiment (Versailles, France)

Perry E, Hower J (1970) Burial diagenesis in Gulf Coast sediments Clays Clay Minerals 18:165–177

Perry E, Hower J (1970) Burial diagenesis in Gulf Coast pelitic sediments. Clays and Clay Minerals 18:165–178

Pevear DR (1992) Illite age analysis, a new tool for basin thermal history analysis. In: Kharaka YK, Maest AS (eds) Water-Rock Interaction. Balkema, Rotterdam pp 1251–1254

Plançon A, Besson G, Gaultier JP, Mamy J, Tchoubar C (1978) Qualitative and quantitative study of a structural reorganization in montmorillonite after potassium fixation. In: Mortland MM, Farmer VC (eds) Proc VI Clay Conf Oxford, pp 45–54

Poinssot C, Baeyens B, Bradbury MH (1999) Experimental and modelling study of caesium sorption on illite. Geochim Cosmochim Acta 63:3217–3227

Pourcelot L, Gauthier-Lafaye F (1999) Hydrothermal and supergene clays of the Oklo natural reactors: conditions of radionuclide release, migration and retention. Chem Geol 157:155–174

Price L (1983) Geologic time as a parameter in organic metamorphism and vitrinite reflectance as an absolute paleo-geothermometer. J Petrol Geol 6:5–38

Price KL, MacDowell SD (1993) Illite/smectite geothermometry if the Proterozoic Oronto Group, Midcontinent rift system. Clays Clay Minerals 41:134–147

Proust D, Léchelle J, Meunier A, Lajudie A (1990) Hydrothermal reactivity of mixed-layer kaolinite/smectite and implications for radioactive waste disposal. Eur J Miner 2:313–325

Purdy JW, Jäger E (1976) K-Ar ages of rock forming minerals from the central Alps. Mem Inst Geol Min Univ Padova, p 30

Pusch R, Madsen FT (1995) Aspects on the illitization of the Kinnekulle bentonites. Clays Clay Miner 43:261–270

Pytte AM, Reynolds RC (1989) The thermal transformation of smectite to illite. In: Naesser ND, MCCulloh TH (eds) The thermal history of sedimentary basin: methods and case history. Springer, Berlin Heidelberg New York, pp 133–140

Pytte AM (1982) The kinetics of the smectite to illite reaction in contact metamorphic shales. MS thesis, Dartmouth College, Hanover, NH p 78

Quirt DH (2002) Clay minerals: Host-rock alteration related to uranium mineralization in the Athabasca Group. In: Delaney G, Jefferson C(ed) The Athabasca Basin and its uranium deposits. Geological Association of Canada – Mineralogical Association of Canada joint annual meeting. Special Session 18, abstracts vol 27

Raffenperger JP, Garven G (1995b) The formation of unconformity-type uranium ore deposits. 2. Coupled hydrochemical modeling. Amer J Sci 295:639–696

Raffensperger JP, Garven G (1995a) The formation of unconformity-type uranium ore deposits. 1. Coupled groundwater flow and heat transport modeling. Amer J Sci 295:581–636

Ramseyer K, Boles JR (1986) Mixed-layer illite/smectite minerals in Tertiary sandstones and shales, San Joaquin basin, California. Clays Clay Miner 34:115–124

Reichenbach H Graf von, Rich CI (1975) Fine grained micas in soils; pp 59–95. In: Gieseking JE (ed) Soil components, vol 2. Inorganic Components. Springer, Berlin Heidelberg New York, pp 684

Renac C, Meunier A (1995) Reconstruction of paleothermal conditions in a passive margin using illite/smectite mixed-layered series (BA1 scientific drill-hole, Ardèche, France). Clay Min 30:107–118

Rex RW (1966) Authigenic kaolinite and mica as evidence for phase equilibria at low temperatures. Clays Clay Miner 13:95–104

Reynolds RC, Thomson CH (1993) Illite from the Postdam sandstone of New York: A probable noncentrosymmetric mica structure. Clays Clay Miner 41:66–72

Reynolds RC (1994) WILDFIRE: a computer program for the calculation of three dimensional X-ray diffraction patterns of mica polytypes and their disordered variations Hanover, NH, Reynolds RC Jr, 8 Brook Rd

Reynolds RC Jr (1992) X-ray diffraction studies of illite/smectite from rocks, <1 μm randomly oriented powders, and <1 μm oriented powder aggregates: the absence of laboratory-induced artifacts. Clays Clay Miner 40:387–396

Reynolds RC Jr (1992) X-ray diffraction studies of illite/smectite from rocks, <1 μm randomly oriented powders, and <1 μm oriented powder aggregates: the absence of laboratory-induced artifacts. Clays Clay Miner 40:387–396

Reynolds RC Jr (1993) Three-dimensional X-ray powder diffraction from disordered illite: Simulation and interpretation of the diffraction patterns. In: Reynolds RC, Walker JR (eds) Computer application to X-ray powder diffraction analysis of clay minerals. Clay Minerals Society workshop lectures, Boulder CO 5:43–78

Rich CI, Obenshain SS (1955) Chemical and clay mineral properties of a red-yellow podzolic soil derived from muscovite schist. Soil Sci Soc Amer Proc 19:334–339

Rich CI (1958) Muscovite weathering in a soil developed in the Virginia Piedmont. Proc Fifth Nat Conf Clays Clay Miner pp 203–212

Righi D, Meunier A (1995) Origin of clays by rock weathering and soil formation. In: Velde B (ed) Origin and Mineralogy of clays. Clays and the environment. Springer, Berlin Heidelberg New York, pp 43–161

Righi D, Huber K, Keller C (1999) Clay formation and podzol development from postglacial moraines in Switzerland. Clay Minerals 34:319–332

Robertson HE, Lahann RW (1981) Smectite to illite conversion rates: effects of solution chemistry. Clays Clay Minerals 29:129–135

Rosenberg PE (2002) The nature, formation, and stability of end-member illite: A hypothesis. Amer Miner 87:103–107

Sainz-Diaz CI, Hernandez-Laguna A, Dove MT (2001) Theoretical modelling of cis-vacant and trans-vacant configurations in the octahedral sheet of illites and smectites. Phys Chem Minerals 28:322–331

Sass BM, Rosenberg PE, Kittrick JA (1987) The stability of illite/smectite during diagenesis: an experimental study. Geochim Cosmochim Acta 51:2103–2115

Sato T, Murakami T, Watanabe T (1996) Change in layer charge of smectites and smectite layers in illite/smectite during diagenetic alteration. Clays Clay Minerals v 44:460–469

Savary V (1995) Histoire thermique et conditions redox des zones de réactions d'Oklo. Gabon Thèse Univ Nancy I, p 269
Schmidt G (1973) Interstitial water composition and geochemistry of deep Gulf Coast shales and sandstones Ame Assoc Petrol Geol Bull 57:321–337
Schultz LG (1969) Lithium and potassium absorption, dehydroxylation temperature, and structural water content in aluminous smectites. Clays Clay Miner 17:115–149
Scott AD, Reed MG (1966) Expansion of potassium-depleted muscovite. Proc 13th Nat Conf Clays Clay Miner 247–261
Serres O de (1600) Le théâtre d'Agriculture et Mesnages des champs d'Olivier de Serres, seigneur du Pradel, dans lequel est représenté tout ce qui est requis et nécessaire pour bien dresser, gouverner, enrichir et embellir la maison rustique. Coll Thesaurus, Acte Sud (2001), pp 1545
Shawney BL (1970) Selective sorption and fixation of cations by clay minerals: a review. Clays Clay Miner 20:93–100
Shutov VD, Drits VA, Sakharov BA (1969) On the mechanism of a postsedimentary transformation of montmorillonite to hydromica. In: Heller L (ed) Proc Int Clay Conf, Tokyo, 1, Israel University Press, Jerusalem, pp 523–531
Singer A, Stoffers P (1980) Clay mineral diagenesis in two East African lake sediments. Clay Miner 15:291–307
Small JS (1993) Experimental determination of the rates of precipitation of authigenic illite and kaolinite in the presence of aqueous oxalate and comparison to the K–Ar ages of authigenic illite in reservoir sandstones. Clays Clay Miner 41:191–208
Small JS, Hamilton DL, Habesch S (1992) Experimental simulation of clay precipitation within reservoir sandstones. 2: Mechanism of illite formation and controls on morphology. J Sedim Petrol 62:520–529
Smith JV, Yoder HS (1956) Experimental and theoretical studies of the mica polymorphs. Miner Mag 31:209–231
Srodon J (1981) X-ray identification of randomly interstratified illite/smectite in mixtures with discrete illite. Clay Miner 16:297–304
Srodon J (1984) X-ray powder diffraction identification of illitic materials. Clays Clay Miner 32:337–349
Srodon J (1999) Extracting K-Ar ages from shales: a theoretial test. Clay Miner 33:375–378
Srodon J, Clauer N, Eberl DD (2002) Interpretation of K-Ar dates of illitic clays from sedimentary rocks aided by modeling. Amer Miner 87:1528–1535
Srodon J, Morgan DJ, Eslinger EV, Eberl DD, Karlinger M (1986) Chemistry of illite/smectite and end-member illite. Clays Clay Miner 34:368–378
Srodon J, Eberl DD (1984) Illite. In: Bailey SW (ed) Reviews in Mineralogy 13, Micas, Mineralogical Society, Washington D DC, pp 495–544
Srodon J, Elsass F, McHardy WJ, Morgan DJ (1992) Chemistry of illite-smectite inferred from TEM measurements of fundamental particles. Clay Minerals v 27:137–158
Steiner A (1968) Clay minerals in hydrothermally altered rocks at Wairakei, New Zealand. Clays Clay Miner 16:193–213
Stern W, Mullis J, Rahn M, Frey M (1991) Deconvolution of the first "illite" basal reflection. Schweiz Min Perog Mitt 71:453–462
Sucha V, Elsass F, Eberl D, Kuchita L, Madejova J, Gates WP, Komadel P (1998) Hydrothermal synthesis of ammonium illite. Amer. Mineral. 83:58–67
Sucha V, Kraus I, Gerthofferova H, Petes J, Serekova M (1993) Smectite to illite conversion in bentonites and shales of the East Slovak basin. Clay Miner 28:243–253

Sucha V, Srodon J, Elsass F, McHardy WJ (1996) Particle shape versus coherent scattering domain of illite/smectite: Evidence from HRTEM of Dolna Ves clays. Clays Clay Minerals 44:665–671

Sweeney J, Burnham AK (1990) Evaluation of a simple model of vitrinite reflectance based upon chemical kinetics. Ame Assoc Petrol Geol Bull 74:1559–1570

Talibudeen O, Goulding KWT (1983) Charge heterogeneities in smectites. Clays Clay Miner 31:37–42

Tamura T, Jacobs DG (1960) Structural implications in cesium sorption. Health Physics 2:391–398

Tardy Y, Garrels RM (1974) Method for estimating the Gibbs energies of formation of layer silicates. Geochim Cosmochim Acta 38:1101–1116

Tardy Y, Garrels RM (1976) Prediction of Gibbs energies of formation. I – Relationships among Gibbs energies of formation of hydroxides, oxides and aqueous ions. Geochim Cosmochim Acta 40:1051–1056

Tatard L (2003) Les échanges K^+/NH_4^+ dans les illites (undergraduate report, Univ Paris IX)

Tillick DA, Peacor DR, Mauk JL (2001) Genesis of dioctahedral phyllosilicates during hydrothermal alteration volcanic roks: I. The Golden Cross epithermal ore deposit, New Zealand. Clays Clay Miner 49:126–140

Tissot B, Welte DH (1978) Petroleum Formation and Occurrence, Springer Verlag, pp 662

Titley SR (1982) The style and progress of mineralization and alteration in porphyry copper systems. In: Titley SR (ed) Advances in Geology of the Porphyry Copper Deposits, Southwestern North America. University Arizona Press, Tucson, AZ, pp 93–116

Turpault MP, Berger G, Meunier A (1992a) Dissolution-precipitation processes induced by hot water in a fractured granite. Part 1: Wall-rock alteration and vein deposition processes. Eur J Mineral 4:1457–1475

Turpault MP, Meunier A, Guilhaumou N, Touchard G (1992b) Analysis of hot fluid infiltration in fractured granite by fluid inclusion study. Appl Geochim Suppl 1:269–276

Utada M (1980) Hydrothermal alteration related to igneous acidity in Cretaceous and Neogene formations of Japan. Mining Geol Jpn Spec Issue 8:67–83

Utada M, Aoki M, Inoue A, Kusakabe H (1988) A hydrothermal alteration envelope and alteration minerals in the Shinzan mineralized area, Akita Prefecture, northeast Japan. Mining Geology Japan Special Issue 12:67–77

Varajao A, Meunier A (1995) Particle morphological evolution during the conversion of I/S to illite in lower Cretaceous shales from Sergipe-Alagoas basin, Brazil. Clays Clay Miner 43:35–59

Veblen DR, Guthrie G, Livi KJT, Reynolds RC Jr (1990) High-resolution transmission electron microscopy and electron diffraction of mixed-layer illite/smectite: experimental results. Clays and Clay Minerals v 38:1–13

Velde B, Nicot E (1985) Diagenetic clay mineral composition as a function of pressure, temperature, and chemical activity. J Sedim Petrol 55:541–547

Velde B, Church T (1999) Rapid clay transformations in Delaware salt marshes. Appl Geochem 14:559–568

Velde B, Espitalié J (1989) Comparison of kerogen maturation and illite/smectite composition in diagenesis. Jour Petrol Geology 12:103–110

Velde B, Renac C (1996) Smectite to illite conversion and K-Ar ages. Clay Miner 31:25–32

Velde B, Weir A (1979) Synthetic illite in the chemical system K_2O-Al_2O_3-SiO_2-H_2O at 300 °C and 2 kb. Proc Sixth Int Clay Conf, Elsevier, pp 395–404

Velde B (1965) Experimental determination of muscovite polymorph stabilities. Amer Miner 50:436–449

Velde B (2001) Clay minerals in the agricultural surface soils in the Central United States. Clay Minerals 36:277–294

Velde B, Goffe B, Hoellard A (2003) Evolution of clay minerals in a chronosequence of poldered sediments under the influence of natural pasture development. Clays Clay Minerals 51:205–217

Velde B, Suzuki T, Nicot E (1986) Pressure-Temperature-Composition of illite/smectite mixed-layer minerals: Niger delta mudstones and other examples. Clays Clay Miner 34:435–441

Velde B, Peck T (2002) Clay mineral changes in the Morrow experimental plots, University of Illinois. Clays Clay Minerals 50:364–370

Velde B (1969) The compositional join muscovite-pyrophyllite at moderate temperatures and pressures. Bulletin de la Société Française de Minéralogie et de Cristallographie 92:360–368

Velde B (1985) Clay Minerals: A physico-chemical Explanation of their Occurence. Elsevier Ed, pp 426

Velde B, Brusewitz A-M (1982) Metasomatic and non-metasomatic low-grade metamorphism of Ordovician meta-bentonites in Sweden. Geochim Cosmochim Acta 46:447–452

Velde B, Iijima A (1988) Comparison of clay and zeolite mineral occurrences in Neogene Age sediments for several deep wells. Clays Clay Minerals 36:337–342

Velde B, Lanson B (1993) Comparison of I-S transformations and maturity of organic matter at elevated temperatures. Clays Clay Miner. 41:178–183

Velde B, Vasseur G (1992) Estimation of the diagenetic smectite to illite transformation in time-temperature space. Amer. Miner. 34:651–976

Velde B, Brusewitz AM (1986) Compositional variation in component layers in natural illite/smectite. Clays Clay Miner. 34:651–657

Vieillard P (1994) Prediction of enthalpy of formation based on refined crystal structures of multisite compounds. 1. Theories and examples. Geochim Cosmochim Acta 58:4049–4063

Vieillard P (2000) A new method for the prediction of Gibbs free energies of formation of hydrated clay minerals based on the electronegativity scale. Clays Clay Miner 48:459–473

Wang H, Sterni W, Frey M (1995) Deconvolution of the X-ray "illite" 10A complex: A case study of Helvetic sediments from eastern Switzerland. Schweiz Min Petr Min 75:187–199

Waples D (1980) Time and temperature in petroleum formation – application of Lopatin's method to petroleum exploration. Amer Assoc Petrol Geol Bull 64:916–926

Warr L, Rice A (1994) Interlaboratory standardization of calibration of clay mineral crystallinity and crystalite size data. J Metamorph Geol 12:141–152

Warr L, Primmer T, Robinson D (1991) Varsican very low-grade metamorphism in southwest England: A diastathermal and thrust-related origin. J Metamoph Geol 9:751–764

Watanabe T (1988) The structural model of illite/smectite interstratified minerals and the diagram for their identification. Clay Science v 7:97–114

Weaver CE, Pollard LD (1973) The chemistry of clay minerals. Developments in Sedimentology no. 15, pp 213

Weaver CE (1967) Potassium, illite and the ocean. Geochim Cosmochim Acta 31:2181–2196

Weaver CE (1959) The clay petrology of sediments. Clays Clay Minerals 6:154–187

Whitney G, Northrop HR (1988) Experimental investigation of the smectite to illite reaction: Dual reaction mechanisms and oxygen isotope systematics. Amer Miner 73:77–90

Whitney G, Velde B (1993) Changes in particle morphology during illitization: an experimental study. Clays Clay Miner 41:209–218

Wilke BM, Mishra KE, Rehfuess KE (1984) Clay mineralogy of a soil sequence in slope deposits derived from hauptdolomit (dolomite) in the Bavarian Alps. Geoderma 32:103–116

Wilkinson M, Haszeldine RS (2002) Fibrous illite in oilfield sandstones - a nucleation kinetic theory of growth. Terra Nova 14:56–60

Wilson MJ, Bain DC, Duthie DML (1984) The soil clays of Great Britain II Scotland Clay Minerals 19:709–735

Xinhua D, Youngchuan S, Xinrong L, Qi L (1996) Illite/smectite diagenesis in the NanXiang, Yitong and North China Permian-Carboniferous basins: Application to petroleum exploration. Amer Assoc Petrol Geol Bull 80:157–173

Yamada H, Nakasawa H (1993) Isothermal treatments of regularly interstratified montmorillonite-beidellite at hydrothermal conditions. Clays Clay Miner 41:726–730

Yamada H, Nakasawa H, Yoshioka K, Fujita T (1991) Smectites in the montmorillonite series. Clay Miner 26:359–369

Yates DM, Rosenberg PE (1996) Formation and stability of endmember illite: I. Solution equilibration experiments at 100–250 °C and $P_{v,soln}$. Geochim Cosmochim Acta 60:1873–1883

Yates DM, Rosenberg PE (1997) Formation and stability of endmember illite: II. Solution equilibration experiments at 100–250 °C and $P_{v,soln}$. Geochim Cosmochim Acta 61:3135–3144

Yates DM, Rosenberg PE (1998) Characterization of neoformed illite from hydrothermal experiments at 250 °C and $P_{v,soln}$: An HRTEM/ATEM study. Amer Miner 83:1199–1208

Yau Y-C, Peacor DR, McDowell SD (1987) Smectite to illite reactions in Salton Sea shales: A transmission and analytical electron microscope study Jour Sed Petrol 57:335–342

Ylagan RE, Altaner SP, Pozzioli A (2000) Reaction mechanisms of smectite illitization associated with hydrothermal alteration from Ponza Island, Italy. Clays Clay Miner 48:610–631

Ylagan RF, Pevear DR, Vrolijk PJ (2000) Discussion of "Extracting K-Ar ages from shales: a theoretical test". Clay Miner 35:599–604

Yoder HS, Eugster HP (1955) Synthetic and natural muscovites. Geochim Cosmochim Acta 8:225–280

Zöller M, Brockamp O (1997) 1M- and $2M_1$-illites: different minerals and not polytypes. Eur J Mineral 9:821–827

Index

T

U

V

W

X

Z

Printing: Mercedes-Druck, Berlin
Binding: Stein+Lehmann, Berlin

GPSR Compliance

The European Union's (EU) General Product Safety Regulation (GPSR) is a set of rules that requires consumer products to be safe and our obligations to ensure this.

If you have any concerns about our products, you can contact us on ProductSafety@springernature.com

In case Publisher is established outside the EU, the EU authorized representative is:

Springer Nature Customer Service Center GmbH
Europaplatz 3
69115 Heidelberg, Germany

Batch number: 10620056

Printed by Printforce, the Netherlands